Stefano Selleri
Luca Vincetti
Annamaria Cucinotta

Optical and Photonic Components

Società Editrice
ESCULAPIO

In Copertina:
Helmholtzian Landscape, 1946 - David Smith

ISBN 978-88-7488-924-2

Società Editrice Esculapio s.r.l.
Via Terracini, 30 - 40131 Bologna
www.editrice-esculapio.com - info@editrice-esculapio.it

Editorial Staff: Giancarla Panigali, Carlotta Lenzi

Società Editrice
ESCULAPIO

Preface

The aim of this work is to collect the contents of the numerous courses on photonics and optical components held at the Faculty of Engineering of the University of Modena and Reggio Emilia and the University of Parma. As this manual is intended for future telecommunications and electrical engineers, who will design the components and subsystems of modern optical communication systems, we shall necessarily adopt a more didactic approach – while also allowing the reader to delve deeper into various subjects, such as the guided propagation of light or the most common components used in photonics.

Photonics emerged as a research discipline in the 1960s with the fabrication of the first lasers. In the 1970s the development of these sources was followed by the large-scale deployment of optical fibers as a way of processing information through light beams. Subsequently, scientific research and technological progresses led to a revolution in the field of telecommunications through, for example, the birth of the Internet network – which thus owes part of its success and global expansion to photonics, too.

Today, this discipline has grown far beyond the most optimistic expectations, exceeding its original field of research (*i.e.* telecommunications) to encompass a wide range of scientific and technological applications such as *int. al.* laser processing, sensors, medical diagnosis and treatments, as well as lighting technologies and optical computing. The convergence between photonics and other disciplines has also opened up new horizons which were simply unimaginable until a few decades ago. Biophotonics, a multidisciplinary research area that employs photonic technologies in medicine and biology, is perhaps the most interesting example in this respect. The potential of photonics are such that it has been acknowledged by the European Commission itself as one of the key technologies for addressing future challenges in terms of environment, health and social policies. Albeit of great interest, our work will not tackle these important issues as we shall focus on the field of telecommunications only; nevertheless, we are confident that the concepts here presented may provide a sound basis for approaching different disciplines that can interact with photonics practically.

Before concluding, we wish to express our gratitude to all our students for their precious feedback and suggestions all along these years. We are also happy to acknowledge the generous and competent support of our colleagues.

Stefano Selleri
Department of Information Engineering
University of Parma

Luca Vincetti
Department of Engineering "Enzo Ferrari"
University of Modena and Reggio Emilia

Annamaria Cucinotta
Department of Information Engineering
University of Parma

Index

Introduction

Optical and photonic components are widely used in modern fiber-optic telecommunication systems. They have been gradually implemented and have become more specialised over the years, and mature technologies as well as new approaches have been able to turn mere objects of study into high-performance commercial devices. Optical components in fiber, microfiber, integrated optics, and *photonic crystals* can perform all or most of the functionalities required by the different channels in an optical network: from signal generation and relay, to its detection and processing – thus minimising the possible bottleneck risk linked to opto-electronic conversions. The development of such components has kept pace with the progress of telecommunication systems: on one hand, the former have supported and stimulated the latter, since they have fostered innovations and paved the way to new opportunities and applications; on the other hand, the need for new performance and technical specifications has encouraged the development of the components necessary to upgrade the ever-expanding network, with a particular focus on speed and bandwidth coverage. Therefore, the use and improvement of optical components are strictly connected to the systems in which they are utilised, especially in terms of their operating characteristics. Modern optical communication systems are moving towards a more and more efficient and *dense* use of the available bandwidth, with obvious consequences on the spectral characteristics of each device.

One of the main figures of merit of a communication system is the product of the bit rate and the repeaters distance. In optical systems, this figure is mainly limited by optical fiber attenuation and dispersion. In this respect, if years ago the maximum distance allowed was only a few dozens of kilometers, the advent of optical amplifiers – mostly, erbium-doped fiber amplifiers (EDFAs) – and the development of dispersion compensation devices have managed to extend such limits to hundreds of kilometers. Beyond this distance, the inevitable noise generated by amplifiers degrades the signal-to-noise ratio (SNR) to such an extent that the signal must be regenerated, often electronically, resulting in bit rate limitation and loss of network transparency – though researchers are working hard to build all-optical networks and reduce the said impairments. The implementation of optical amplifiers has not only benefited transmission lengths, but channel capacity as well, since these devices can boost optical signals on a bandwidth of a few dozens of nanometers. Dense wavelength division multiplexing (DWDM) is the most efficient technology to exploit this higher bandwidth and increase the existing network capacity without laying new expensive cables, which significantly reduces the costs of the system update. Such a technology offers new opportunities in terms of optical network planning, provided that the potentials linked to the *add/drop* operations of the different wavelength channels are exploited efficiently. Nevertheless, such advantages come at a price. In fact, if optical component properties and cable characteristics could once be neglected – with no performance loss – in systems using simpler transmission techniques based on one or few wavelengths, they now must be addressed and accurately designed. Secondly, DWDM systems have introduced the new spectral dimension which requires new criteria for network design and for the choice of components with much tighter specifications than *e.g.* single-wavelength systems.

DWDM systems can be simply represented as a collection of optical channels simultaneously working at different wavelengths on the same fiber, as shown in figure 0.1. It is important to point out that all channels must be handled similarly along the optical path from the transmitter to the receiver. Therefore, optical sources, multiplexers, demultiplexers, optical amplifiers, filters, receivers, and even the fiber itself must all be carefully selected so as to achieve and maintain the desired spectral performance throughout the whole network. So, in order to ensure that cross-talk is minimised, all possible interactions between closely spaced channels have to be thoroughly studied. Parameters such as insertion loss, optical return loss (ORL), polarisation effects, passband bandwidth, spectral stability, SNR, *et al.* must all be controlled as the wavelength varies through the whole spectral bandwidth used by the system.

The International Telecommunication Union (ITU) has determined the standard optical frequency grid through the Recommendation ITU-T G.692, which defines channel spacing of $100\,GHz$ ($0.8\,nm$) and $50\,GHz$ ($0.4\,nm$). Hence, even a small spectral drift in DFB lasers commonly used as transmission sources can have serious effects on signal levels at the receiving end due to the narrow spacing of the passband. Moreover, spectral side lobes in the radiation pattern may contribute noise to adjacent channels. Stability and spectral parity are thus of paramount importance.

Multiplexers and demultiplexers are other critical components of a DWDM system. These devices combine signals into a single fiber and then split them on the receiving end. Their passband has to adapt to the wavelength chosen with extreme precision, and, to avoid cross-talk, their rejection ratio has to be high enough so as to eliminate adjacent channels and their side lobes.

Some characteristics of the optical fiber itself are much more important in DWDM systems than they were in conventional single-wavelength networks. For instance, the effects of chromatic dispersion at $1550\,nm$ are well known in single-wavelength systems, but these have to be taken into greater consideration in DWDM networks. In addition to that, single-wavelength systems try to completely eliminate chromatic dispersion, while in DWDM networks a small amount of dispersion is usually present to prevent the impairment due to fiber nonlinearity. Such impairments are mainly caused by the scattering of high-power optical signals and are particularly annoying as they can generate

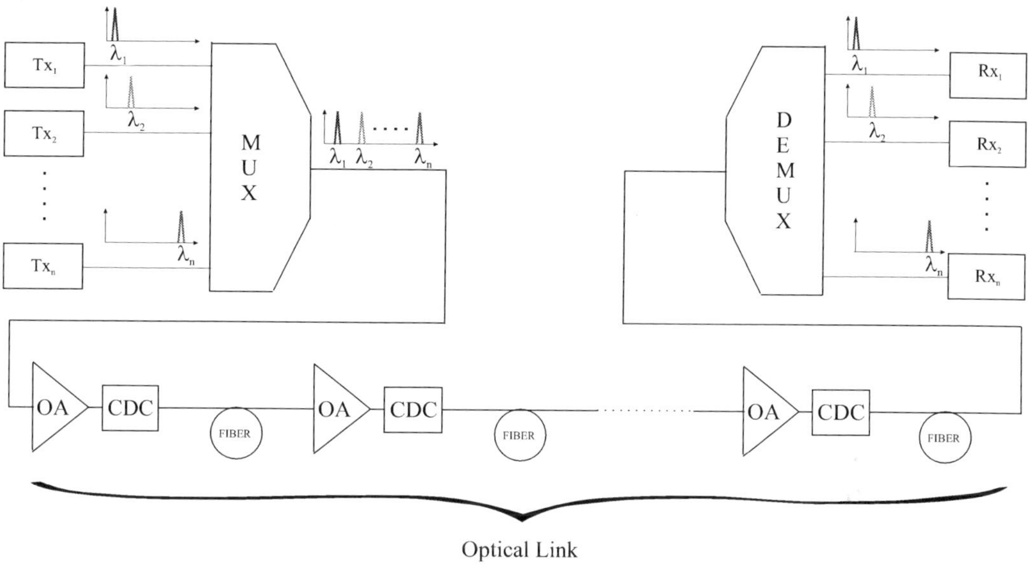

Fig. 0.1: DWDM System Structure.

cross-phase modulation, four-wave-mixing, and signal power loss. Moreover, channel spacing must be greater than the bandwidth of each channel to minimise negative side effects. Time-division multiplexing (TDM) is an effective method to expand the overall capacity of the system by increasing the rate of a single optical carrier. However, due to the high bit rate, such an expansion is only possible by optically multiplexing channels. The Optical Time Domain Multiplexing (OTDM) technique has already made it possible to develop multiplexers and demultiplexers based on an efficient combination between interferometric structures and highly nonlinear devices, such as semiconductor optical amplifiers.

What has been said so far reflects the strong need for an in-depth knowledge of physics and the mechanisms regulating the propagation of an electromagnetic field in optical components – not to mention the importance of appropriate techniques for the analysis and, especially, the design of devices satisfying specific system requirements. All these aspects will be thoroughly discussed in this work, starting from sources such as fibers and dielectric waveguides, and then taking into account the numerous components – either in fiber, integrated optics or microfiber – used in signal detecting and processing.

Chapter 1

Dielectric Waveguides

The term "guided propagation" is used in cases where the electromagnetic field is propagated in a particular direction and confined to a limited portion of the plane transverse to the propagation direction itself. Since such a propagation cannot be obtained with homogenous media, it is necessary to impose specific boundary or continuity conditions along the whole path of the wave. There are essentially two possible solutions in this respect: the first employs electrical conductors and follows the principle that the electromagnetic field has a Poynting vector oriented towards the longitudinal axis, as it cannot be propagated in them. The second solution resorts to total reflection and dielectric materials with different values of permittivity. In this case, we mean *dielectric waveguides*, while the others are called *metallic waveguides*. The only choice, in terms of optical communications, is represented by dielectric waveguides, since there are no perfect conductive materials for frequencies of hundreds of terahertz. The simplest and most common example is the optical fiber, a cylinder made of homogeneous dielectric material with a circular cross-section and surrounded by other material with lower permittivity. In dielectric waveguides, the field is indefinitely extended on the transverse plane, even though, in practical applications, it is supposed to be non-zero only where permittivity is higher.

Guided propagation is best represented using a *cylindrical structure*, where we assume that the electromagnetic characteristics of the material do not vary along the propagation direction of the field. The solutions of Maxwell's equations applied to these particular structures are called *modes*. The aim of this chapter is to review the main properties of cylindrical structures and their modes.

1.1 Waveguide Modes

Consider a generic structure with electromagnetic characteristics, dielectric permittivity ε, magnetic permeability μ and conductivity σ invariant in a specific direction: with these elements, we mean a *cylindrical* structure. Unless otherwise stated, henceforth we will refer to dielectric media – therefore, $\sigma = 0$ and $\mu = \mu_0$. Also, consider an orthogonal reference frame (u_1, u_2, z) oriented in such a way that the z-axis coincides with the invariance direction of the structure, as shown in figure 1.1.

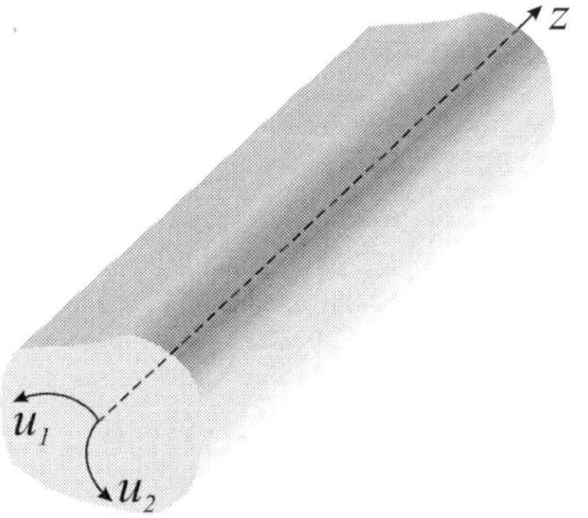

Fig. 1.1: Cylindrical Structure. The z-axis determines the invariance direction; coordinates u_1 and u_2 determine the transverse plane.

In this chapter, we always presume to operate in a harmonic regime, leaving considerations on aperiodic signal propagation to the next chapters. Thus, vector and scalar quantities are represented by complex vectors and numbers, respectively. Considering the invariance of the cylindrical structure, particular solutions of Maxwell's equations can be

$$\begin{aligned} \overline{e}(u_1, u_2, z) &= \overline{E}(u_1, u_2)e^{-\gamma z} \\ \overline{h}(u_1, u_2, z) &= \overline{H}(u_1, u_2)e^{-\gamma z}, \end{aligned} \tag{1.1}$$

characterised by the three quantities $\overline{E}$, $\overline{H}$, γ. Note that, in general, these are functions of the angular frequency ω. The propagation constant γ identifies the evolution of both the phase and the field amplitude along the propagation direction. In presence of losses, $\gamma = \alpha + j\beta$, where α is the *attenuation constant* and β the *phase constant*. The *mode functions* $\overline{E}$ and $\overline{H}$ define the evolution of the electromagnetic field on the transverse plane, but they are not solutions of Maxwell's equations *per se*. Figure 1.2 shows an example of electromagnetic field with no losses (*i.e.* $\alpha = 0$), which is assumed to have only one component, and with the mode functions depending on just one transverse coordinate.

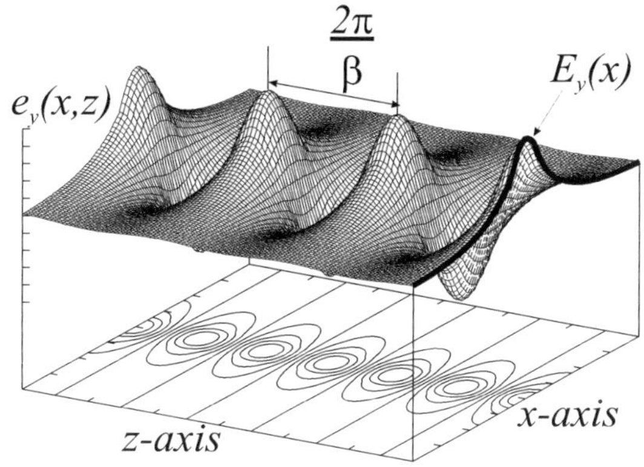

Fig. 1.2: Example of the real part of the electromagnetic field of a mode. Transverse mode function in bold. The z-axis covers three periods of the wave.

On the x-axis, the electric field follows the evolution of the mode function (in bold), while a sine wave with a period of $2\pi/\beta$ is represented on the z-axis.

As we shall see, a deep knowledge of both the mode functions and the propagation constant is essential in order to study and design not only dielectric waveguides, but for other optical devices as well, such as couplers, power dividers and DBRs (distributed Bragg reflectors). The methods adopted to calculate these values that are applied to specific types of dielectric waveguides will be dealt with later in this work. For the moment, we shall briefly review the general procedure for calculating mode functions in a dielectric waveguide.

1.2 Mode Function Calculation

If $\alpha = 0$ (no loss case), the invariance of the structure along the direction of propagation of the electromagnetic field determines an important relation between the longitudinal and the transverse components of both the electric and magnetic field. According to the theory of cylindrical structures, it is possible to express the transverse components $\overline{E}_t$, $\overline{H}_t$ of mode functions as follows:

$$\begin{aligned} \overline{E}_t(u_1,u_2) &= \frac{j}{\left(\frac{\omega}{c}n\right)^2 - \beta^2}\Big(- \beta\overline{\nabla}_t E_z - \omega\mu_0\overline{\nabla}_t H_z \times \hat{z}\Big); \\ \overline{H}_t(u_1,u_2) &= \frac{j}{\left(\frac{\omega}{c}n\right)^2 - \beta^2}\Big(- \beta\overline{\nabla}_t H_z + \omega\varepsilon_0 n^2\overline{\nabla}_t E_z \times \hat{z}\Big), \end{aligned} \tag{1.2}$$

where E_z and H_z are the transverse components of the mode functions, n the refractive index of the medium, c the speed of light in vacuum, and ω the angular frequency of the field. Due to the imaginary unit j in the second member of (1.2), a phase shift of $\pi/2$ is always present between the longitudinal and the transverse components. (1.2) also shows that, if the longitudinal components are known, it then becomes possible to determine both the transverse components and the mode functions. Furthermore, said longitudinal components are solutions to the following differential equations:

$$\begin{aligned}\{\nabla_t^2 E_z + \left[\left(\frac{\omega}{c}n\right)^2 - \beta^2\right] E_z\}\hat{z} &+ \frac{\beta^2}{\left(\frac{\omega}{c}n\right)^2 - \beta^2}\frac{\overline{\nabla}n}{n} \times \left(\overline{\nabla}_t E_z \times \hat{z}\right) \\ &= \frac{\beta\omega\mu_0}{\left(\frac{\omega}{c}n\right)^2 - \beta^2}\frac{\overline{\nabla}n}{n} \times \overline{\nabla}_t H_z;\end{aligned}$$

$$\begin{aligned}\{\nabla_t^2 H_z + \left[\left(\frac{\omega}{c}n\right)^2 - \beta^2\right] H_z\}\hat{z} &+ \frac{\left(\frac{\omega}{c}n\right)^2}{\left(\frac{\omega}{c}n\right)^2 - \beta^2}\frac{\overline{\nabla}n}{n} \times \left(\overline{\nabla}_t H_z \times \hat{z}\right) \\ &= -\frac{\beta\omega\varepsilon_0 n^2}{\left(\frac{\omega}{c}n\right)^2 - \beta^2}\frac{\overline{\nabla}n}{n} \times \overline{\nabla}_t E_z.\end{aligned}$$

The main problem in solving these equations lies in the fact that the refractive index n depends on the transverse coordinates, thus producing coupled equations. Fortunately, the cylindrical structure is often like the one presented in figure 1.3, where the medium is only piecewise homogeneous – non-homogeneity being a necessary condition for field confinement.

Where the material is homogeneous, the equations previously seen can be traced back to the well-known Helmholtz equation:

$$\nabla_t^2\phi + \left[\left(\frac{\omega}{c}n\right)^2 - \beta^2\right]\phi = 0,$$

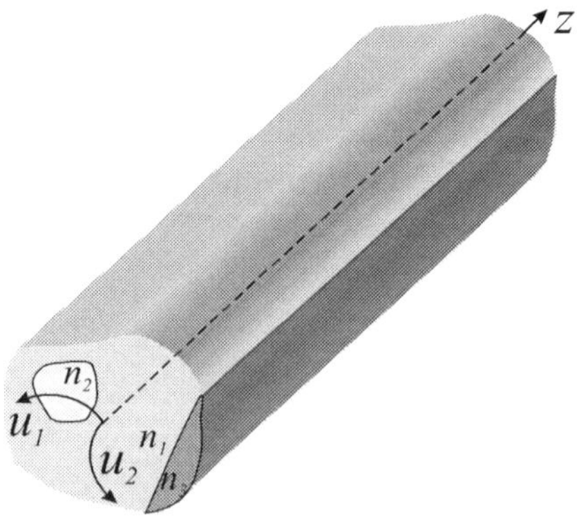

Fig. 1.3: Piecewise homogeneous cylindrical structure. The continuity conditions are applied to the interfaces, here delimited by black lines.

where ϕ is the longitudinal component of either the mode function, the electric or the magnetic field. After the equation is solved in the different areas, the transverse components are determined by the expressions from (1.2); subsequently, continuity and boundary conditions are defined and applied. As the dielectric medium is extended indefinitely on the transverse plane, the boundary conditions coincide with the regularity conditions at infinity of the solution.

1.3 Power Carried by a Mode

The power carried by a mode is equal to the flux of the Poynting vector through the plane S orthogonal to the propagation direction. Consequently, once the boundary and continuity conditions have been satisfied, the mode functions are defined up to a multiplicative constant common to both $\overline{E}$ and $\overline{H}$. In fact, if we multiply a generic solution of the Helmholtz equation $\phi(u_1, u_2)$ by a constant A, we obtain the new function $\phi'(u_1, u_2) = A\phi(u_1, u_2)$, which is still a solution of the Helmholtz equation. This complies with the uniqueness theorem, as no boundary condition has been applied to the input and output ends of the waveguide yet. The mode functions can thus be *normalised* by imposing that

$$\frac{1}{2}\int_S \left(\overline{E} \times \overline{H}^*\right) \cdot \hat{z} dS = 1.$$

Consequently, it is possible to represent a mode with generic amplitude as

$$\begin{aligned} \overline{e}(u_1, u_2, z) &= A\overline{E}(u_1, u_2)e^{-\alpha z}e^{-j\beta z} \\ \overline{h}(u_1, u_2, z) &= A\overline{H}(u_1, u_2)e^{-\alpha z}e^{-j\beta z} \end{aligned} \tag{1.3}$$

and the power carried by the mode is equal to:

$$P = \int_S \frac{1}{2}\left(\overline{e} \times \overline{h}^*\right) \cdot \hat{z} dS = \int_S \frac{1}{2}|A|^2 e^{-2\alpha z}\left(\overline{E} \times \overline{H}^*\right) \cdot \hat{z} dS = |A|^2 e^{-2\alpha z}.$$

If P_0 is the input power in $z = 0$ carried by the mode under consideration, it follows that:

$$P(z) = P_0 e^{-2\alpha z}$$

where $P_0 = |A|^2$. Therefore, α is linked to the power lost in propagation due to the losses. The attenuation $P_0/P(z)$ is often expressed in dB; in this case, $10\log(P_0/P(z)) = \alpha_{dB} z$, where:

$$\alpha_{dB} = 20\log(e)\alpha.$$

If α_{dB} is expressed in dB/km and α in m^{-1}, then:

$$\alpha_{dB} = 8.6858\, 10^3\alpha.$$

Typical values of an optical fiber used in the third window of telecommunications (around 1550 nm) are $\alpha = 0.15 - 0.2$ dB/km. However, these are extremely low compared to coaxial cables, where α depends both on the characteristics of the cable itself and on the operating frequency, thus varying between $4 \div 20$ dB/km at 100 MHz and hundreds of dB/km at 1 GHz.

1.4 Modes Completeness and Orthogonality

The waveguide modes are a complete set, and they are orthogonal to each other, thus constituting a *basis* of solutions for cylindrical structures. Hence, any field distribution propagated in the waveguide analysed can be expressed as a linear combination of the same waveguide modes:

$$\begin{aligned} \overline{e}_t(u_1, u_2, z) &= \sum_\nu (A_\nu e^{-j\beta_\nu z} + B_\nu e^{j\beta_\nu z})\overline{E}_{t\nu} \\ e_z(u_1, u_2, z) &= \sum_\nu (A_\nu e^{-j\beta_\nu z} - B_\nu e^{j\beta_\nu z})E_{z\nu}; \\ \\ \overline{h}_t(u_1, u_2, z) &= \sum_\nu (A_\nu e^{-j\beta_\nu z} - B_\nu e^{j\beta_\nu z})\overline{H}_{t\nu} \\ h_z(u_1, u_2, z) &= \sum_\nu (A_\nu e^{-j\beta_\nu z} + B_\nu e^{j\beta_\nu z})H_{z\nu}; \end{aligned} \tag{1.4}$$

A_ν refers to direct modes, while B_ν to reflected modes. The different signs associated to B_ν are necessary to satisfy Maxwell's equations and keep the same mode functions for both direct and reflected modes. In case of absence of losses, and with negligible reflected mode amplitude – by virtue of orthogonality – the power carried by the field $\overline{e}$, $\overline{h}$ is equal to:

$$P_0 = \sum_\nu |A_\nu|^2.$$

The width of a generic mode contributing to field representation – and therefore the power carried by the mode itself – is determined by the *excitation conditions* of the waveguide, *i.e.* by the field distribution on the input end $z = 0$ of the waveguide:

$$A_\nu = \frac{1}{2}\int_S \left(\overline{e}_{ting} \times \overline{H}^*_{t\nu}\right) \cdot \hat{z}dS \tag{1.5}$$

where $\overline{e}_{ting} = \overline{e}_t(u_1, u_2, 0)$ is the transverse component of the input field, also known as *exciting field.* The integral in the second member of the equation is generally known as *overlap integral.* Note that the condition of excitation can be compared to the setting of appropriate boundary conditions on the waveguide input end; on the output end, instead, the boundary conditions were automatically defined when we assumed that the reflected modes are negligible – that is to say, the waveguide is perfectly matched to the load.

1.5 Waveguide Transfer Function

The different modes of an ideally flawless waveguide propagate independently of each other thanks to their orthogonality, which allows the principle of superposition to be applied. Thus, let us assume we excite a waveguide of length L with a field distribution corresponding to one of the modes of the

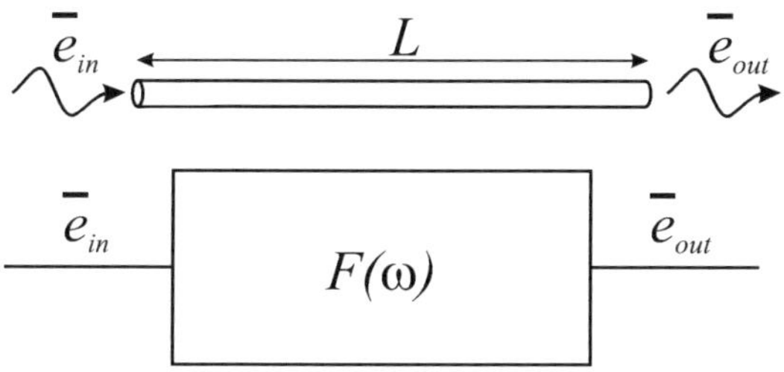

Fig. 1.4: Block diagram of a single-mode waveguide.

guide and characterised by $\overline{E}$, $\overline{H}$ and γ:

$$\overline{e}_{in} = A\overline{E}(u_1, u_2).$$

According to (1.1), the output field is equal to:

$$\overline{e}_{out} = A\overline{E}(u_1, u_2)e^{-\alpha L}e^{-j\beta L}.$$

Hence, it is clear that the quantity

$$F(\omega) = e^{-\alpha(\omega)L}e^{-j\beta(\omega)L} \tag{1.6}$$

represents the waveguide transfer function of the mode considered in figure 1.4. The trend of both β and α as a function of ω respectively indicate the *dispersion* and *attenuation curve* of the mode; figure 1.5 shows a possible qualitative trend. Every mode is generally characterised by the mode functions, as well as the attenuation and phase constants; therefore the the waveguide transfer function varies from one mode to another. It should be remembered that the non-distortion conditions of a transmission system require the transfer function to have constant magnitude, while its phase varies linearly with frequency. Suppose now that N modes propagate in the waveguide simultaneously as a result of the excitation field – each mode being described by its own amplitude A_i given by (1.5), and by the transfer function:

$$F_i(\omega) = e^{-\alpha_i(\omega)L}e^{-j\beta_i(\omega)L}$$

with $i = 1, 2, ..., N$. In this case, we can represent the waveguide as a block diagram in which N linear modes with transfer function $F_i(\omega)$ are connected in parallel (cf. figure 1.6).

1.6 The Electric Field of a Mode in the Time Domain

It is possible to calculate the electric field of a mode in the time domain by making the three constituents of $\overline{E}$ explicit, and writing the complex components in polar form:

$$\overline{E}(u_1, u_2) = E_{u_1}(u_1, u_2)e^{j\theta_{u_1}}\hat{u_1} + E_{u_2}(u_1, u_2)e^{j\theta_{u_2}}\hat{u_2} + E_z(u_1, u_2)e^{j\theta_z}\hat{z},$$

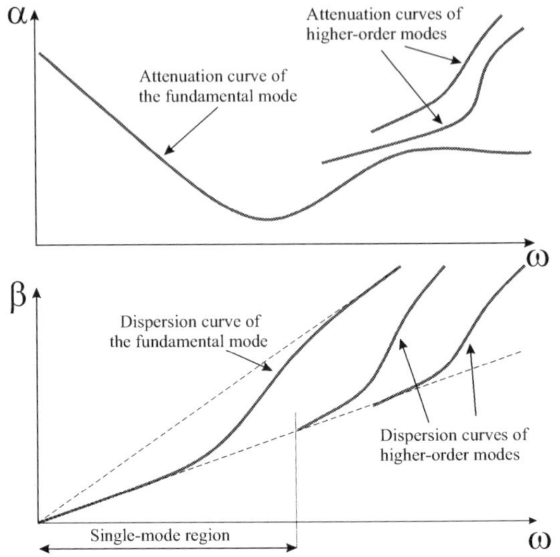

Fig. 1.5: Attenuation and dispersion curves of the first three modes of a generic dielectric waveguide.

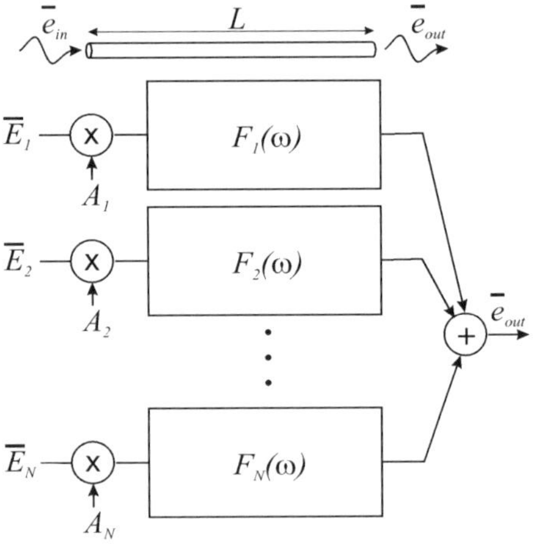

Fig. 1.6: Block diagram of an N-mode waveguide.

where θ_i – with $i = u_1, u_2, z$ – is the phase shift of E_i at the instant $t = 0$, from which:

$$\begin{aligned}
\overline{\mathcal{E}}(u_1, u_2, z, t) &= Re\left\{\overline{e}(u_1, u_2, z, \omega_0)e^{j\omega_0 t}\right\} = Re\left\{A\overline{E}(u_1, u_2)e^{-\gamma z}e^{j\omega_0 t}\right\} \\
&= Ae^{-\alpha z}\left\{E_{u_1}(u_1, u_2)cos(\omega_0 t - \beta z + \theta_{u_1}(u_1, u_2))\hat{u_1}\right. \\
&+ E_{u_2}(u_1, u_2)cos(\omega_0 t - \beta z + \theta_{u_2}(u_1, u_2))\hat{u_2} \\
&+ \left. E_z(u_1, u_2)cos(\omega_0 t - \beta z + \theta_z(u_1, u_2))\hat{z}\right\}; \qquad (1.7)
\end{aligned}$$

where ω_0 is the operating angular frequency. A similar expression can be obtained for the magnetic field. The equation (1.7) offers a general, though complicated, representation of the electric field of a mode in the time domain and in the harmonic regime. However, the same expression can be considerably simplified when it comes to actual practice. (1.2) shows in fact that in the absence of losses, or with $\alpha << \beta$ at least, there is a phase shift of $\frac{\pi}{2}$ between the transverse and longitudinal components. For this reason, henceforth we shall designate the origin of the time reference so that $\overline{E}_t$ and $\overline{H}_t$ are real, while E_z and H_z are purely imaginary; thus, if $\theta_{u_1} = \theta_{u_2} = 0$ and $\theta_z = \pi/2$, the equation above can be reformulated as follows:

$$\overline{\mathcal{E}}(u_1, u_2, z, t) = Ae^{-\alpha z}\Big(\overline{E}_t(u_1, u_2)\cos(\omega_0 t - \beta z) + E_z(u_1, u_2)\sin(\omega_0 t - \beta z)\hat{z}\Big). \tag{1.8}$$

The equiphase surfaces are planes orthogonal to the direction of invariance of the cylindrical structure and are propagated at the phase velocity

$$v_p = \frac{\omega_0}{\beta}. \tag{1.9}$$

The phase delay

$$\tau_p = \frac{L}{v_f} = \frac{\beta L}{\omega_0}. \tag{1.10}$$

is the time an equiphase surface requires to cover the whole length of the waveguide.

1.7 Mode Coupling

As previously stated, the property of orthogonality characteristic of waveguide modes indicates that, in practical terms, there is no exchange of power between two modes as they propagate.

A different case is depicted in figure 1.7, where the dielectric permittivity also depends on the longitudinal z-axis, as we are confronted with a perturbed waveguide. In this case, the continuity conditions of the field are altered, with consequent energy transfer between different modes – a phenomenon known as *mode coupling.*

If the perturbation $\Delta\epsilon$ on the z-axis is of modest extent, we can resort to the *mode-coupling theory* (cf. Appendix), which holds that it is possible to define the overall electric and magnetic field thanks to the equations (1.4), yet with the coefficients varying along the propagation direction:

$$\begin{aligned}
\overline{e}_t(u_1, u_2, z) &= \sum_\nu (A_\nu(z)e^{-j\beta_\nu z} + B_\nu(z)e^{j\beta_\nu z})\overline{E}_{t\nu} \\
e_z(u_1, u_2, z) &= \sum_\nu (A_\nu(z)e^{-j\beta_\nu z} - B_\nu(z)e^{j\beta_\nu z})E_{z\nu}; \\
\overline{h}_t(u_1, u_2, z) &= \sum_\nu (A_\nu(z)e^{-j\beta_\nu z} - B_\nu(z)e^{j\beta_\nu z})\overline{H}_{t\nu} \\
h_z(u_1, u_2, z) &= \sum_\nu (A_\nu(z)e^{-j\beta_\nu z} + B_\nu(z)e^{j\beta_\nu z})H_{z\nu}.
\end{aligned} \tag{1.11}$$

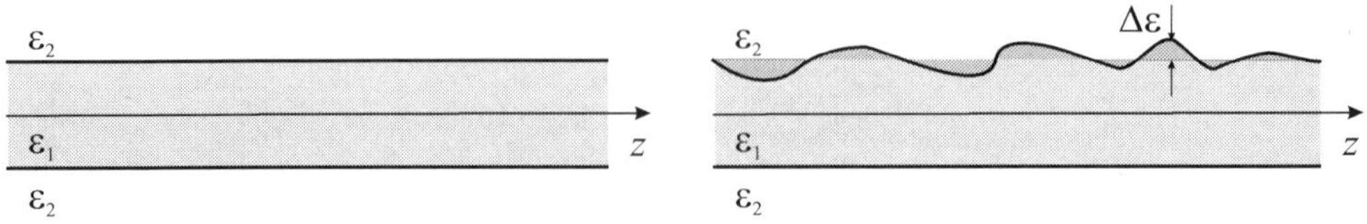

Fig. 1.7: On the left, an ideal unperturbed waveguide. On the right, a perturbed waveguide.

In the most common situation, two modes i and j propagate in the same direction: the power exchange, which is responsible for A_ν depending on z, is proportional to the difference between the propagation constant of the two modes, $\Delta\beta = \beta_i - \beta_j$; additionally, the power exchange is also proportional to the coupling coefficient K_{ij} – which depends both on the perturbation $\Delta\epsilon$ and on the superposition of the two electric fields at the level of the perturbation itself:

$$K_{ij} = \frac{\omega}{4} \int_{S_\infty} \Delta\epsilon \overline{E}_{t_i} \cdot \overline{E}_{t_j} dS.$$

Chapter 2

Dielectric Slab Waveguides

Slabs are amongst the most common examples of dielectric waveguides. Figure 2.1 shows a symmetric slab consisting of different dielectric layers: the central region, delimiting the field, is called the *core*, or the *film*, and it is made of dielectric material with refractive index n_{co} surrounded by another kind of material with refractive index n_{cl} lower than n_{co}; this second layer is called *cladding*. The core and the cladding are indefinitely extended along both the propagation direction and one of the transverse directions. This, in particular, prevents the field from being confined to a limited portion of the transverse plane. The slab cannot be considered a dielectric waveguide of any particular practical interest. However, apart from the fact that some waveguides in integrated optics may sometimes be effectively analysed according to the slab model, slab waveguides allow for considerable mathematical simplification which permits focusing on the numerous general aspects of electromagnetic field propagation in dielectric waveguides.

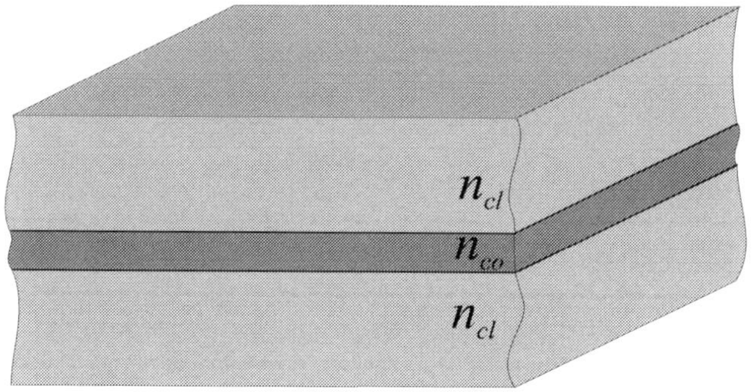

Fig. 2.1: Dielectric Slab Waveguide. The darker central layer represents the core, while the surrounding region is the cladding.

2.1 TE and TM Modes

As the core is separated from the two claddings by flat interfaces, the mathematical analysis is easier if we choose a Cartesian frame of reference oriented in such a way that the separating interfaces coincide with coordinate planes. In this work, the system introduced is oriented so that the z-axis corresponds to the propagation direction, while the y-axis coincides with the transverse invariance direction (see figure 2.2). The thickness of the core is $2d$, ranging from $x = -d$ to $x = d$. As our main aim is to define the characteristics of modes propagating at a distance, the propagation constant γ is initially presumed to be purely imaginary:

$$\gamma = j\beta.$$

The invariance of the structure on the y-axis is a necessary but not sufficient condition for constant solutions to exist with respect to y. These particular solutions are thus present in slabs too, and – though being indefinitely extended – they are dependent to a certain extent on the y-coordinate. However, since the aim of this chapter is to analyse the mechanism of electromagnetic field confinement in dielectric waveguides, we shall focus our attention on the constant solutions along y. It follows that, in the course of this chapter, we will assume the following to be true in all equations:

$$\frac{\partial}{\partial y} = 0. \tag{2.1}$$

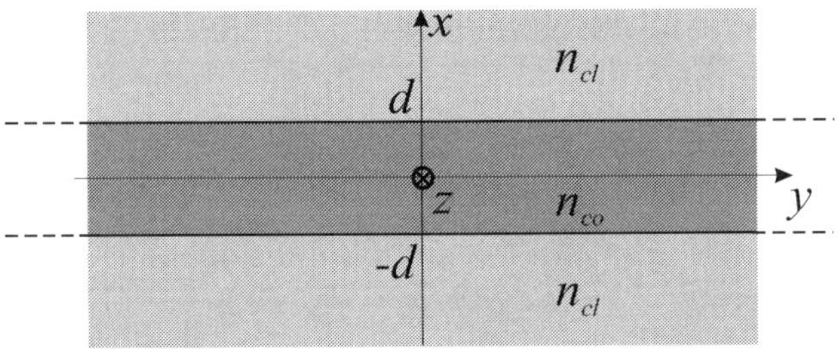

Fig. 2.2: Transverse section of a slab. The separating interfaces between the core and the cladding are at $x = \pm d$.

Considering (2.1), we can return to (1.2), which proved useful to define the transverse components as a function of the longitudinal ones:

$$\begin{aligned} E_x &= -j\frac{\beta}{n^2\left(\frac{\omega}{c}\right)^2-\beta^2}\frac{dE_z}{dx} \\ E_y &= j\eta_0\frac{\frac{\omega}{c}}{n^2\left(\frac{\omega}{c}\right)^2-\beta^2}\frac{dH_z}{dx} \\ H_x &= -j\frac{\beta}{n^2\left(\frac{\omega}{c}\right)^2-\beta^2}\frac{dH_z}{dx} \\ H_y &= -j\frac{n^2}{\eta_0}\frac{\frac{\omega}{c}}{n^2\left(\frac{\omega}{c}\right)^2-\beta^2}\frac{dE_z}{dx} \end{aligned} \tag{2.2}$$

where $n = \sqrt{\varepsilon_r}$ is the refractive index of the medium, $k_0 = \omega/c$ the wavenumber, and $\eta_0 = \sqrt{\mu_0/\varepsilon_0}$ the intrinsic impedance of free space. Thanks to (2.1), the transverse components depend on the two longitudinal components, which permits the separation of the solutions into two groups: with $E_z = 0$, only E_y, H_x and H_z are non-zero in the solutions; with $H_z = 0$, instead, H_y, E_x and E_z are non-zero. The first group produces *transverse electric (TE) modes* as there is no electric field in the direction of propagation, while in the second case we refer to *transverse magnetic (TM) modes* since the magnetic field is completely transverse along the direction of propagation.

	Components	
Mode	Transverse	Longitudinal
TE	E_y, H_x	H_z
TM	H_y, E_x	E_z

We will use $\phi(x)$ to refer to the sole non-zero longitudinal component, *i.e.* H_z for TE modes and E_z for TM modes. $\phi(x)$ is the solution of the Helmholtz equation, which in a piecewise homogeneous medium – and according to (2.1) – can be thus expressed:

$$\begin{cases} \dfrac{d^2\phi}{dx^2} + \left[\left(n_{co}\frac{\omega}{c}\right)^2 - \beta^2\right]\phi = 0 & \text{se} \quad |x| \leq d \\ \dfrac{d^2\phi}{dx^2} + \left[\left(n_{cl}\frac{\omega}{c}\right)^2 - \beta^2\right]\phi = 0 & \text{se} \quad |x| > d \end{cases} \tag{2.3}$$

2.2 Plane Waves and Field Confinement

Before going on to calculate the mode functions and the phase constant through the integration of the Helmholtz equation associated to the continuity conditions of tangent components, we shall describe the mechanism of field confinement in the transverse plane and the process of propagation along the longitudinal direction by means of plane waves. Plane waves are in fact solutions to Maxwell's equations both in the core and in the cladding, as these are homogeneous media.

Let us reconsider the Helmholtz equation (2.3), rewritten as follows:

$$\frac{d^2\phi}{dx^2} + \tilde{k}_x^2\phi = 0, \tag{2.4}$$

where

$$\tilde{k}_x = \sqrt{n^2\left(\frac{\omega}{c}\right)^2 - \beta^2} = \sqrt{n^2k_0^2 - \beta^2}, \tag{2.5}$$

from which:

$$\tilde{k}_x^2 + \beta^2 = n^2k_0^2 = n^2\left(\frac{\omega}{c}\right)^2.$$

The solution to (2.4) can be written either as a linear combination of sines and cosines, or as a combination of imaginary exponentials; the latter will serve the purpose of this paragraph:

$$\phi(x) = \tilde{A}e^{-j\tilde{k}_x x} + \tilde{B}e^{j\tilde{k}_x x}.$$

If the longitudinal component of the electric or magnetic field can be calculated by multiplying ϕ by the phase term $e^{-j\beta z}$, it is clear that the field in the three homogeneous sections of the slab is described by a pair of plane waves having the following propagation vector:

$$\overline{S} = j(\pm\tilde{k}_x\hat{x} - \beta\hat{z}).$$

For the time being, we will assume that the constant β is always real and, for the sake of solution continuity, we will also consider its value as constant throughout the whole waveguide. If $\tilde{k}_x$ is real too, plane waves are uniform; otherwise, if $\tilde{k}_x$ is imaginary, we will have evanescent waves along the x-axis. Hence, according to (2.5), the waves are uniform in the core when $\beta < n_{co}\omega/c$, but they are evanescent when $\beta > n_{co}\omega/c$; as for the cladding, if $\beta < n_{cl}\omega/c$ the waves are uniform, while if $\beta > n_{cl}\omega/c$ we will have evanescent waves (cf. figure 2.3). The confinement is in the transverse direction if plane waves are uniform in the core and evanescent in the cladding; therefore:

$$n_{cl}\frac{\omega}{c} < \beta < n_{co}\frac{\omega}{c}.$$

The solutions satisfying the relation above are the *guided modes* of the slab. They can be represented by a pair of plane waves propagating through the core by multiple reflections, as outlined in figure 2.4. The inequality $\beta < n_{co}\omega/c$ is due to the fact that β is the projection of the phase vector $\overline{k}$ along the z-axis; if we consider figure 2.4, in fact, since $|\overline{k}| = n_{co}\omega/c$, it follows that $\beta = |\overline{k}|\sin\theta =$

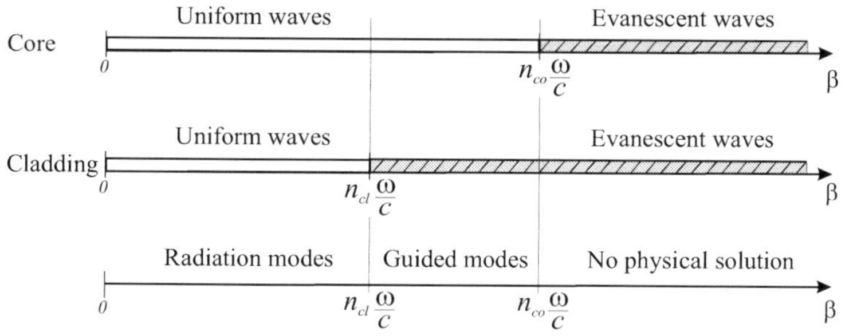

Fig. 2.3: The relationship between the propagation constant and the type of plane wave.

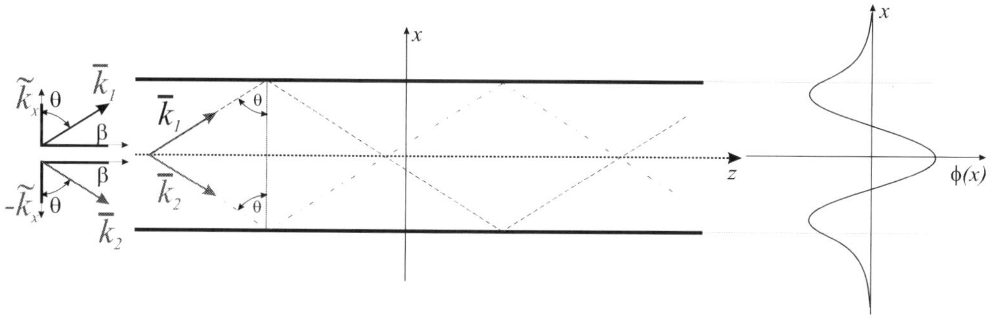

Fig. 2.4: A pair of plane waves propagating by total reflection through the core of a slab. The waves are evanescent in the cladding, with the attenuation vector oriented along the x-axis.

$n_{co}\omega/c \sin\theta < n_{co}\omega/c$. The second inequality represents the condition of total reflection: indeed, if $\beta = n_{co}\omega/c \sin\theta > n_{cl}\omega/c$, we obtain $\theta > \arcsin(n_{cl}/n_{co}) = \theta_c$. Figure 2.4 also shows the qualitative field profile of a guided mode on the transverse plane. The superposition of two uniform plane waves in the core generates a standing wave with a sinusoidal pattern along the transverse direction; on the other hand, the presence of evanescent waves in the cladding causes exponential decay.

However, if

$$\beta < n_{cl}\frac{\omega}{c}$$

there is no total reflection, and waves are uniform in the cladding. The solution is not limited to the core and, moreover, a flux of the Poynting vector is present in the transverse direction too – see figure 2.5 for the qualitative profile. These solutions are called *radiation modes* and, though they do not allow power transmission at a distance, they are nonetheless important in order to ensure that the set of modes is complete.

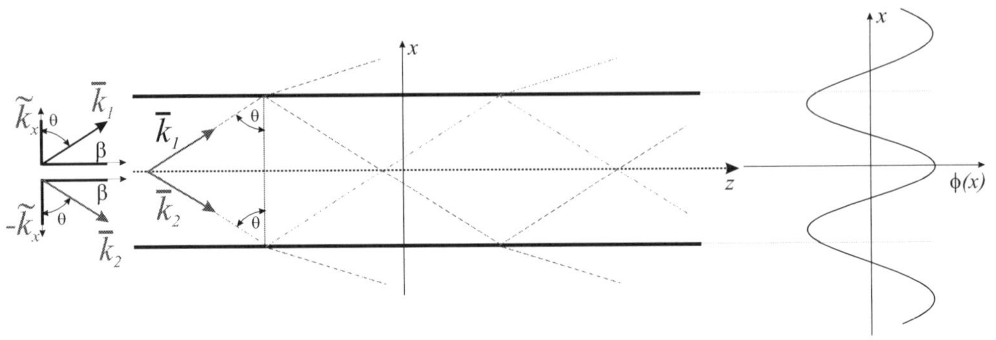

Fig. 2.5: A pair of plane waves propagating through the core of a slab. Due to the lack of total reflection, uniform waves are present in the cladding.

If then,

$$\beta > n_{co}\omega/c,$$

we have evanescent waves both in the core and in the cladding, and the profile of ϕ is that shown in figure 2.6. However, this solution cannot be accepted since the function, though being continuous as imposed by the continuity conditions, is not regular enough. According to (2.2) in fact, the cusps corresponding to the core-cladding interface provoke the discontinuity of E_y and H_y – which, instead, should always be continuous, as they are tangential to the interface. As a consequence, β must always be less than $n_{co}k_0 = n_{co}\omega/c$.

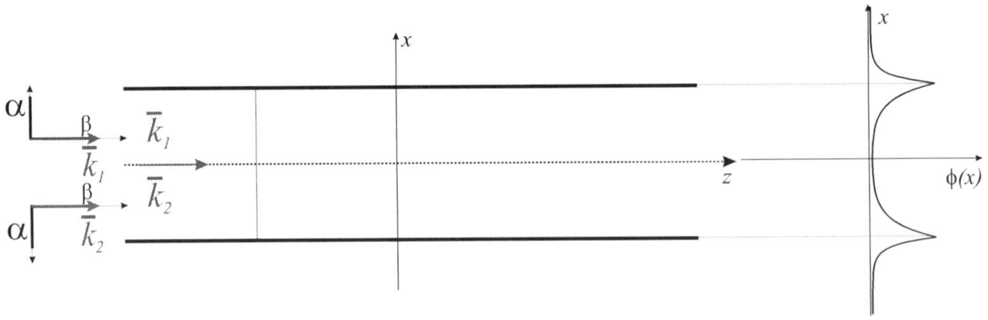

Fig. 2.6: A pair of evanescent plane waves propagating through the core of a slab. Since $\beta > n_{cl}\omega/c$, we have evanescent waves in the cladding too.

For the sake of completeness, consider the case of the following completely real propagation constant:

$$\gamma = \alpha.$$

The propagation vector results:

$$\overline{S} = \pm j\tilde{k}_x\hat{x} + \alpha\hat{z}.$$

In terms of plane waves, this refers to evanescent plane waves in the propagation direction transmitted transversely, as the phase vector is orthogonal to the core-cladding interface. These solutions are named *reactive modes*, but they do not generate any power flux along the propagation direction. The possible values of the phase constant and the different modes associated to it as a function of frequency are schematised in figure 2.7. Besides, it is important to point out that what has been said so far applies to any dielectric waveguide composed of two homogeneous media having indices n_{co} and n_{cl} – though we specifically considered the case of a slab with flat surfaces separating the core from the cladding.

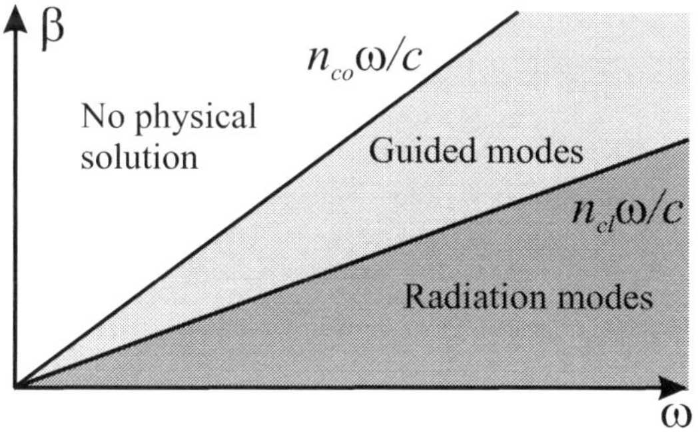

Fig. 2.7: The solutions of β according to the different modes of a symmetric slab.

2.3 Guided Modes Calculation

We shall now calculate the mode functions through the integration of the Helmholtz equation (2.3). The solution ϕ represents the component H_z for TE modes, and the component E_z for TM modes. Other components are obtained by means of (2.2). In order to simplify the analysis, we will introduce the following quantities:

$$k_x = \sqrt{\left(\frac{\omega}{c}\right)^2 n_{co}^2 - \beta^2};$$

$$\gamma = \sqrt{\beta^2 - \left(\frac{\omega}{c}\right)^2 n_{cl}^2}$$

where γ should not be mistaken for the mode propagation constant. As far as guided modes are concerned, k_x and γ prove to be totally real since $n_{cl}\omega/c < \beta < n_{co}\omega/c$. In addition, we suppose that

$$(k_x d)^2 + (\gamma d)^2 = v^2 \tag{2.6}$$

where

$$v = \frac{\omega}{c} d \sqrt{n_{co}^2 - n_{cl}^2} = 2\pi \frac{d}{\lambda} \sqrt{n_{co}^2 - n_{cl}^2} \tag{2.7}$$

is called *normalised frequency* and only depends on the operating wavelength and on the characteristics of the waveguide, *i.e.* core dimensions and refractive indices. That said, we can reinterpret (2.3) as follows:

$$\begin{cases} \dfrac{d^2\phi}{dx^2} - \gamma^2 \phi = 0 & \text{se} \quad x > d \\ \\ \dfrac{d^2\phi}{dx^2} + k_x^2 \phi = 0 & \text{se} \quad |x| \leq d \\ \\ \dfrac{d^2\phi}{dx^2} - \gamma^2 \phi = 0 & \text{se} \quad x < -d \end{cases} \tag{2.8}$$

its solution being:

$$\phi(x) = \begin{cases} b_1 e^{-\gamma x} + b_3 e^{\gamma x} & \text{se} \quad x > d \\ a_1 \sin(k_x x) + a_2 \cos(k_x x) & \text{se} \quad |x| \leq d \\ b_2 e^{\gamma x} + b_4 e^{-\gamma x} & \text{se} \quad x < -d \end{cases}$$

where the coefficients a_1, a_2, b_1, b_2, b_3 and b_4 must all be determined. Since the solution cannot physically diverge at infinity, we immediately obtain:

$$b_3 = b_4 = 0.$$

Hence, the solution also satisfies the Sommerfeld radiation conditions. As the general expression of the longitudinal components is known, we shall now determine the transverse components and impose the continuity conditions. To do so, let us tackle the TE and the TM modes separately.

2.3.1 *TE* Modes

In this case, $E_z = 0$ and $\phi(x) = H_z(x)$, therefore:

$$H_z(x) = \begin{cases} b_1 e^{-\gamma x} & \text{if} \quad x > d \\ a_1 \sin(k_x x) + a_2 \cos(k_x x) & \text{if} \quad |x| \leq d \\ b_2 e^{\gamma x} & \text{if} \quad x < -d \end{cases}$$

From (2.2) it follows that:

$$E_y(x) = j\eta_0 \frac{\omega}{c} \begin{cases} \frac{1}{\gamma} b_1 e^{-\gamma x} & \text{if} \quad x > d \\ \frac{1}{k_x} \left[a_1 \cos(k_x x) - a_2 \sin(k_x x)\right] & \text{if} \quad |x| \leq d \\ -\frac{1}{\gamma} b_2 e^{\gamma x} & \text{if} \quad x < -d \end{cases} \tag{2.9}$$

and finally:

$$H_x = -\frac{\beta c}{\eta_0 \omega} E_y. \tag{2.10}$$

It is now necessary to impose the continuity of the components tangential to the interfaces in order to determine the value of the integration constants a_1, a_2 and b_1, b_2. Before doing so, a further distinction is necessary: considering E_y as the only component of the electric field, we observe that this is defined by a linear combination of sines and cosines in the core – which are linearly independent functions. Therefore, TE modes can be further split into *even* and *odd TE modes* according to the symmetry of the trigonometric function defining the profile of E_y in the core. A similar distinction will be made for TM modes too.

Even TE Modes

In this case, $a_2 = 0$. The expressions of the components tangent to the interface become the following:

$$H_z(x) = \begin{cases} b_1 e^{-\gamma x} & \text{if} \quad x > d \\ a_1 \sin(k_x x) & \text{if} \quad |x| \leq d \\ b_2 e^{\gamma x} & \text{if} \quad x < -d \end{cases}$$

$$E_y(x) = j\eta_0 \frac{\omega}{c} \begin{cases} \frac{1}{\gamma} b_1 e^{-\gamma x} & \text{if} \quad x > d \\ a_1 \frac{1}{k_x} \cos(k_x x) & \text{if} \quad |x| \leq d \\ -\frac{1}{\gamma} b_2 e^{\gamma x} & \text{if} \quad x < -d \end{cases}$$

for H_z and E_y respectively. By imposing the continuity conditions at $x = d$, we obtain:

$$\begin{cases} a_1 \sin(k_x d) - b_1 e^{-\gamma d} = 0 \\ j a_1 \eta_0 \frac{k_0}{k_x} \cos(k_x d) - j\eta_0 \frac{k_0}{\gamma} b_1 e^{-\gamma d} = 0. \end{cases} \tag{2.11}$$

If the same conditions are applied at $x = -d$, we will have the same equations, except for b_1, substituted with $-b_2$; therefore:

$$b_2 = -b_1.$$

The system of equations (2.11) is a homogeneous system of two equations in the variables a_1 and b_1. If the two equations are linearly independent, *i.e.* the determinant of the system is non-zero, the solution is then $a_1 = b_1 = 0$ and it is both common and null. Such a solution has obviously no practical interest, and the only way to obtain non-zero solutions is to ensure that the determinant is null. With

$$Det \begin{vmatrix} \sin(k_x d) & -e^{-\gamma d} \\ \frac{1}{k_x} \cos(k_x d) & -\frac{1}{\gamma} e^{-\gamma d} \end{vmatrix} = 0,$$

the result is

$$\gamma = k_x \tan(k_x d)$$

which is the *characteristic equation* of TE modes. Basically, the values of γ and k_x – and therefore, the profile of the solution on the transverse plane – cannot be arbitrary, yet they are linked to the characteristic equation in order to satisfy all the continuity conditions of the components. As γ and k_x have to comply with (2.6) as well, their value can be calculated by solving the following system:

$$\begin{cases} \gamma = k_x \tan(k_x d) \\ \gamma^2 + k_x^2 = \left(\frac{\omega}{c}\right)^2 \left(n_{co}^2 - n_{cl}^2\right). \end{cases} \tag{2.12}$$

Since the system is composed of transcendental equations, analytic solutions are not an option. Therefore, we shall continue graphically. Figure 2.8 shows the trends of the two equations in the (k_x, γ) plane. As both k_x and γ are positive quantities, we will restrict our analysis to the first quadrant. The equation (2.6) describes the arc of a circle of radius $(\omega/c)\sqrt{n_{co}^2 - n_{cl}^2}$, while the characteristic equation has vertical asymptotes in $k_x = (2m+1)\pi/2d$ with integer m. As we can see, there are generally different intersections between the curves, *i.e.* several solutions are accepted for the system (2.12). By using γ_ν and $k_{x\nu}$ to indicate the values of the νth solution of the (2.12) system, the value of the phase constant of the corresponding mode is

$$\beta_\nu = \sqrt{\left(\frac{\omega}{c}\right)^2 n_{co}^2 - k_{x\nu}^2}.$$

As we have already said, the determinant has to be null to have non-zero solutions; moreover, this reduces the number of equations with respect to the variables. Consequently, the (2.11) system is reduced to a two-variable equation:

$$a_1 \sin(k_{x\nu} d) = b_1 e^{-\gamma_\nu d} = A \neq 0,$$

whose solution

$$\begin{aligned} a_1 &= \frac{A}{\sin(k_{x\nu} d)} \\ b_1 &= \frac{A}{e^{-\gamma_\nu d}} \\ b_2 &= -\frac{A}{e^{-\gamma_\nu d}}; \end{aligned}$$

is not unique, but can be expressed as a function of the parameter A – its value being still arbitrary. This last degree of freedom allows us to normalise the solution, that is to say, the power carried by the mode:

$$\int_S \frac{\overline{E} \times \overline{H}^*}{2} \cdot \hat{z}\, dS = 1. \tag{2.13}$$

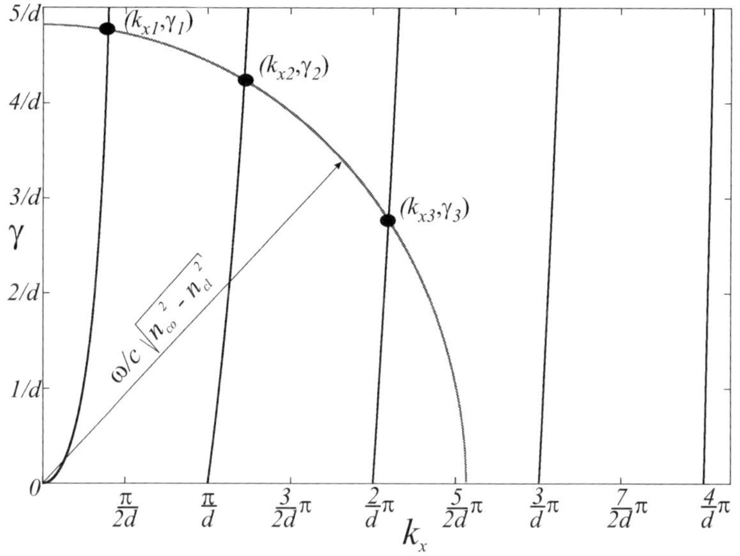

Fig. 2.8: Graphical representation of the (2.12) system. Intersections represent the solutions to the system.

In our case, only E_y, H_x and H_z are non-zero, thus – by resorting to (2.10) and (2.12) – the expression (2.13) becomes:

$$\begin{aligned}
&\int_{-\infty}^{\infty} \frac{\overline{E}\times\overline{H}^*}{2}\cdot\hat{z}\,dx = -\frac{1}{2}\int_{-\infty}^{\infty} E_y H_x^*\,dx = \frac{1}{2}\frac{\beta c}{\eta_0\omega}\int_{-\infty}^{\infty}|E_y|^2\,dx \\
&= \frac{1}{2}\beta\eta_0\frac{\omega}{c}A^2\left[\int_{-\infty}^{-d}\frac{e^{2\gamma_\nu(x+d)}}{\gamma_\nu^2}\,dx + \int_{-d}^{d}\frac{\cos^2(K_{x\nu}x)}{k_{x\nu}^2\sin^2(k_{x\nu}d)}\,dx + \int_{d}^{\infty}\frac{e^{-2\gamma_\nu(x-d)}}{\gamma_\nu^2}\,dx\right] \\
&= \frac{1}{2}\beta\eta_0\frac{\omega}{c}A^2\left[\frac{1}{\gamma_\nu^3} + \frac{1}{k_{x\nu}^2}\left(\frac{d}{\sin^2(k_{x\nu}d)} + \frac{1}{\gamma_\nu}\right)\right] = 1.
\end{aligned}$$

From this we can obtain:

$$A = \frac{\sqrt{2}}{\sqrt{\beta\eta_0\frac{\omega}{c}\left[\frac{1}{\gamma_\nu^3} + \frac{1}{k_{x\nu}^2}\left(\frac{d}{\sin^2(k_{x\nu}d)} + \frac{1}{\gamma_\nu}\right)\right]}}.$$

Odd TE Modes

For E_y to have odd symmetry with respect to the centre of the core, we must have $a_1 = 0$. By imposing the continuity conditions of the tangent components H_z and E_y we obtain:

$$\begin{cases} a_2 \cos(k_x d) = b_1 e^{-\gamma d} \\ -a_2 \frac{1}{k_x} \sin(k_x d) = \frac{1}{\gamma} b_1 e^{-\gamma d}. \end{cases}$$

If the determinant is null, we have the characteristic equation of odd TE modes:

$$\gamma = -k_x \cot(k_x d).$$

We can determine the values of the constants of integration by adopting a methodological approach similar to the one proposed for even TE modes:

$$\begin{aligned} a_2 &= \frac{A}{\cos(k_{x\nu} d)} \\ b_1 &= \frac{A}{e^{-\gamma_\nu d}} \\ b_2 &= -\frac{A}{e^{-\gamma_\nu d}}. \end{aligned}$$

with

$$A = \frac{\sqrt{2}}{\sqrt{\beta \eta_0 \frac{\omega}{c} \left[\frac{1}{\gamma_\nu^3} + \frac{1}{k_{x\nu}^2} \left(\frac{d}{\cos^2(k_{x\nu} d)} + \frac{1}{\gamma_\nu} \right) \right]}}.$$

2.3.2 TM Modes

It is possible to handle the solution of TM modes by following almost the same method used for TE modes. For this reason, the detailed calculations will be omitted and we shall only highlight the most significant results. Since $H_z = 0$ and $\phi(x) = E_z(x)$, we have:

$$E_z(x) = \begin{cases} b_1 e^{-\gamma z} & \text{if } x > d \\ a_1 \sin(k_x x) + a_2 \cos(k_x x) & \text{if } |x| \le d \\ b_2 e^{\gamma z} & \text{if } x < -d. \end{cases}$$

From (2.2) it follows that:

$$H_y(x) = -j \frac{n_{cl}^2}{\eta_0} \frac{\omega}{c} \begin{cases} \frac{1}{\gamma} b_1 e^{-\gamma z} & \text{if } x > d \\ \frac{n_{co}^2}{n_{cl}^2} \frac{1}{k_x} (a_1 \cos(k_x x) - a_2 \sin(k_x x)) & \text{if } |x| \le d \\ -\frac{1}{\gamma} b_2 e^{\gamma z} & \text{if } x < -d \end{cases} \tag{2.14}$$

and finally:

$$E_x = \frac{\beta \eta_0 c}{\omega} \begin{cases} \frac{1}{n_{co}^2} H_y & \text{if } |x| \le d \\ \frac{1}{n_{cl}^2} H_y & \text{if } |x| > d \end{cases}$$

According to the symmetry of H_y, the solutions can be either even or odd, but – contrary to what has been said for TE modes earlier – the multiplicative coefficients also depend on the refractive index of the medium; hence, the characteristic equations are basically the same as before, except for the multiplicative factor $(n_{cl}/n_{co})^2 < 1$:

$$\gamma = \frac{n_{cl}^2}{n_{co}^2} k_x \tan(k_x d) \quad \text{for even TM modes}$$

$$\gamma = -\frac{n_{cl}^2}{n_{co}^2} k_x \cot(k_x d) \quad \text{for odd TM modes}$$

By imposing the normalisation condition on the power, it then becomes possible to calculate the multiplicative constant which defines the amplitude of the mode function taken into account.

2.4 Cutoff Conditions

It is possible to define some important properties of the guided modes from the characteristic equations. Figure 2.9 shows the trend of the characteristic equations and of (2.6) in the $(k_x d, \gamma d)$ plane. The latter is the arc of a circle of radius v centred at the origin, and the points indicate the intersections between (2.9) and the characteristic curves – *i.e.* the pairs k_x, γ – corresponding to the various guided modes.

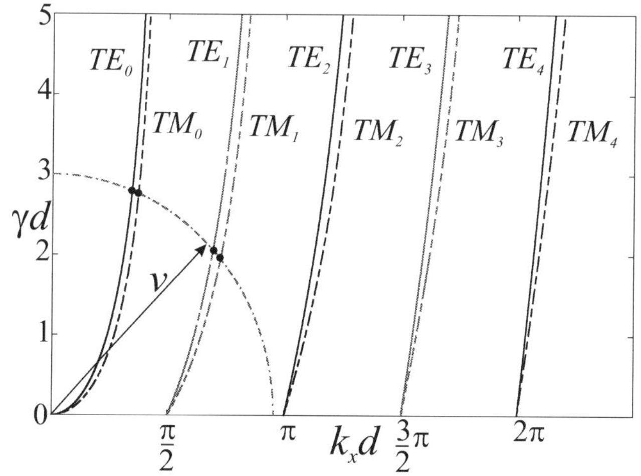

Fig. 2.9: The trend of TE (solid lines) and TM (dashed lines) modes in the $(k_x d$, γd plane. The dispersion curves of even modes are shown in black, those of odd modes in grey. The dash-dot line indicates the equation (2.6).

The set of solutions is discrete and countable. For each TE and TM polarisation, the modes are numbered according to the integer index ν, which is zero for the solution with the lowest k_x. As the normalised frequency decreases, the intersections are shifted towards lower values of γ: hence, the field decays more slowly in the cladding and is less confined. When $\gamma = 0$, the mode is no longer confined and this is called the cutoff frequency of the waveguide. If the normalised frequency is further reduced, the mode can no longer exist as the continuity conditions are no longer satisfied. This is a considerable difference with metallic waveguides, where the mode – without being able to carry the power at a distance – continues to exist even in cutoff conditions. Another major difference between dielectric and metallic waveguides lies in the cutoff frequency of the fundamental mode, which can be equal to zero in the former, while, in the latter, this is only possible when it comes to transverse electromagnetic (TEM) modes – never for TE or TM modes. The slab cannot be strictly single-mode because there are always two modes, the TE_0 and the TM_0 modes with cutoff frequency equal to zero. The same happens in optical fibers too, where the two modes, at least ideally, are degenerate – that is to say, they have the same phase constant. On the contrary, TE_0 and TM_0 always have different phase constants in slabs, and this is due to the factor $(n_{cl}/n_{co})^2 < 1$ in the characteristic equation of TM modes, which reduces the value of the phase constant: the higher the step index between the core and the cladding, the greater the difference.

Thanks to the value of the normalised frequency, it then becomes possible to define the number of modes above cutoff. The cutoff frequency of the νth mode is:

$$v_{c\nu} = \nu \frac{\pi}{2}.$$

If we consider we can both have a TE and a TM mode, the number N of modes that can propagate is:

$$N = 2\,\text{INT}(\frac{v}{\pi/2}).$$

Even if improper in the case of a slab, the single-mode condition turns out to be:

$$\lambda > 2d\sqrt{n_{co}^2 - n_{cl}^2}.$$

Once λ is known, if the product $d\sqrt{n_{co}^2 - n_{cl}^2}$ is too high, the single-mode regime is lost. It is not possible to reduce the dimensions of the core too much in order to avoid imposing overly strict conditions of coupling optimisation between the source and the waveguide: this is why it is necessary to work on low step-index waveguides.

2.5 Mode Functions Profiles

Figure 2.10 shows the profiles of the component E_y of the first four TE modes on the transverse plane. As a result of the interference caused by two uniform plane waves, the functions have a sinusoidal profile inside the core, yet they decay exponentially in the cladding – which is typical of evanescent plane waves. As we can see in figure 2.9, $k_x < \pi/(2d)$ for TE_0: this means that the function $\cos(k_x x)$, defining the profile of E_y inside the core, is never null. For TE_1, $\pi/(2d) < k_x < \pi/d$, therefore in $\sin(k_x x)$ there is only one zero-crossing point in the core. A further analysis shows that the value of ν is equal to the number of zero-crossing points of E_y for TE modes, and of H_y for TM modes. In order to indicate the trend of the electric field in time, we need to multiply the mode function by the propagation factor, and pass from phasors to real vectors:

$$\overline{\mathcal{E}}(x, z, t) = \mathcal{E}_y(x, z, t)\hat{\jmath} = E_y(x)\cos(\omega_0 t - \beta z)\hat{\jmath}.$$

The trend of the component $\mathcal{E}_y$ of the TE_0 mode in the x, z plane is presented in figure 2.11.

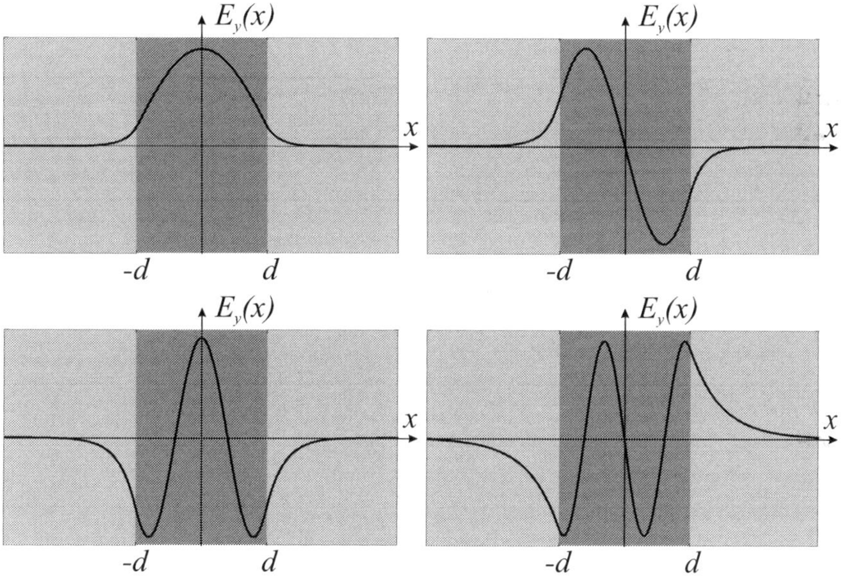

Fig. 2.10: From left to right: the profile of the component E_y of the modes TE_0, TE_1 (top), and TE_2, TE_3 (bottom) of a slab. The darker sections represent the core, surrounded by the cladding.

2.5.1 Field Confinement

As already stated, the constant γ determines the field decay rate in the cladding: the higher the value of γ, the more confined the mode is in the core. With reference to figure 2.9, for a particular mode, field confinement becomes stronger as the normalised frequency ν increases; according to (2.7), the index difference, the core dimensions and the wavelength are other important parameters: field confinement, in fact, becomes stronger as the first two grow or as the latter is reduced. We shall analyse the profile of the component E_y of the TE_0 mode by considering two different wavelengths,

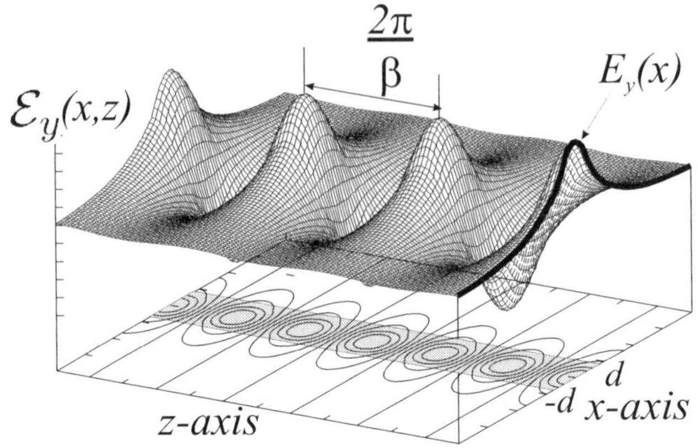

Fig. 2.11: The trend of the component $\mathcal{E}_y$ of the TE_0 mode of a slab with $t = 0$. The core-cladding interfaces correspond to the planes $x = d$ and $x = -d$. The z-axis plots three periods of the wave. Transverse mode function in bold.

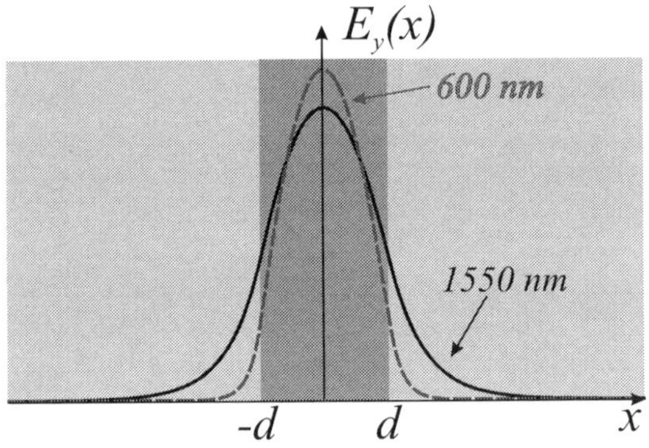

Fig. 2.12: The profile of the component E_y of the TE_0 mode at different wavelengths: $\lambda = 600nm$ and $\lambda = 1550nm$.

as reported in figure 2.12. The higher the wavelength, the closer the normalised frequency is to cutoff conditions and the less confined the mode is.

If we compare different modes at the same normalised frequency, the confinement becomes weaker as the mode order increases (cf. figure 2.10). Hence, after determining the operating frequency, the fundamental mode will always be the one with the highest level of confinement.

2.6 Dispersion Curves

If we consider figure 2.9 once more, we notice that the variation of the angular frequency ω engenders the displacement of the point of intersection between the circumference and the characteristic curves, which involves the variation of the phase constant β, as

$$\beta = \sqrt{\left(n_{co}\frac{\omega}{c}\right)^2 - k_x^2} = \sqrt{\left(n_{cl}\frac{\omega}{c}\right)^2 + \gamma^2}. \tag{2.15}$$

Figure 2.13 shows the trend of β as a function of the angular frequency $\omega = 2\pi\nu$. In this respect, we refer to a *dispersion curve* because – as we shall see in chapter 14 – the non-linear dependence of β from ω causes the so-called *time dispersion*, *i.e.* an increase in the duration of electromagnetic pulses propagating in the waveguide. Note that, in the case of cutoff, $\gamma = 0$, and therefore $\beta = n_{cl}k_0$. As the angular frequency increases, k_x tends to a finite value; hence, according to (2.15):

$$\lim_{\omega\to\infty} \beta(\omega) = n_{co}\frac{\omega}{c}.$$

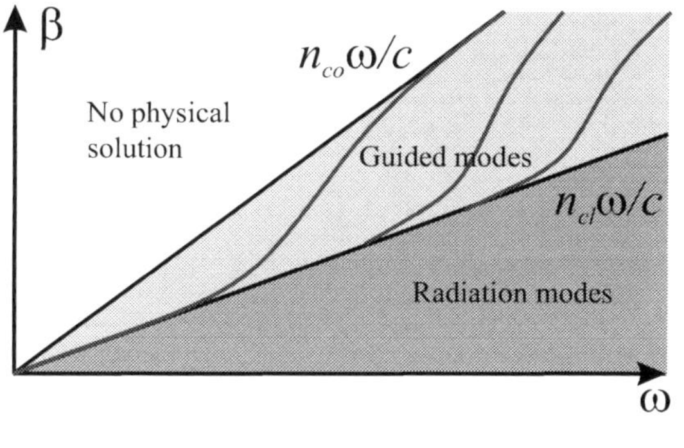

Fig. 2.13: The dispersion curve of the first three TE modes of a slab as a function of the angular frequency ω.

2.6.1 High Confinement and Paraxial Approximation

When the value of the normalised frequency and of the guided modes is particularly high, it is possible to obtain an approximate expression of β which will be particularly useful later on when we analyse important components such as MMI (Multi Mode Interference) couplers and laser cavities.

Let us return to figure 2.9 and observe that, when $v >> 1$, the following relations apply to the νth mode – be it TE or TM:

$$\begin{aligned}\gamma_\nu &>> 1/d;\\ k_{x_\nu} &<< n_{co}\frac{2\pi}{\lambda};\\ k_{x_\nu} &\simeq \frac{\pi}{2}\frac{\nu+1}{d}.\end{aligned}$$

From the first inequality we can infer that, in these conditions, the mode propagates in the cladding over a negligible distance. The second inequality and the last equation are a result of the fact that, in correspondence to the points $k_x d = (\nu+1)\pi/2$, the dispersion curves have vertical asymptotes. This said – and without forgetting that $\sqrt{1-x^2} \simeq 1 - x^2/2$ if $x << 1$ – it follows that:

$$\begin{aligned}\beta_\nu &= \sqrt{\left(n_{co}\frac{2\pi}{\lambda}\right)^2 - k_{x_\nu}^2}\\ &= n_{co}\frac{2\pi}{\lambda}\sqrt{1-\left(\frac{k_{x_\nu}}{2\pi n_{co}}\lambda\right)^2}\\ &\simeq n_{co}\frac{2\pi}{\lambda}\left[1-\frac{1}{2}\left(k_{x_\nu}\frac{\lambda}{2\pi n_{co}}\right)^2\right]\\ &= n_{co}\frac{2\pi}{\lambda} - (\nu+1)^2\frac{\pi\lambda}{4n_{co}(2d)^2}.\end{aligned} \tag{2.16}$$

2.6.2 The Effective Index

Rather than the phase constant, the effective index n_{eff} is preferably employed when it comes to dielectric waveguides. We define n_{eff} as follows:

$$n_{eff} = \frac{\beta}{k_0}.$$

Since

$$\beta = n_{eff}k_0$$

as far as the phase characteristics are concerned, it is as if the field were propagated in a homogeneous dielectric medium with refractive index n_{eff}. According to what has been said before, the following must be true for the guided modes:

$$n_{cl} < n_{eff} < n_{co};$$

as for the radiation modes, on the other hand:

$$n_{eff} < n_{cl}.$$

The dispersion curves often account for the trend of the effective index as a function of the wavelength. Figure 2.14 shows the curves of the first three TE and TM modes of a slab. The condition

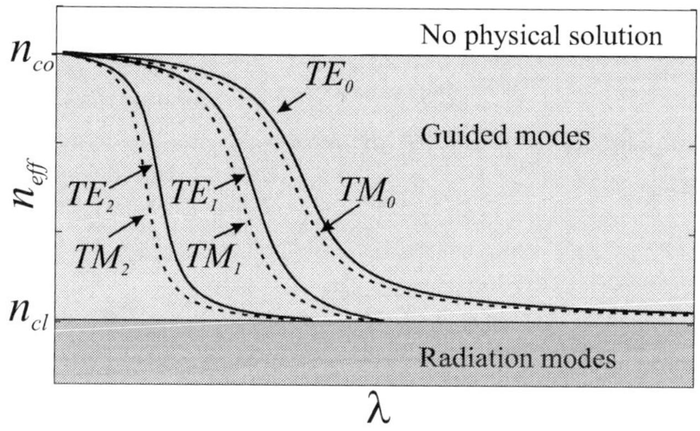

Fig. 2.14: The trend of the effective index of the first three modes of a slab as a function of the wavelength λ. The solid lines describe the TE modes, while TM modes are indicated by dashed lines.

$n_{eff} = n_{cl}$ determines the mode cutoff wavelength, while $n_{eff} = n_{co}$ in $\lambda = 0$, since it corresponds to the following condition:

$$\lim_{\omega \to \infty} \beta(\omega) = n_{co}\frac{\omega}{c}.$$

The effective index is useful to determine how close the cutoff condition can be and to retrieve valuable information on the level of field confinement. In fact, we can prove that the effective index is basically a weighted average of the core and cladding indices, whose weights depend on the power propagated in the two regions. When the mode is close to the cutoff wavelength, a considerable portion of the field is in the cladding, while the part in the core is negligible, and thus $n_{eff} \simeq n_{cl}$. As λ becomes smaller then, the field is mainly confined in the core and, consequently, the effective index becomes higher until it reaches the value of the core for $\lambda = 0$.

2.7 Radiation Modes

If $\beta < n_{cl}\omega/c$, the solution is represented by plane waves not only in the core, but in the cladding too. Instead of guided modes, with the field being confined along the core of the waveguide, in this case the distribution of the field is completely delocalised, thus we talk about *radiation modes*, a peculiar characteristic of open waveguides – dielectric waveguides being a particular case of open waveguides. No transmission at a distance is possible with radiation modes because the power they carry tends to progressively diverge from the core and, due to its finite extension on the transverse plane, the receiver is not able to utilize it.

It is possible to apply the same mathematical processes used for guided modes in order to calculate the mode functions, the characteristic equations and the dispersion curves of radiation modes, yet with a substantial difference: in this case, it is not possible to reduce the number of coefficients by imposing the regularity conditions at infinity. If we consider the TE modes, for example, the sole component of the electric field is:

$$E_y(x) = \begin{cases} j\eta_0 \frac{k_0}{k_{xcl}} \left(b_1 \sin(k_{xcl}x) + b_3 \cos(k_{xcl}x)\right) & \text{if} \quad x > d \\ j\eta_0 \frac{k_0}{k_x} \left(a_1 \sin(k_x x) + a_2 \cos(k_x x)\right) & \text{if} \quad |x| \leq d \\ j\eta_0 \frac{k_0}{k_{xcl}} \left(b_2 \sin(k_{xcl}x) + b_4 \cos(k_{xcl}x)\right) & \text{if} \quad x < -d \end{cases}$$

where

$$k_{xcl} = \sqrt{\left(n_{cl}\frac{\omega}{c}\right)^2 - \beta^2}$$

is a real quantity. As no function diverges to infinity, no coefficient can be null. Nevertheless, it is still possible to determine whether the solutions are even or odd by alternately imposing $a_1 = 0$ and $a_2 = 0$. In both cases, however, if we impose the continuity conditions at $x = \pm d$, the result is a homogeneous system of four equations in five unknowns with non-zero solutions, with the determinant non necessarily equal to zero. The same discourse applies to TM modes. The phase constant

$$\beta = \sqrt{\left(n_{cl}\frac{\omega}{c}\right)^2 - k_{xcl}^2}$$

may vary continuously as k_{xcl} need not satisfy any extra condition. Hence, radiation modes constitute a *continuous spectrum* of solutions – which is instead discrete in the case of guided modes.

2.8 Reactive Modes

We have a purely imaginary β when $k_{xcl} > n_{cl}\omega/c$. The corresponding modes are called *reactive modes* and the propagation constant is completely real, that is to say $\gamma = \alpha$. For this reason, the solution is constituted by evanescent plane waves propagating in a direction which is transverse to the attenuation vector oriented towards the z-axis:

$$\overline{a} = \alpha\hat{z}.$$

It is evident that such a particular solution cannot be propagated, yet it is often useful to describe the configuration of the field in case of a perturbed waveguide, or at the input end of the waveguide itself. However, the solution seen above will be neglected in the course of our work to avoid weighing down the text excessively.

2.9 Mode Completeness and Coupling Loss

Figure 2.15 shows the possible values that the effective index n_{eff} of the different modes of a symmetric slab can have. Once the operating wavelength $\overline{\lambda}$ is known, there exists a finite number N

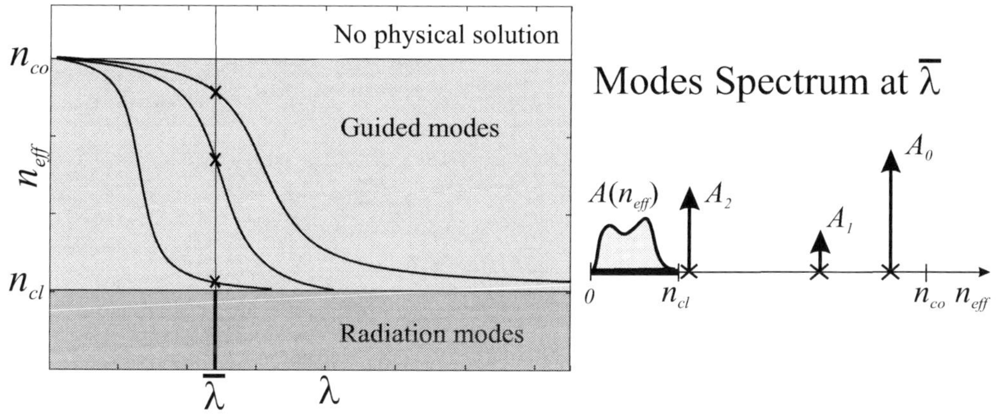

Fig. 2.15: The values of n_{eff} and the series expansion coefficients of the different modes contributing to the overall field representation.

of guided modes and a continuum of radiation modes. These modes form a complete set, which is a *basis* for the solutions to Maxwell's equations in the dielectric waveguide. This is valid along the whole waveguide, and particularly in the input area, where the transverse component of the input field $\overline{e}_{t_0}$ – often known as *exciting field* – can be written in the following form:

$$\overline{e}_{t_0}(x) = \sum_{\nu=0}^{N-1} A_\nu \overline{E}_{t_\nu}(x) + \int_0^{n_{cl}} A(n_{eff})\overline{E}(n_{eff}, x) dn_{eff},$$

where A_ν, $A(n_{eff})$ are the amplitudes of the mode functions $\overline{E}_{t_\nu}(x)$, $\overline{E}_t(n_{eff}, x)$ – of the N guided modes and infinite radiation modes respectively – to contribute to the expansion. As for the TE modes, for example, with reference to (1.5) and (2.10), we obtain:

$$\begin{aligned} A_\nu &= \frac{\int_{-\infty}^{\infty} \left(\overline{e}_{t_0} \times \overline{H}^*_{t_\nu}\right) \cdot \hat{z} dx}{\int_{-\infty}^{\infty} \left(\overline{E}_{t_\nu} \times \overline{H}^*_{t_\nu}\right) \cdot \hat{z} dx} \\ &= \frac{\int_{-\infty}^{\infty} -e_{y_0} H^*_{x_\nu} dx}{\int_{-\infty}^{\infty} -E_{y_\nu} H^*_{x_\nu} dx} \\ &= \frac{\int_{-\infty}^{\infty} e_{y_0} E_{y_\nu} dx}{\int_{-\infty}^{\infty} E^2_{y_\nu} dx}. \end{aligned} \qquad (2.17)$$

Note that, according to (2.17) and since

$$E_\nu(-y) = \begin{cases} E_\nu(y) & \text{if} \quad \nu \quad \text{is even} \\ -E_\nu(y) & \text{if} \quad \nu \quad \text{is odd,} \end{cases}$$

we can assert that, if the exciting field has an even symmetry with respect to the waveguide axis, then $A_\nu = 0$ for odd ν; conversely, if the exciting field has odd symmetry, $A_\nu = 0$ for even ν. In both cases, in fact, the numerator of (2.17) would be the product of an even function and an odd function whose integral from $-\infty$ to ∞ is evidently null.

By virtue of modes orthogonality, the input power is

$$P_0 = \sum_{\nu=0}^{N-1} |A_\nu|^2 + \int_0^{n_{cl}} |A(n_{eff})|^2 dn_{eff}.$$

The quantity

$$\Delta P = \int_0^{n_{cl}} |A(n_{eff})|^2 dn_{eff}$$

represents the power lost due to coupling, *i.e.* the amount of power carried by radiation modes and therefore useless on the receiving end. To minimise ΔP, it is important that the distribution of the exciting field is as similar as possible to that of a mode or combination of guided modes above cutoff. This is why the exciting field in single-mode waveguides must necessarily be consistent with the fundamental mode. Figure 2.16 presents the distribution of the field along a single-mode slab excited with three different field distributions. In the first case, the distribution is identical to that of the fundamental mode, and the input power is completely transferred to the fundamental mode propagated with unaltered distribution. In the second case, the field is different from the fundamental mode: the contribution of the radiation modes is evident and, as z increases, the field

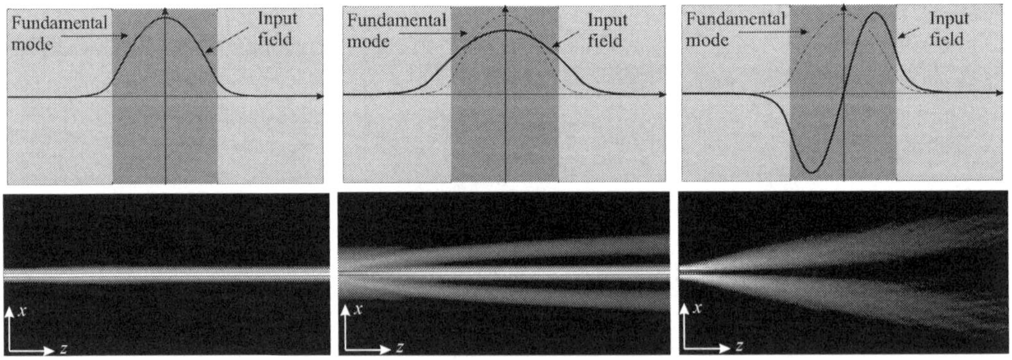

Fig. 2.16: The trend of the field in a single-mode slab according to three different exciting field. The field is more intense in the light-coloured sections.

tends to move away from the core to such an extent that the receiver cannot utilize it. In the last example, the waveguide is excited by a field which is symmetrically different to the fundamental mode: in respect to the waveguide axis, the fundamental mode has an even symmetry, while the exciting field has an odd one. Hence, the fundamental mode is not excited and the input power is completely lost.

2.10 Birefringence

In figure 2.17 we can see the dispersion curves of the TE_0 and TM_0 modes in detail. Due to the different characteristic equations, the two curves do not overlap – that is to say, the modes are not degenerate. The gap between the two curves is called *birefringence* and it determines the different propagation characteristics of the two fundamental modes:

$$B = \left| n^x_{eff} - n^y_{eff} \right| \tag{2.18}$$

n^x_{eff} and n^y_{eff} being the effective indices of the TE_0 and TM_0 modes respectively.

Since the different continuity conditions are at the heart of birefringence, its value depends on the core geometry, on the wavelength and on the refractive indices of the materials. In this respect, we refer to *geometrical birefringence* – not to be confused with *intrinsic birefringence*, which is linked to the anisotropy of the materials chosen. Figure 2.18 displays the spectral trend of birefringence as the index difference between the core and the cladding varies; as expected, all other parameters being equal, the value of birefringence tends to diminish as the index difference decreases. Let us remember that $n_{cl} < n_{eff} < n_{co}$, hence:

$$0 \leq B \leq n_{co} - n_{cl}.$$

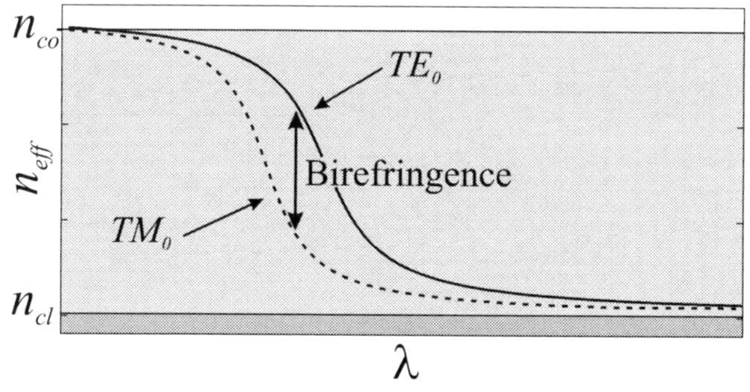

Fig. 2.17: The dispersion curves of the two fundamental modes of a slab.

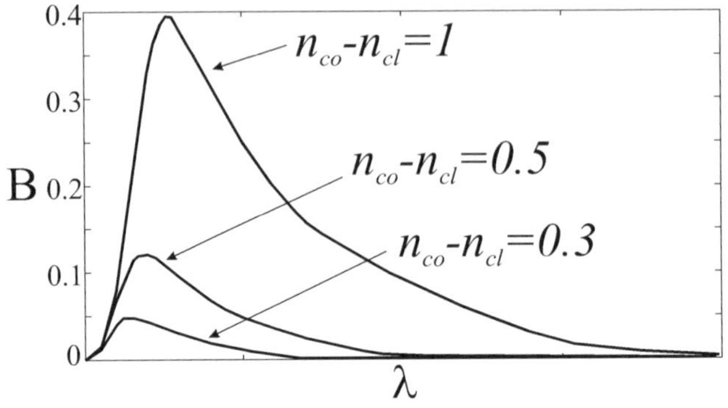

Fig. 2.18: The spectral trend of birefringence according to different index differences in a slab.

The trends for λ tending to zero and to infinity are not due to the characteristics of the waveguide, yet they depend on the general properties of the effective indices of the guided modes of dielectric waveguides. For a generic dielectric waveguide $n_{eff}(\lambda = 0) = n_{co}$ and $\lim_{\lambda \to \infty} n_{eff}(\lambda) = n_{cl}$; in fact, we have:

$$B(\lambda = 0) = 0, \qquad \lim_{\lambda \to \infty} B(\lambda) = 0.$$

Birefringence is present in almost all integrated waveguides and it is an important parameter to evaluate the quality of a waveguide, or of a component more generally. The importance lies in the fact that the degree of polarisation of the field coming out of a fiber is not known *a priori* and, furthermore, it varies randomly as time and frequency change; for this reason, it is necessary for the propagation characteristics of the two waveguide modes to be as similar as possible in order to avoid dramatic temporal and spectral variations in terms of characteristics of the waveguide.

2.11 Anisotropic Slabs

As we have seen in the previous paragraph, birefringence may also be induced by the characteristics of the materials constituting the waveguide. To delve deeper into this aspect, we shall solve Maxwell's equations in an anisotropic medium, described by the tensor $\hat{\varepsilon}$, with different values of dielectric permittivity along the three orthogonal space directions:

$$\hat{\varepsilon} = \begin{bmatrix} \varepsilon_x & 0 & 0 \\ 0 & \varepsilon_y & 0 \\ 0 & 0 & \varepsilon_z \end{bmatrix} = \varepsilon_0 \begin{bmatrix} \varepsilon_{r_x} & 0 & 0 \\ 0 & \varepsilon_{r_y} & 0 \\ 0 & 0 & \varepsilon_{r_z} \end{bmatrix}.$$

By substituting Maxwell's second equation in the first, we obtain the *curl-curl* equation in the electric field $\overline{e}$:

$$\overline{\nabla} \times \overline{\nabla} \times \overline{e} - \omega^2 \mu \hat{\varepsilon} \overline{e} = 0. \tag{2.19}$$

We shall consider a flat structure invariant with respect to y, for which we assume $\partial/\partial y = 0$, and look for modal solutions such as:

$$\bar{e}(x) = \overline{E}(x)e^{-j\beta z},$$

where the mode function only depends on the variable x, and the exponential limits the dependence on z. From this we have:

$$\overline{\nabla} \times \bar{e} = \left[j\beta E_y \hat{\imath} + (-j\beta E_x - \frac{\partial E_z}{\partial x})\hat{\jmath} + \frac{\partial E_y}{\partial x}\hat{z} \right] e^{-j\beta z}$$

and

$$\overline{\nabla} \times \overline{\nabla} \times \bar{e} = \left[\left(\beta^2 E_x - j\beta \frac{\partial E_z}{\partial x} \right) \hat{\imath} - \left(\frac{\partial^2 E_y}{\partial x^2} - \beta^2 E_y \right) \hat{\jmath} - \left(j\beta \frac{\partial E_x}{\partial x} + \frac{\partial^2 E_z}{\partial x^2} \right) \hat{z} \right] e^{-j\beta z}.$$

If we substitute the previous equation in (2.19) and project it along the three Cartesian axes, we have:

$$\begin{aligned} j\beta \frac{\partial E_z}{\partial x} + (\omega^2 \mu \varepsilon_x - \beta^2) E_x &= 0, \\ \frac{\partial^2 E_y}{\partial x^2} + (\omega^2 \mu \varepsilon_y - \beta^2) E_y &= 0, \\ \frac{\partial^2 E_z}{\partial x^2} + j\beta \frac{\partial E_x}{\partial x} + \omega^2 \mu \varepsilon_z E_z &= 0. \end{aligned} \tag{2.20}$$

In order to determine the field profiles and the relative value of the constant β, it is necessary to solve the previous system as many times as the number of homogeneous regions forming the slab; secondly, we must impose the continuity conditions as usual to obtain, and then solve, the characteristic equation. As the reader is left with the task of developing what has just been explained, for the time being we shall focus on other aspects of particular interest. First and foremost, it is worth noting that both the variables E_x and E_z are present in the first and third equation of the system (2.20), while the second equation can only be solved in E_y: for this reason, the two coupled equations describe TM modes, while the second describes TE modes. Secondly, it is worth pointing out that the TM modes rest on ε_x and ε_z, while the TE modes only depend on the value of permittivity along y. It follows that, by using anisotropic materials, the TE and TM modes can be propagated in geometrically identical waveguides which are perceived as different due to contrasting dielectric permittivity values. As a consequence, the dispersion curves of β will necessarily be different and will be responsible for the presence of birefringent phenomena.

Consider *e.g.* a slab made of a medium with anisotropic core – characterised by elements of the dielectric permittivity tensor $\varepsilon_x = \varepsilon_z$ and ε_y – and isotropic cladding. As we mention the relationship between dielectric permittivity and refractive index n,

$$\hat{n}^2 = \begin{bmatrix} n_x^2 & 0 & 0 \\ 0 & n_y^2 & 0 \\ 0 & 0 & n_z^2 \end{bmatrix} = \hat{\varepsilon}/\varepsilon_0 = \begin{bmatrix} \varepsilon_{r_x} & 0 & 0 \\ 0 & \varepsilon_{r_y} & 0 \\ 0 & 0 & \varepsilon_{r_z} \end{bmatrix},$$

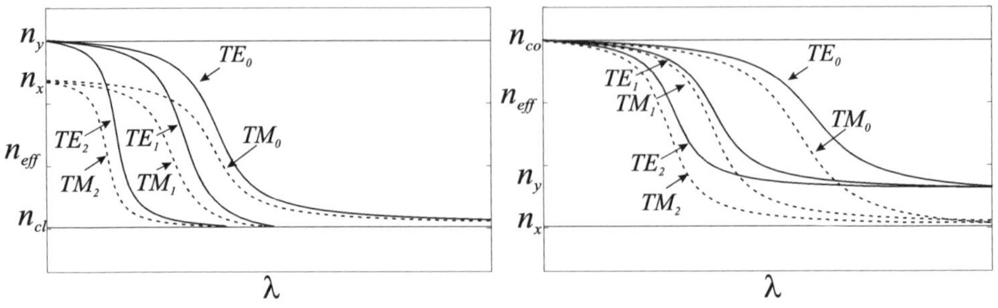

Fig. 2.19: The dispersion curves of the TE and TM modes of a slab: on the left, a slab with anisotropic core – characterised by elements of the dielectric permittivity tensor $n_x = n_z < n_y$ – and isotropic cladding; on the right, the slab has isotropic core and anisotropic cladding, defined by elements of the dielectric permittivity tensor $n_x = n_z < n_y$.

we can observe the qualitative trend of the dispersion curves as reported in figure 2.19(a). As the frequency increases, in fact, the value of the effective index always tends to the value of the core refractive index, which is different here for TE and TM modes. If the cladding is anisotropic then, the dispersion curves diverge and tend to the inferior limit, which corresponds to the value of the refractive index of the cladding – cf. figure 2.19(b).

2.12 Asymmetric Slabs

Although we have focused our attention so far on the symmetric slab, the use of asymmetrical slabs is nonetheless much more common in practical terms. This type of waveguide has two claddings with different refractive indices due to the fact that, in integrated optics, the waveguides are fabricated starting from a homogeneous substrate with refractive index n_{sub}, above which we find the core – produced either by diffusion, deposition or epitaxial growth – with index n_{co}. A protective layer around the core is added only in special cases and, most of the time, the structure is the one depicted in figure 2.20, where the upper cladding layer is air with $n_{sup} = 1$. The same mathematical process used for the symmetric slab can be applied in this case as well and, since the equations (2.2) are still valid, once again we can classify the solutions obtained either as TE or TM modes. However, a further differentiation between even and odd modes is not possible this time due to the asymmetrical properties of the slab.

Figure 2.21 accounts for the different modes of an asymmetric slab according to the solutions of β. When $\beta > n_{co}\omega/c$, no physical solution is possible and no mode exists since evanescent waves would be present in all three regions – which does not satisfy the continuity conditions. If $n_{sub}\omega/c \leq \beta \leq n_{co}\omega/c$, we have guided modes as there are uniform plane waves in the core and

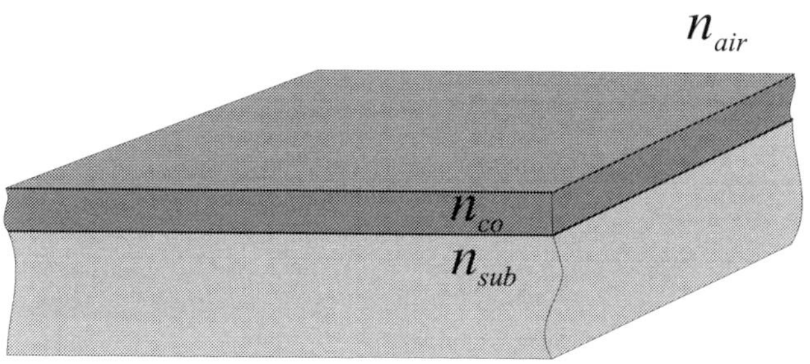

Fig. 2.20: Asymmetric Slab Waveguide. The darker central layer represents the core, the light-coloured region is the substrate.

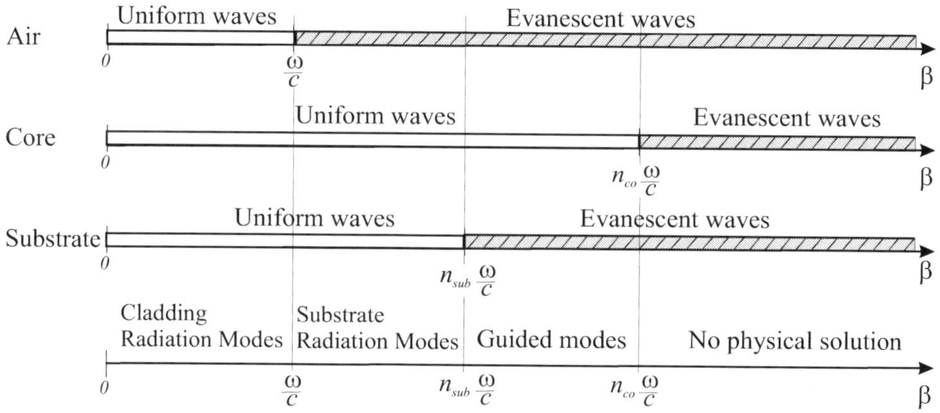

Fig. 2.21: The solutions of β according to the different modes of an asymmetric slab.

evanescent waves in both the substrate and the air cladding. With $\omega/c \leq \beta \leq n_{sub}\omega/c$, we refer to *substrate modes* because the mode is no longer guided as the waves in the substrate become uniform, while they are still evanescent in the air region. Finally, when $\beta < \omega/c$, the waves are uniform in all three layers and the situation corresponds to *cladding radiation modes*.

2.12.1 Mode Functions

Figure 2.22 shows the profile of the component E_y of the fundamental mode TE_0 of an asymmetric slab. Note that, due to the higher index difference, E_y decays more rapidly in the air than in the substrate. For the same reason, the maximum of the function is not situated in a symmetrical position, yet it is shifted more towards the substrate region. The same figure also reports the qualitative profile of a substrate mode: an exponential decay typical of an evanescent wave is present in the air cladding, while the presence of uniform waves in the substrate leads to an oscillatory trend.

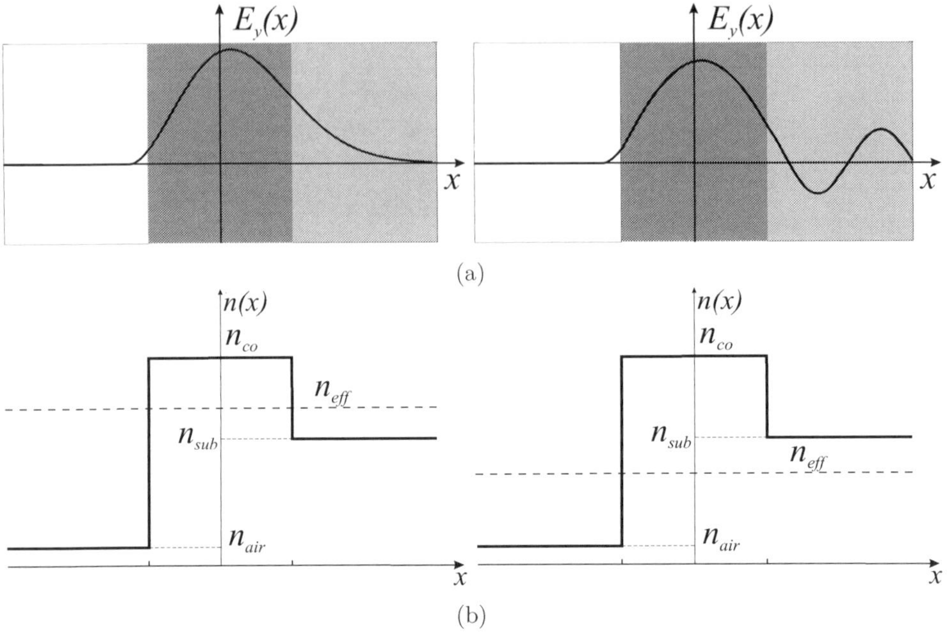

Fig. 2.22: (a) On the left, the profile of the component E_y of the TE_0 mode of an asymmetric slab; on the right, the qualitative profile of a substrate mode. The dark region represents the core while the cladding is in the light-coloured area and the air cladding layer is in white. (b) The comparison between the refractive index profile and the effective index of the mode.

Chapter 3

Integrated Optical Waveguides

Dielectric waveguides in integrated optics are the basic component of modern optical circuits in that they are utilised not only to interconnect the different parts of the circuit, but also as a basic element for more complex components. There are many different types of integrated waveguides, and their characteristics and geometries are strictly related to their design techniques, the chemical-physical characteristics of the materials chosen and the specific function of the components themselves. Furthermore, the rapid evolution of technological processes directly affects integrated waveguides as they become an extremely dynamic and ever-changing technology: for this reason, an all-encompassing analysis of all types of waveguides would be impossible. The aim of this chapter, instead, is to analyse the simplest – conceptually speaking – or the most common waveguides in order to highlight some of the major properties of the modes propagating in them.

3.1 Different Types of Integrated Optical Waveguides

Field confinement in real waveguides must be achieved not only along the vertical direction, as is the case for slabs, but also along the transverse direction. Some examples of waveguides in integrated optics are reported in figure 3.1. The technological processes currently employed in integrated optics seem to reflect the considerable development that characterise microelectronics. Rib, ridge and buried waveguides are mainly fabricated via chemical etching and molecular-beam or vapour-phase epitaxy of material with a different refractive index; in these waveguides, the refractive index is nearly always a piecewise constant function. As for graded-index (in-diffused) waveguides, these are fabricated by insertion – either diffusion, or ionic implantation and subsequent annealing – of dopants modifying the refractive index in the substrate. In this case, the refractive index, which depends on the dopant concentration, is a continuous function of the geometrical coordinates, with the exception of the interface between the substrate and the air where a discontinuity occurs.

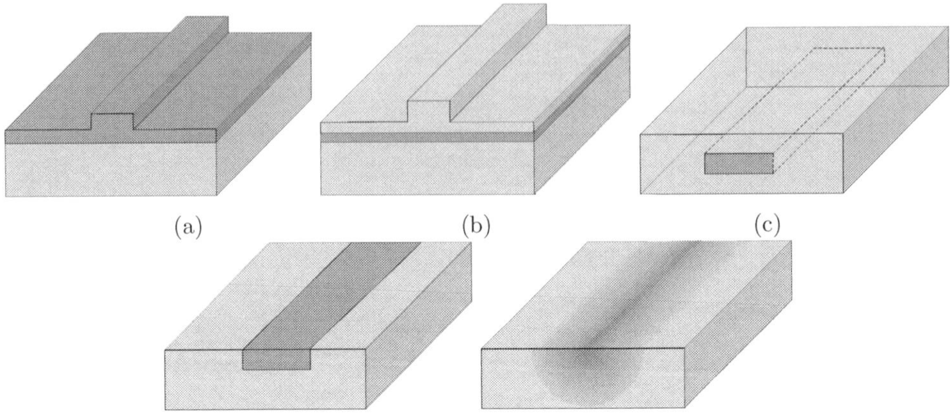

Fig. 3.1: Different types of integrated optical waveguides. (a) rib waveguide, (b) ridge waveguide, (c) buried waveguide, (d) semi-buried waveguide, (e) graded-index waveguide. The darker section represents the guiding region with higher refractive index.

However, there is no analytical solution to the modes of integrated optical waveguides; for this reason, several semi-analytical techniques – such as the effective index method, or Marcatili's method – have been developed over the years, despite their being approximate and having a rather limited field of application. Furthermore, the development of increasingly sophisticated electronic calculators has made it possible to extensively employ numerical simulators, to the benefit of accurate waveguide design. Amongst the most common numerical methods it is worth mentioning the Finite Difference Method (FDM), the Finite Element Method (FEM), and the Method of Lines (MoL).

The materials adopted are mainly semiconductors – such as gallium arsenide ($GaAs$), indium phosphide (InP) and silicon (Si), which may be combined with silica (SiO_2) – and insulating materials such as lithium niobate ($LiNbO_3$). $GaAS$ and InP are good direct band gap semiconductors since both passive and active components can be easily combined on the same chip: as a matter of fact, the use of $Al_xGa_{1-x}As$ and $In_xGa_{1-x}As_yP_{1-y}$ alloys on $GaAs$ and InP substrates respectively allows the modulation of the band gap width by changing the value of the parameters x and y. Were we to build laser sources, optical amplifiers or photodetectors, the energy gap values must be less than the energy of the photons of the propagating electromagnetic field; otherwise, if we were to fabricate passive components, the gap would have to be greater for the material to be transparent to photons. Another advantage these materials have is their high refractive index, as well as the possibility of changing the value of this index by adjusting the alloy composition. Moreover, these particular waveguides are characterised by a high index difference, which guarantees strong field confinement and allows the size of integrated components to be reduced. The Silica-on-insulator (SOI) technology – based on crystalline silicon deposited on an insulating layer of silica – is extremely interesting because, being based on silicon, it easily enables electronic integration and allows important technological processes to be exploited. Furthermore, the high index contrast ($n_{silicon} = 3.5$, $n_{silica} = 1.45$ at $\lambda = 1550\,nm$) allows for a considerable reduction in the size of the components. As far as lithium niobate is concerned, this insulating material is extremely difficult to handle and does not allow the creation of active components – even though researchers have already developed optical amplifiers and erbium-doped $LiNbO_3$ waveguide lasers. Due to its high resistance against chemical etching, it then becomes possible to employ in-diffused waveguides only, and their low degree of field confinement results in very large components. However, lithium niobate has interesting electromagnetic properties: its anisotropy and superior electro- and acousto-optical properties make it perfect for fabricating high-performance polarization converters, EOMs (electro-optic modulators) and optical switches.

3.2 Buried Waveguides

The buried waveguide, also known as channel waveguide, is certainly the simplest form of waveguide not only from the geometrical viewpoint, but also for its mechanism of field confinement. With reference to figure 3.2, this particular type of waveguide is formed by a (usually) rectangular core made of a homogeneous material with index n_{co} completely surrounded by a second homogeneous material having lower refractive index n_{cl} than that of the core. The index difference between the core and the cladding ensures field confinement in both transverse directions.

The distributions of the three components of the magnetic field mode functions of the first four waveguide modes are shown in figures 3.3 and 3.4. Due to the confinement in both transverse directions, the modes are no longer divided between TE or TM modes, but become hybrid, *i.e.* they have non-zero electric and magnetic fields. As we can see nonetheless, there is always a transverse component called *dominant* for it is much larger than the other two components. The amplitude ratio is typically of the order of 1 to $50 \div 1000$ as far as the secondary transverse component is concerned, and of 1 to $5 \div 50$ for the longitudinal component. For this reason, we can approximately assume that the magnetic field vector is mainly oriented along one of the two axes of the transverse plane: thanks to the duality theorem, it is possible to demonstrate that, if the magnetic field has

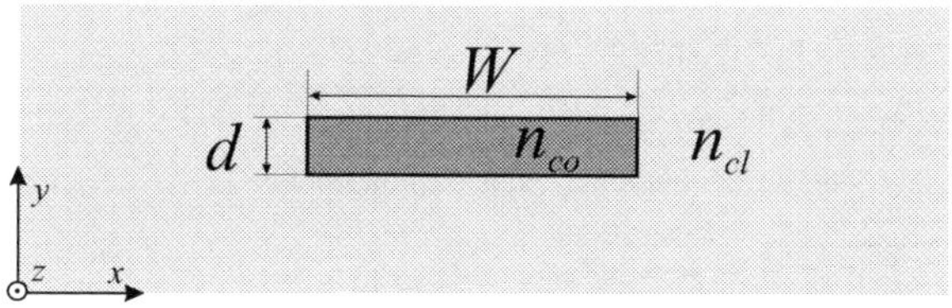

Fig. 3.2: Transverse section of a buried waveguide with rectangular core.

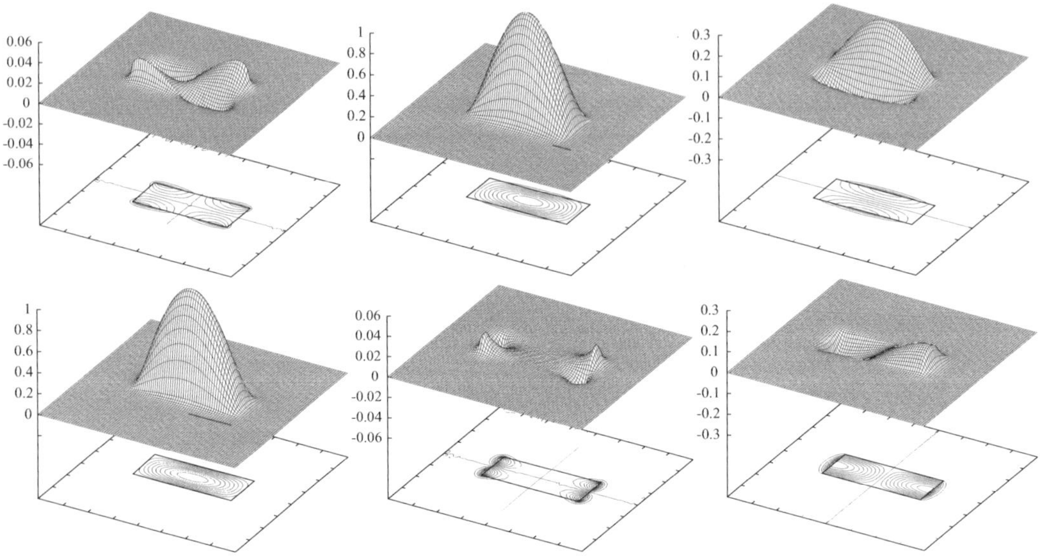

Fig. 3.3: The three columns show the three components H_x, H_y and H_z of the first two modes (H^y_{11} and H^x_{11}) of a buried waveguide with rectangular core.

this property, then this same property is to be found in the electric field as well, provided the two transverse components are exchanged. The situation is essentially the same as the one shown in figure 3.5, where both the magnetic and the electric field are linearly polarised and orthogonal to each other: this property is used in order to categorise the different modes. The first waveguide mode we shall consider is the H^y_{11} mode. The letter H identifies the magnetic field, while the superscript indicates that the dominant transverse component is along the y-axis; as for subscripts, they are used to denote the number of extrema of the dominant component along the x- and the y-axes respectively. The other modes are the H^x_{11}, the H^y_{21} and the H^x_{21}Hy21. Figure 3.6 displays the contours of the first six modes polarized along the y-axis – that is to say, following the H^y_{ij} structure. Similarly, an analogous norm is used in order to refer to the electric field. By virtue of the intrinsic orthogonality between the two vectors, the H^y_{11} and H^x_{11} modes corresponds to the E^x_{11} and E^y_{11} modes respectively – cf. figure 3.7 for the dispersion curves of the first six modes of the rectangular buried waveguide. The cutoff frequency of the first two modes is equal to zero, and this property is shared by most integrated waveguides – as already noted when we dealt with slabs. For this reason, waveguides supporting only one mode per polarization are defined *single-mode integrated waveguides*.

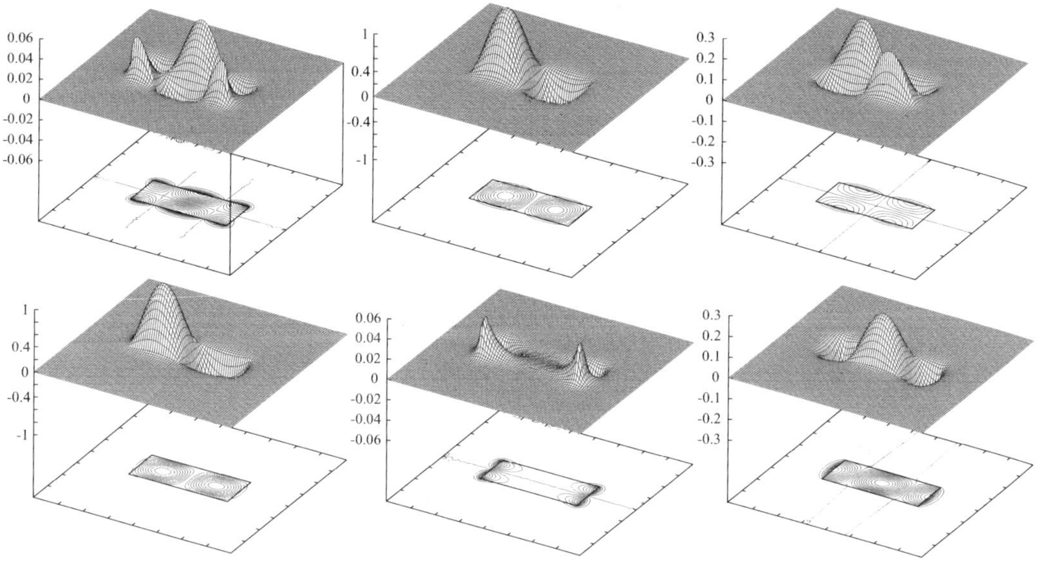

Fig. 3.4: The three columns show the three components H_x, H_y and H_z of the third (H^y_{21}) and the fourth (H^x_{21}) mode of a buried waveguide with rectangular core.

3.2.1 Birefringence

Figure 3.8 presents the contours of the dominant components of the H^y_{11} and H^x_{11} modes with respect to rectangular and square cores. In the first case, we can observe a different profile of the component due to the different continuity conditions that must be satisfied along both the long and the short side of the interface. Such dissimilar distribution ensures that the two modes are not degenerate, as we can evidently infer by observing the dispersion curves in figure 3.7. With a square core, we

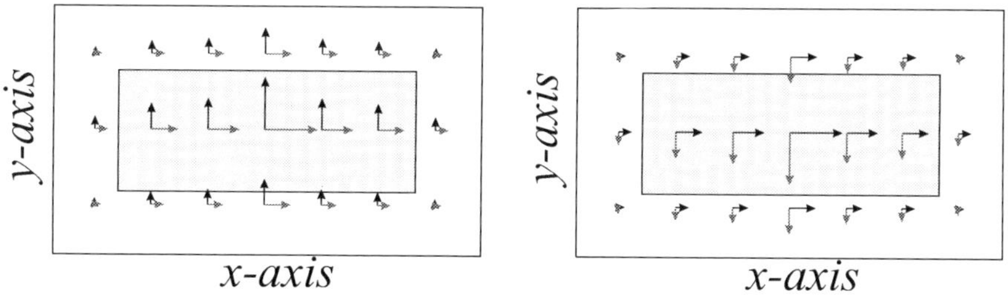

Fig. 3.5: The orientation of the magnetic field vectors (in black) and electric field vectors (in grey) of the H^y_{11} (on the left) and H^x_{11} (on the right) modes of a rectangular buried waveguide.

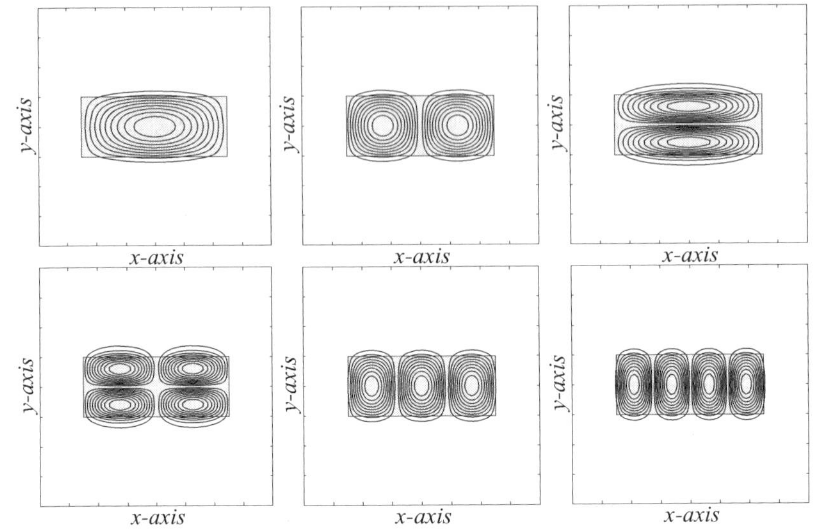

Fig. 3.6: The contours of the fundamental components of the first six polarised modes of a rectangular buried waveguide along y. From right to left: H^y_{11}, H^y_{21}, H^y_{12} (top) and H^y_{22}, H^y_{31}, H^y_{41} (bottom). The contours are spaced at 10% of the peak value.

are still in the presence of dissimilar distributions, but these can actually coincide in case of a $\pi/2$ rotation. Indeed, the continuity conditions must be applied to same-length sides, therefore having the same effect on the distributions of the field. The symmetrical feature, with respect to $\pi/2$ rotations, in the core of a square waveguide ensures that the two modes with different polarisation are degenerate, and, as a consequence, the waveguide is not birefringent. Figure 3.9 presents the

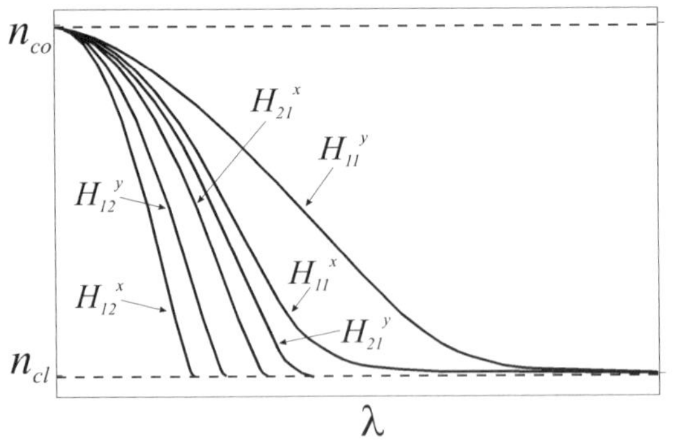

Fig. 3.7: The dispersion curves of the first six modes of a rectangular buried waveguide.

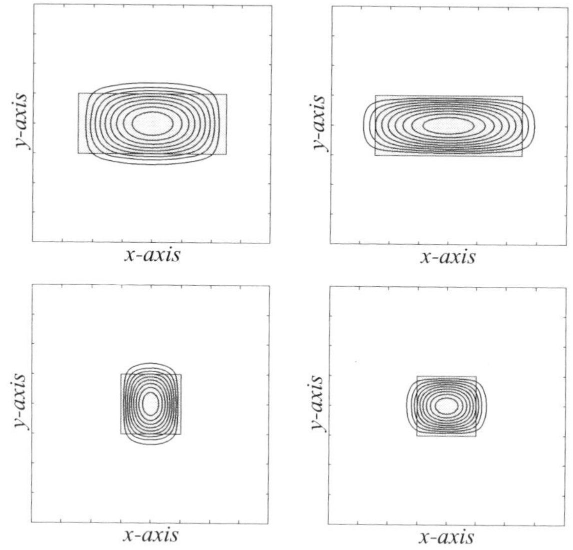

Fig. 3.8: The contours of the dominant components of the H^y_{11} (on the left) and H^x_{11} (on the right) modes with $W/d = 2.5$ (top) and $W/d = 1$ (bottom). The contours are spaced at 10% of the peak value.

spectral trend of birefringence according to the varying ratio between the two sides of the core. As expected, and with all other parameters being equal, the value of birefringence decreases as the dimensions of the two sides tend to coincide, and it finally becomes constant and equal to zero when the sides are equal.

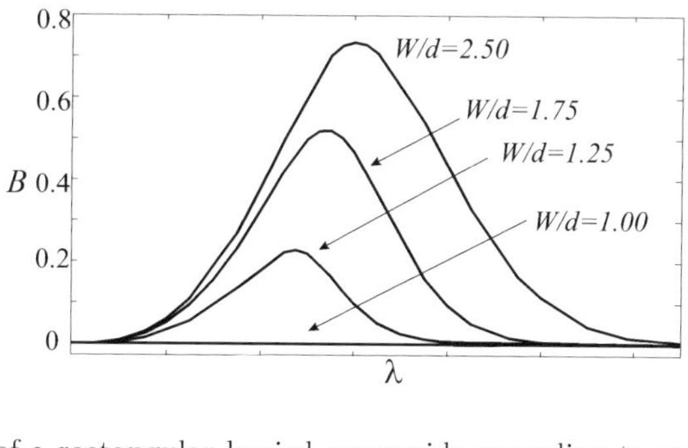

Fig. 3.9: Birefringence of a rectangular buried waveguide according to wavelengths with different W/d ratios.

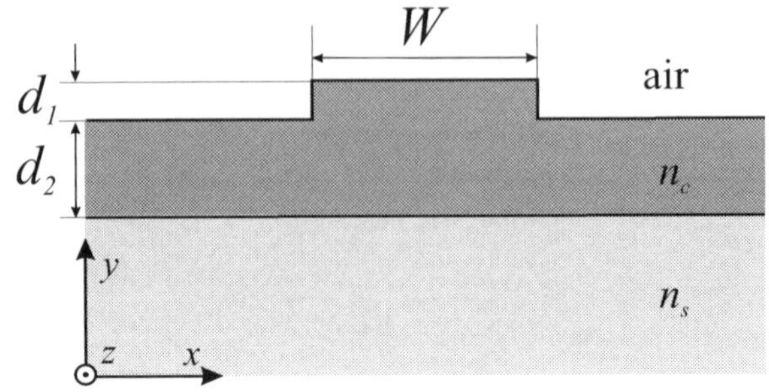

Fig. 3.10: Transverse section of a rib waveguide.

3.3 Rib Waveguides

The rib waveguide, whose transverse section is shown in figure 3.10, is a particularly common choice in integrated optics. In this waveguide, there is no actual index difference that can explain field confinement in the horizontal direction; however, it is still possible to justify field confinement if we fractionate the rib waveguide into three asymmetrical slabs. Since the central slab is characterised by a larger core, the modes associated with it will have greater effective indices compared to the side waveguides modes – as schematised in figure 3.11. Hence, the horizontal confinement can somehow be attributed to the increasing effective index in the rib section with respect to the side regions.

Figure 3.12 displays the profiles of the three components of the first two magnetic field modes of an InP rib waveguide. The rib is in $In_{0.96}Ga_{0.04}As_{0.08}P_{0.92}$ with index $n_c = 3.2$ and has the

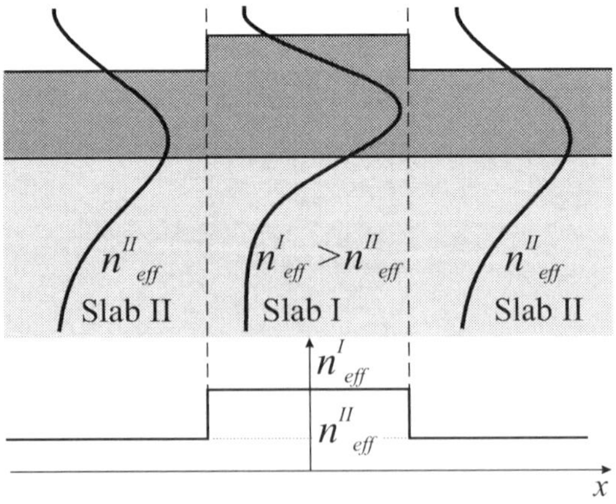

Fig. 3.11: The mechanism of lateral confinement in a rib waveguide.

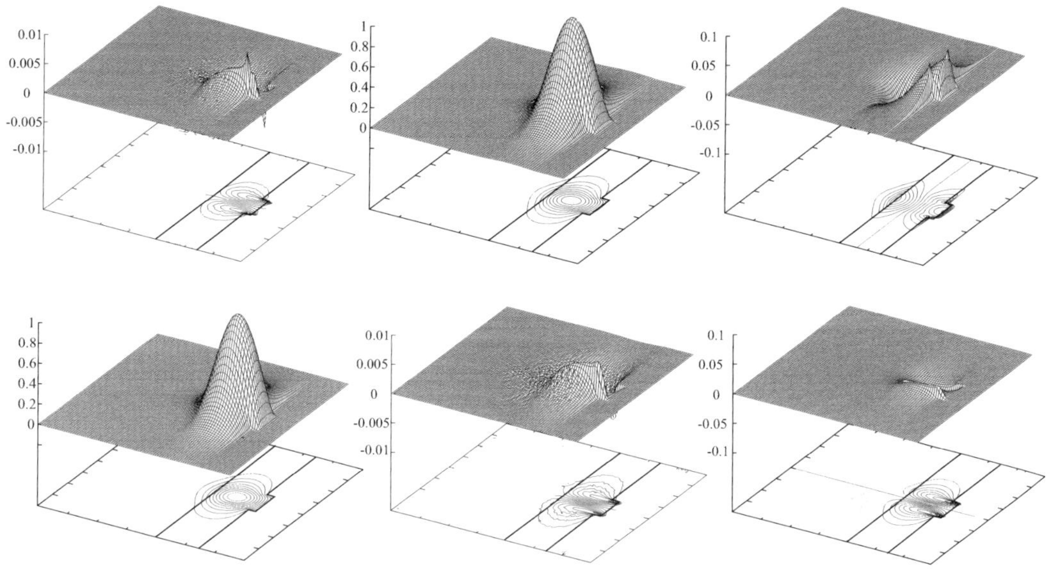

Fig. 3.12: From left to right: the components H_x, H_y and H_z of the H_{11}^y (top) and H_{11}^x (bottom) modes of a rib waveguide with the following physical and geometrical characteristics: $W = 6\mu m$, $d_1 = 0.5\mu m$, $d_2 = 3\mu m$, $n_a = 1$, $n_c = 3.2$ and $n_s = 3.17$ at $\lambda = 1.55\,\mu m$.

following dimensions: $W = 6\mu m$, $d_1 = 0.5\mu m$, $d_2 = 3\mu m$. The InP substrate has index $n_s = 3.17$. The characteristics of the modes are similar to the ones of the buried waveguide: the two modes are hybrid and their cutoff frequency is equal to zero; moreover, they are characterised by a transverse component which is dominant with respect to the other components. The main difference lies in vertical asymmetry, which is due to the different index difference between the rib and the air, on the one hand, and the rib and the substrate, on the other. Figure 3.13 shows that the level of confinement of the fundamental mode depends heavily on the geometrical characteristics of the rib: as the rib height increases, so does the difference between the central and side regions in terms of

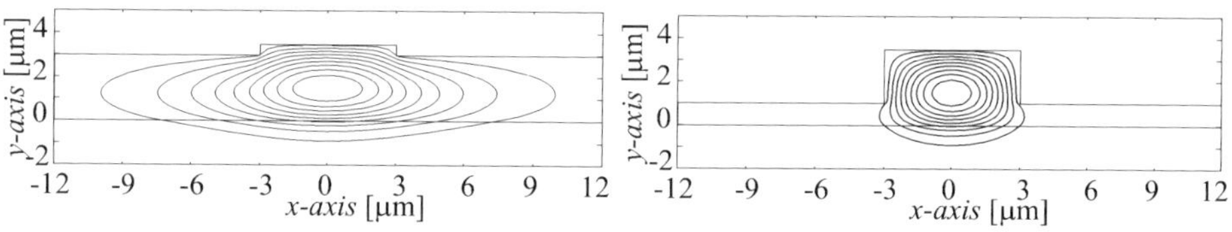

Fig. 3.13: The confinement of the H_{11}^y mode depends on the rib width. On the left: $d_1 = 0.5\mu m$; on the right: $d_1 = 2.5\mu m$. The contours are spaced at 10% of the peak value.

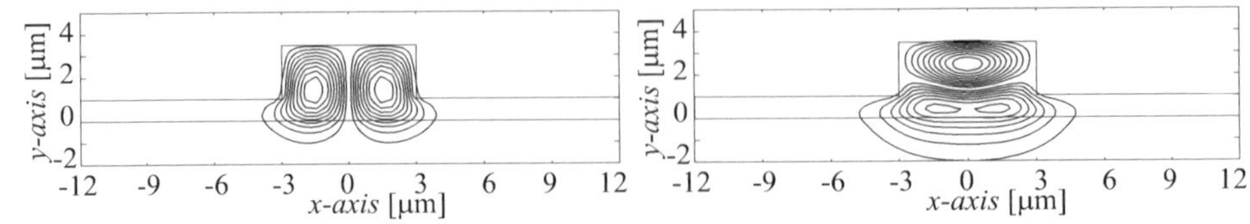

Fig. 3.14: The dominant component of the higher-order modes H_{21}^y and H_{12}^y of a rib waveguide with the following physical and geometrical characteristics: $W = 6\mu m$, $d_1 = 3\mu m$, $d_2 = 0.5\mu m$, $n_a = 1$, $n_c = 3.2$ and $n_s = 3.17$. The contours are spaced at 10% of the peak value.

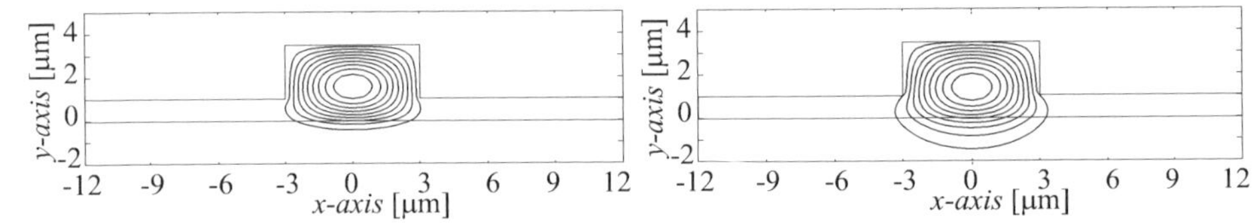

Fig. 3.15: Field confinement depends on the wavelength. The results refer to the dominant component of the H_{11}^y mode at the wavelengths of $1000\,nm$ and $2000\,nm$ – in the simpler hypothesis that the refractive indices of the materials do not depend on λ. The contours are spaced at 10% of the peak value.

the effective index and, consequently, field confinement. An analogous discourse can be applied to the rib width. The contours of the higher-order modes H_{21}^y and H_{12}^y are illustrated in figure 3.14. Finally, figure 3.15 underlines that field confinement depends on the wavelength: as the wavelength increases, in fact, the level of field confinement becomes weaker; in particular, the field tends to move more towards the substrate, with a consequent reduction of the effective index, which tends to that of the substrate.

3.4 Strip and Strip-raised Waveguides

These waveguides are commonly used in semiconductor technologies since they ensure a high degree of field confinement. The transverse cross-sections of these waveguides are illustrated in figure 3.16. As we describe the main properties, we shall refer to a SOI (silicon-on-insulator) strip-raised waveguide with the following features: $n_c = 3.5$, $n_s = 1.45$, $W = 350\,nm$, $d_u = d_d = 200\,nm$ and $d_s = 300\,nm$. Figure 3.17 represents the three magnetic field components of the first two modes H_{11}^y and H_{11}^x at $\lambda = 1550\,nm$. Note that, despite the sub-micrometric size of the waveguiding region, the field is strongly confined. However, the high index difference is the cause of a significant level of birefringence – cf. figure 3.18 for its spectral trend. In this case too, it is possible to eliminate birefringence – over a certain range of wavelengths, at least – by acting on the geometric dimensions of the waveguide.

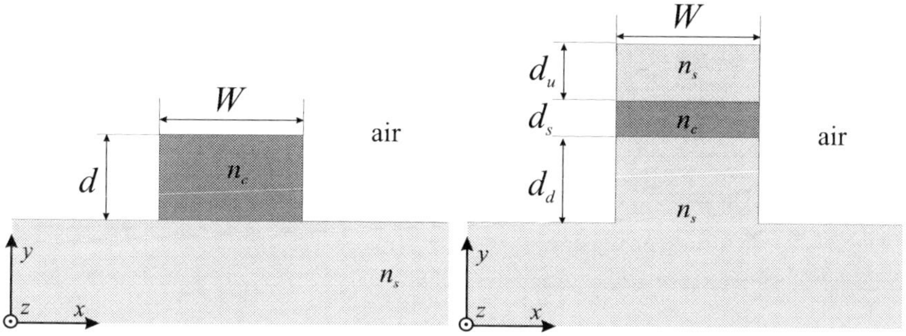

Fig. 3.16: Strip (left) and strip-raised (right) waveguides.

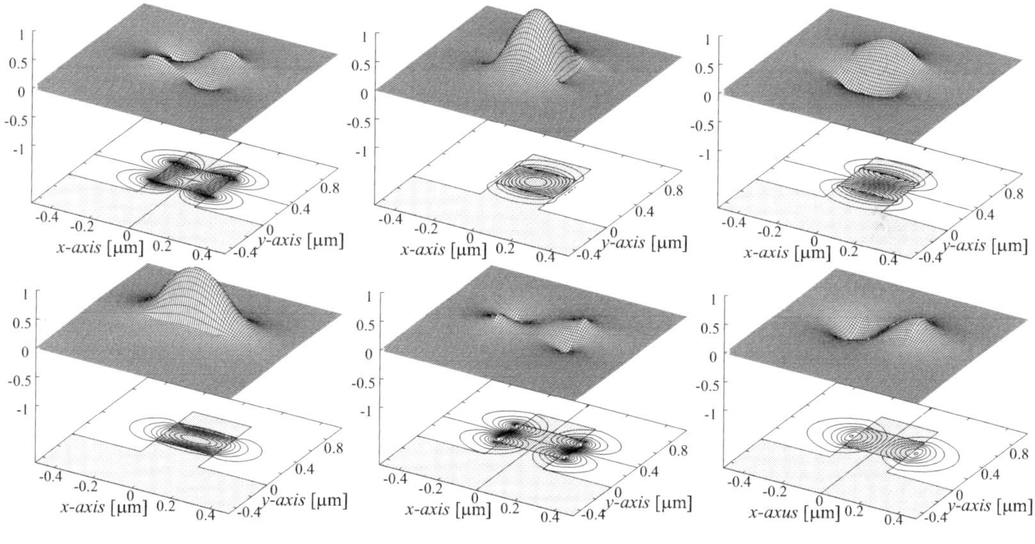

Fig. 3.17: The three components H_x, H_y and H_z of the first two modes (H_{11}^y and H_{11}^x) at the wavelength $\lambda = 1550\,nm$ of a SOI strip-raised waveguide with $n_s = 1.45$, $n_c = 3.5$, $W = 350\,nm$, $d_u = d_d = 200\,nm$ and $d_s = 300\,nm$.

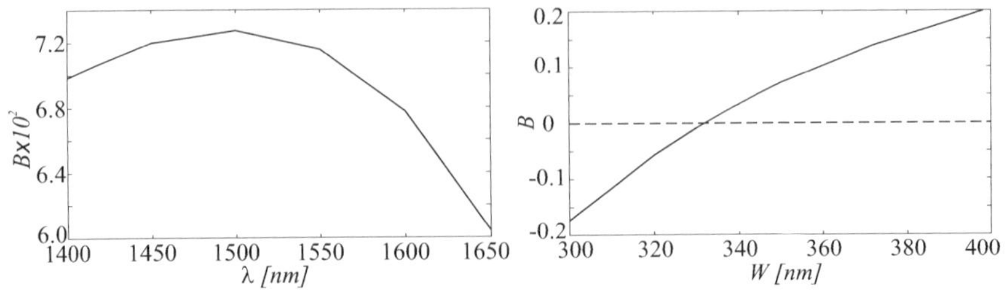

Fig. 3.18: On the left, the birefringence spectral trend of an SOI strip-raised waveguide with $n_s = 1.45$, $n_c = 3.5$, $W = 350\mu m$, $d_u = d_d = 200\,nm$ and $d_s = 300\,nm$. On the right, the birefringence trend at $\lambda = 1550\,nm$ as the width W of the waveguide varies.

3.5 Graded-index (or In-diffused) Waveguides

To give an example, consider a titanium in-diffused $LiNbO_3$ waveguide – where titanium is used to increase the refractive index. Figure 3.19 shows its transverse cross-section as well as the variation of the refractive index Δn with respect to the substrate value $n_{sub} = 2.2$. In figure 3.20 we can observe the components of the magnetic field mode function in relation to the first two waveguide modes H_{11}^x and H_{11}^y at the wavelength $\lambda = 1550nm$. If we consider the waveguides previously analysed, in this case the variation of the refractive index along the horizontal direction and towards the depth of the substrate is considerably lower. As a consequence, the secondary components are considerably reduced, though the same also applies to the level of field confinement, which is the main drawback of this type of waveguide.

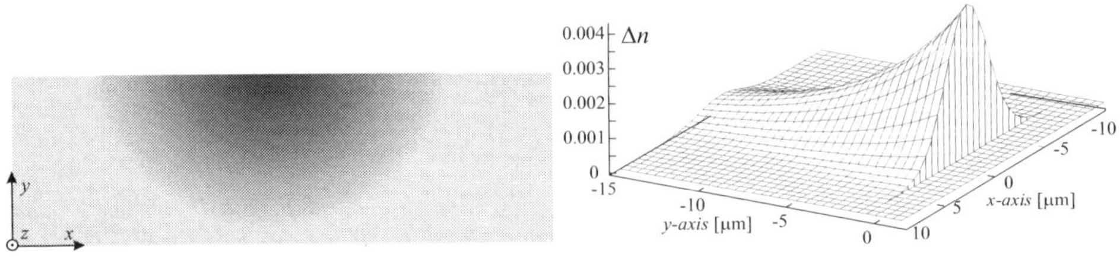

Fig. 3.19: On the left, the transverse cross-section of a graded-index waveguide; the level of grey is proportional to the refractive index value. On the right, the trend of Δn in an in-diffused $LiNbO_3$ waveguide. The interface between lithium niobate and air is equal to $y = 0$.

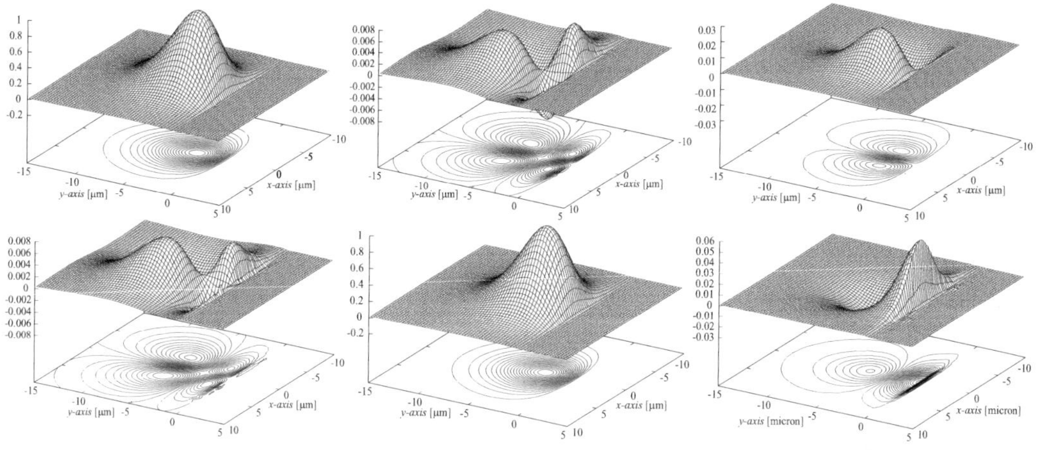

Fig. 3.20: From left to right, the components H_x, H_y and H_z of the magnetic field vector with respect to the first (top) and the second mode (bottom) of an in-diffused $LiNbO_3$ waveguide.

3.6 Waveguide Losses

In this section we shall investigate the main factors causing guided power loss in dielectric waveguides. A first distinction between the various types of losses can be made between *concentrated* and *distributed losses*. In the first category we find spatially well-localised phenomena, to the extent that they can be ascribed to precise cross-sections of the waveguide. In this case, the extent of the loss can be measured by calculating the ratio between the power values before (P^+) and after (P^-) these phenomena occur in a specific area:

$$L = \frac{P^+}{P^-},$$

often presented in dB and as positive values:

$$L_{dB} = P^+_{dB} - P^-_{dB}.$$

The second category, on the other hand, comprises those phenomena which are presumed to be evenly distributed along the waveguide. In this respect, the following coefficient (cf. paragraph 1.3) is used to quantify these losses:

$$\alpha_{dB} = 20log(e)\alpha = 8.6858\alpha\ [dB/m].$$

Thus, the propagation for a certain distance z leads to an attenuation in dB of:

$$L = \alpha_{dB} z.$$

3.6.1 Coupling Losses

Coupling losses are due to the considerably different level of field confinement between the fundamental mode of standard optical fibers and that of integrated waveguides, as figure 3.21 clearly demonstrates. The reason for this lies in the different needs that fibers and integrated waveguides have to meet: as far as fibers are concerned, it is essential that the core is not too small so as to prevent critical interconnections among them in the cable-laying process; when it comes to integrated waveguides, on the other hand, trying to minimise the size of the core is of paramount importance in order to reduce the dimension of the components.

Since the power that can be transferred from one mode to another depends on the overlap integral, it is evident how, due to the excitation of the radiation modes, a direct "fiber-integrated waveguide" connection considerably reduces the guided power. The losses can easily be the order of several dozens of dB. For this reason, the fiber and the integrated waveguides always interconnect by means of specific structures called *tapers*, which gradually turn the mode of one waveguide into the mode of another. This transformation is necessary to adjust the level of field confinement and it is possible as the waveguide characteristics – usually the size of the core – progressively vary along the direction of propagation. An example of taper is depicted in figure 3.22. Thus, it is clear that the more slowly these transformation occur, the lower the power lost due to radiation modes; however, the main downside of tapers is their cumbersome size.

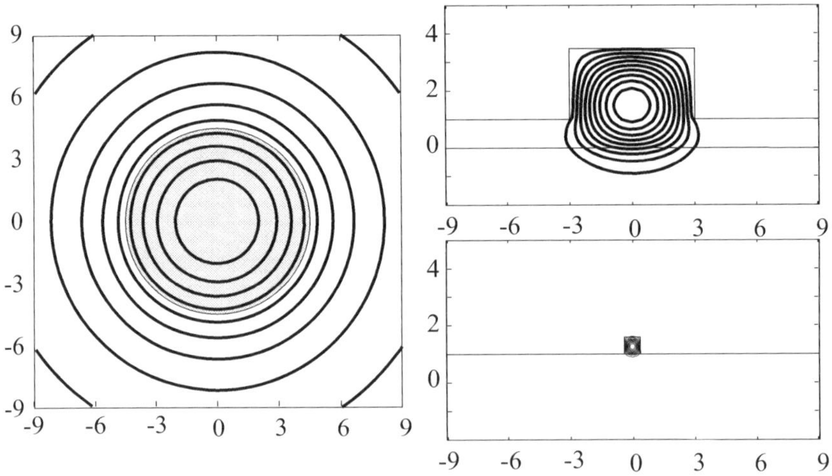

Fig. 3.21: Comparison between the distribution of the fundamental mode in an optical fiber (on the left) and two integrated waveguides (on the right) – namely an InP rib waveguide (top) and a SOI strip waveguide (bottom). All dimensions are in micron.

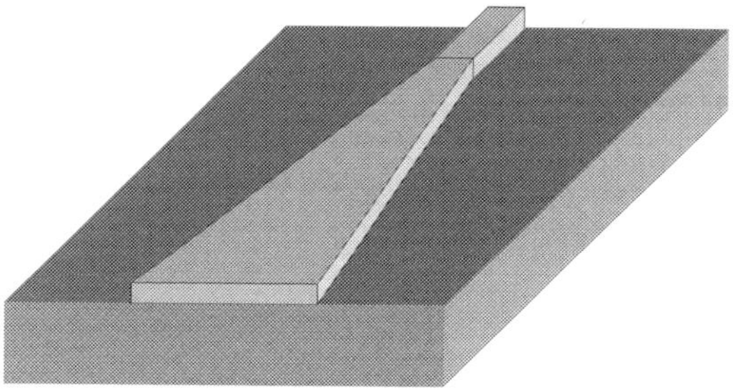

Fig. 3.22: An example of tapered rib waveguide obtained by altering the rib width.

3.6.2 Curved Waveguide Losses

Curved waveguides are particularly common in integrated optics as they have many functions: for example, curved waveguides are useful to change the direction of field propagation, separate parallel waveguides or bring them closer, or again to merge different waveguides in a single waveguide or vice versa. An example of their utility is shown in figure 3.23. Any sudden change in the direction of propagation triggers two different mechanisms of power loss: *bending losses* and *transition losses.* The following will be briefly described below.

Bending Losses

In the case of planar structures with an invariance also along one of the transverse directions, and considering a constant bending radius, the choice of a cylindrical frame of reference allows us to

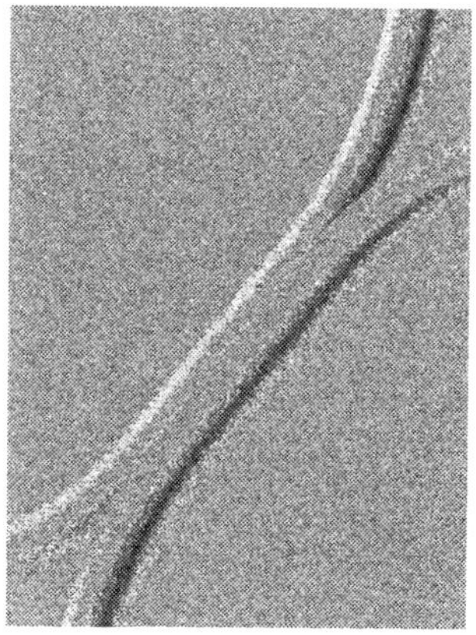

Fig. 3.23: An example of curved waveguide in an integrated optical coupler.

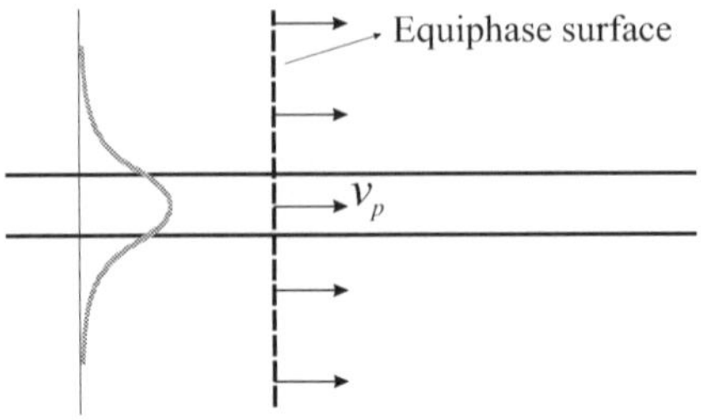

Fig. 3.24: Straight waveguide. The intersection between an equiphase plane and the (x, z) plane is represented by the dotted line. The phase velocity v_p is constant over the whole plane. The field profile of the fundamental mode is in grey.

analyse the curved waveguide as a cylindrical structure according to the properties of its relative modes. However, we deem it more appropriate to adopt a less rigorous procedure, which permits the highlighting of important properties nonetheless. To this end, some observations on straight waveguides are first necessary (cf. figure 3.24). The field has a dependence of the $e^{-j\beta z}$ type, therefore the equiphase surfaces are planes orthogonal to the z-axis, translating at the phase velocity $v_p = \omega/\beta = c/n_{eff}$. Since we are dealing with a guided mode, $n_{cl} < n_{eff} < n_{co}$, and therefore

$$c/n_{co} < v_p < c/n_{cl}$$

where c is the speed of light in vacuum. Inside the core, the phase velocity is higher than the speed of light in the medium. This apparent contradiction can be explained if we recall that the field inside the core can be considered as formed by two plane waves propagating in an oblique direction – along which the phase velocity is measured – with respect to the z-axis (cf. the chapter on slab waveguides). On the contrary, the phase velocity in the cladding is less than the speed of light in the medium, since the plane wave is evanescent, *i.e.* slow.

Now consider the case of a curved waveguide with constant bending radius equal to R, as measured from the centre of the waveguide (fig. 3.25). In this case, the equiphase planes do not propagate in a straight line, rather they tend to follow the curvature of the waveguide. The phase velocity is not constant on the plane and increases linearly as we move away from the centre of curvature. With r denoting the distance from the centre of curvature, the phase velocity now depends on r:

$$v_p(r) = v_{pR}\frac{r}{R}, \tag{3.1}$$

with

$$v_{pR} = \frac{c}{n_{eff}} \tag{3.2}$$

being the phase velocity in correspondence to the waveguide axis, n_{eff} being the effective index of the mode. In the upper cladding, the phase velocity of the evanescent wave increases as we move

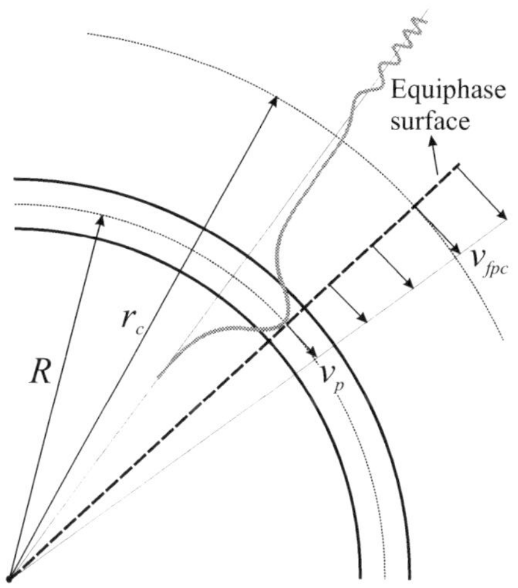

Fig. 3.25: Curved waveguide. The intersection between an equiphase plane and the (x, z) plane is represented by the dotted line. The profile of the field of the fundamental mode is in grey. The phase velocity v_p increases as we move away from the centre of curvature. In correspondence to the critical radius r_c, the phase velocity $v_p(r_c) = v_{pc}$ matches the speed of light in the cladding. Beyond r_c, the wave presents a radial component of the wave vector and has an oscillatory trend.

away from the core. There is definitely a distance called *critical radius* (r_c), where:

$$v_p(r_c) = \frac{c}{n_{cl}}. \tag{3.3}$$

Beyond this distance, the phase velocity necessary for an evanescent wave to propagate in a direction tangent to the waveguide axis becomes excessive. Thus, the wave turns into a uniform wave with $\overline{k}$ wave vector. The radial component of $\overline{k}$ determines a flux of the Poynting vector in the radial direction, thereby leading to losses. The flux is proportional to the width of the uniform wave and, since the decay is exponential according to $r < r_c$, the lower the radial power flux, the greater the critical radius. From (3.1)-(3.3) it follows that:

$$r_c = R\frac{n_{eff}}{n_{cl}}.$$

This expression is valid qualitatively speaking, but it permits the main factors affecting this type of losses to be identified nonetheless. The larger the radius of curvature R, the smaller the losses and the higher the effective index, compared to that of the cladding. In particular, since

$$1 \leq \frac{n_{eff}}{n_{cl}} \leq 1 + \delta$$

$\delta = (n_{co} - n_{cl})/n_{cl}$ being the relative index difference, it is important to maximise both the core-cladding index difference – so as to increase the upper limit – and the level of field confinement – in order to bring the n_{eff} close to the same limit (however, all this must obviously comply with the single-mode condition). This study highlights an important property of the electromagnetic field, known as *optical tunnel effect*: this occurs when the evanescent wave turns into a uniform wave and, from a certain point on, produces a power flux in the transverse direction.

The Conformal Mapping Technique

Besides the approach based on uniform and evanescent plane waves, the problem of curved waveguides can be addressed by resorting to the conformal mapping technique. Without going into too much detail, it is possible to demonstrate that a curved planar waveguide with radius R, and characterised by a general refractive index profile $n(r)$, is equivalent to a straight waveguide having the following profile:

$$n_{eq}(x) = n(x + R)e^{x/R}.$$

If the radius of curvature is high enough to assume that $x/R << 1$, then it is possible to obtain a first-order approximation of the exponential, as shown below:

$$n_{eq}(x) = n(x + R)\left(1 + \frac{x}{R}\right).$$

In figure 3.26 we report the case of a slab waveguide with piecewise constant index. In the point x_c on the x-axis, the effective index of the mode is no longer higher than $n(x)$: from the exponential decay typical of guided modes, the solution changes to an oscillatory trend, which is a characteristic of radiation modes. With this approach too, it is clear how the losses are reduced as the radius of curvature and field confinement both increase: in the first case, because the slope of the index profile falls, and, in the second case, since the field value in x_c grows smaller. The latter, in particular, attributes the phenomenon to an equivalent straight dielectric waveguide, hence being particularly convenient when used in combination with numerical methods for the modal analysis of dielectric waveguides.

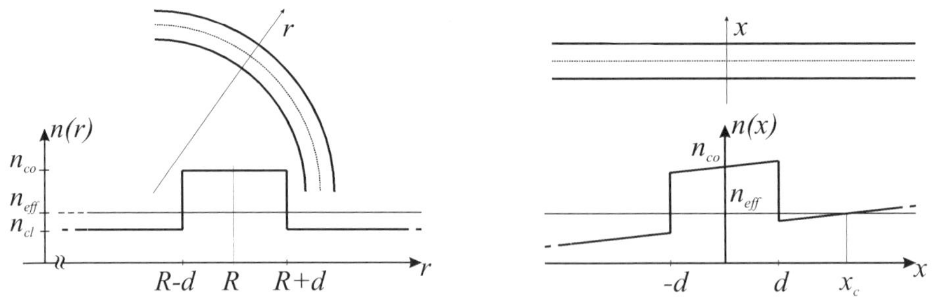

Fig. 3.26: A curved waveguide with step profile and the equivalent straight waveguide with the relative index profile.

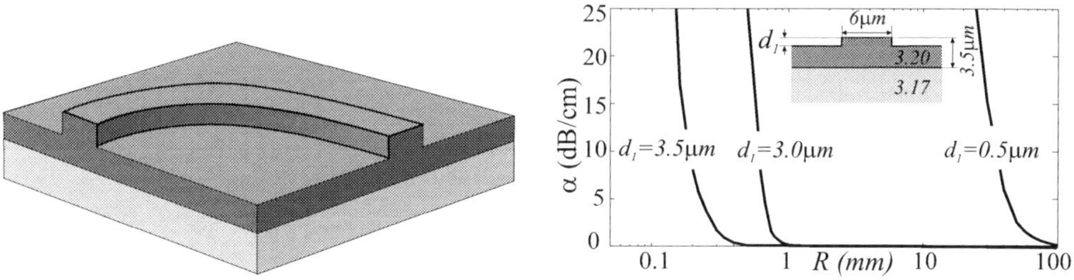

Fig. 3.27: On the left: a curved rib waveguide. On the right: the bending losses of the rib waveguide as the radius of curvature R varies according to different rib width values d_1 ($\lambda = 1550\,nm$). The geometrical and physical parameters of the waveguide are reported in the figure above.

Bending Losses in a Rib Waveguide: A Case Study

To give an example, figure 3.27 displays the bending losses (in dB/cm) of an in InP rib waveguide according to the radius of curvature R, and for different values of the rib height d_1 at the wavelength $\lambda = 1550\,nm$. The losses have been determined through a modal solver based on the finite element method, together with the conformal mapping technique. As we have just seen, the losses rapidly increase as the radius of curvature decreases. Furthermore, an increased rib depth permits the reduction of the radius of curvature significantly, with an increase in the level of field confinement. The latter observation proves that, in order to reduce the size of the components, it would be desirable to opt for high-index materials which allow waveguides with high field confinement to be obtained.

Transition Losses

Besides causing losses in terms of guided power, the curvature of the waveguide also deforms the profile of the mode which is propagated. Figure 3.28 compares the profiles of the fundamental mode of a curved waveguide with that of a straight one, both having the same characteristics. The profile difference gives rise to the so-called *transition losses*, which occur when a curved waveguide is connected to a straight one. To minimise these losses, it is necessary to maximise the overlap integral ((1.5)) between the two modes, so as to make the two profiles as similar as possible. Figure 3.28 shows how this can be achieved by simply creating a Δx offset between the axis of the straight waveguide and the axis of the curved one. The improvement in terms of reduction of power lost for imperfect coupling is illustrated in figure 3.29, which gives an account of the results of a study on an S-curve with and without offsets. The straight appearance of the waveguides in the figure is based on the fact that the simulations were carried out using the conformal mapping technique.

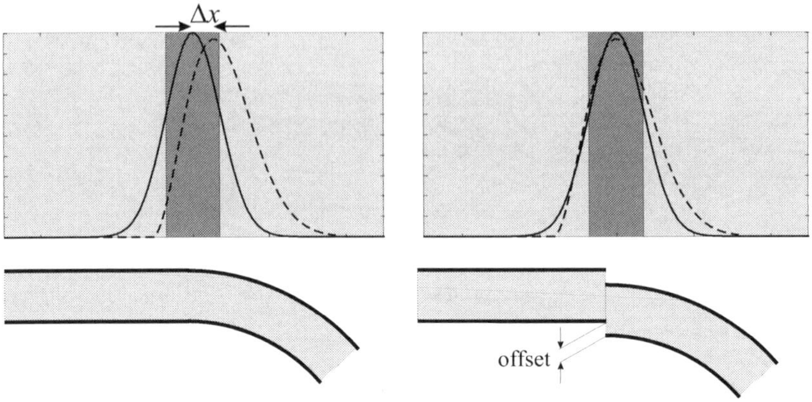

Fig. 3.28: On the left, the modes of a straight and curved waveguide with no offset, respectively described by a solid line and a dotted line. On the right, the same modes in presence of an offset.

3.6.3 Scattering Losses

Power loss can also be triggered by the texture of the surfaces surrounding the waveguide core when these are excessively rough or coarse. Figure 3.30 shows an example of this by presenting the case of an InP rib waveguide. The losses are due to the coupling between the guided mode and the radiation modes caused by the fluctuating surfaces. The latter strongly depend on the quality of the manufacturing process, the type of material chosen and the waveguide dimensions – typical values ranging from 0.1 to $5\,dB/cm$. An approximate theory can show that the scattering loss coefficient

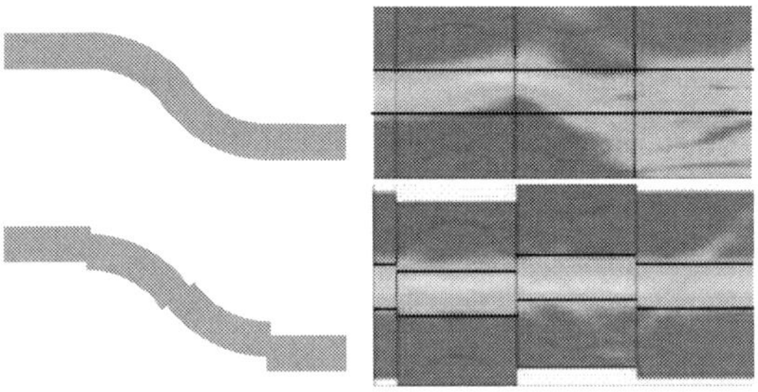

Fig. 3.29: The field intensity distribution along an S-shaped curve with (bottom) and without (top) appropriate offsets.

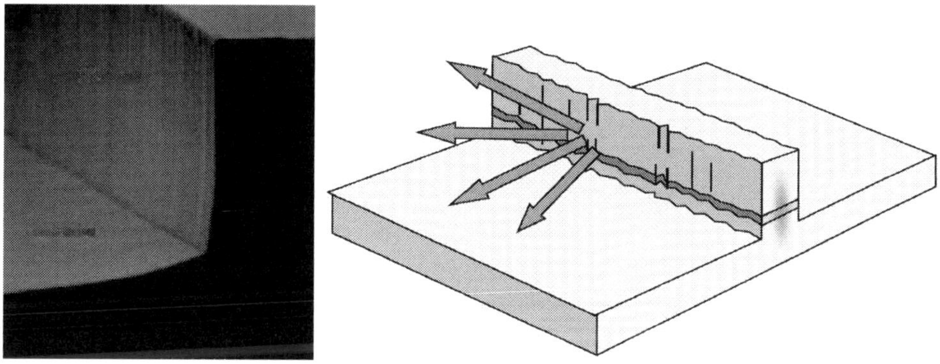

Fig. 3.30: On the left: electron microscope image of an InP rib waveguide showing the rough walls that separate the waveguide core from the air. On the right: scattering losses due to the irregularities of the waveguide.

α_{sc} is proportional to the core-cladding step index Δn, to the level of roughness of the surface, and to the electric field value in proximity of the interface; cf. the following relation:

$$\alpha_{sc} = \left(\frac{2\pi}{\lambda}\right)^2 \frac{\sigma^2 n_{co}}{n_{eff}} \Delta n^2 \frac{\oint |\overline{E}|^2 dl}{P} \tag{3.4}$$

where the line integral is calculated on the line defined by the intersection between the rough surface and the transverse plane, σ is the roughness RMS (root mean square), P the power carried by the mode, n_{eff} the effective index and n_{co} the core index.

In semiconductor waveguides the step index Δn is particularly high, between 1 and 2.5, which is why the technological processes that take place have to ensure as smooth a surface as possible. The fluctuations of the surfaces are typically random, yet if we manage to compare the length of the autocorrelation of the fluctuations with the waveguide wavelength, the waveguide behaves like a Bragg reflector and the losses increase significantly. A proper design of the waveguide can help reduce scattering losses, *e.g.* by minimising the electric field value in correspondence to rough surfaces, thus reducing the line integral in (3.4).

The dependence on the wavelength is not only due to λ explicitly appearing in (3.4), but this is also motivated by the fact that the effective index and the electric field mode function depend on the same λ. The two types of dependence have opposing effects on the value α_{sc}, and cause minimal losses in terms of wavelength. In actual fact, as the wavelength decreases, α_{sc} tends to increase because of λ being in the denominator; on the other hand, as the wavelength increases, the effective index and the level of field confinement of the mode – which is close to cutoff conditions – decrease, consequently reducing the value of α_{sc}.

3.6.4 Polarization-dependent Losses

Polarization-dependent losses (PDLs) are not due to a specific element or condition, and every phenomenon previously analysed can be used to explain such losses. As we went through the different types of losses, we observed that, somehow, they always depend on the effective index and field distribution. As is known, these quantities are different for the two fundamental modes due to birefringence. Therefore, while it propagates, one mode might have different losses from those of the mode being orthogonally polarised. PDLs are an important parameter for assessing the quality of a waveguide or an integrated component: in fact, the state of polarization of the signal coming from a fiber is not known *a priori* and, furthermore, it changes randomly as time and frequency vary. It is therefore essential to ensure as low a PDL as possible to avoid dangerous fluctuations of the signal power which would compromise the transmission system performance. Defining as L_{dB}^{x} the power loss of the polarised mode along the x-axis, and L_{dB}^{y} the loss of the polarised mode along y (both in decibels), the following is defined as *polarization-dependent loss* (PDL):

$$PDL = |L_{dB}^{x} - L_{dB}^{y}|.$$

PDL values in modern transmission systems are usually below $1\,dB$.

Chapter 4

Power Dividers

At a functional level, a power divider is a device consisting of a block with one input port and two output ports: as schematised in figure 4.1, the input power is thus divided equally between the output ports. Typically, we refer to $x : 100 - x$ power dividers, with x indicating the percentage of the input power on one of the two waveguides, and $100 - x$ the percentage on the other waveguide. Since the relation $P = |A|^2$ correlates the mode amplitude $|A|$ with the power carried by it, it follows that the amplitude of the two output waveguides is $|A|\sqrt{x/100}$ and $|A|\sqrt{1-(x/100)}$.

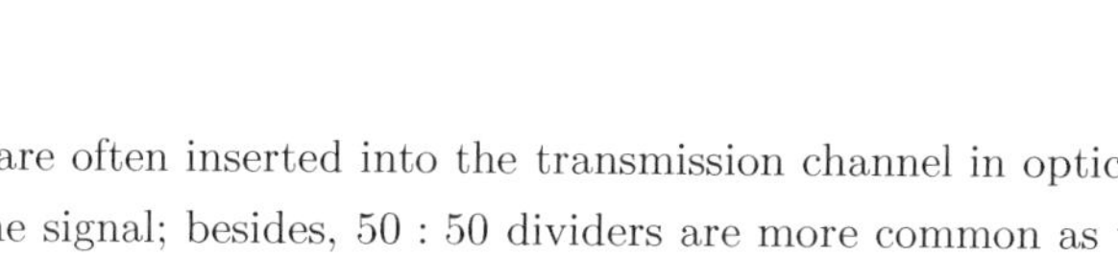

Fig. 4.1: Power divider.

$1 : 99$ power dividers are often inserted into the transmission channel in optical fibers to check the level and quality of the signal; besides, $50 : 50$ dividers are more common as they divide the input power equally between the two output waveguides. In this case, we often mean a "3-dB power divider" because the output power is half the input power, *i.e.* it is 3 dB less, while the field amplitude drops by $1/\sqrt{2}$. Several techniques can be used in this respect: we will now consider the so-called Y-junction as well as the directional coupler.

4.1 *Y*-junction Power Dividers

The Y-junction power divider – cf. figure 4.2 – is a simple device in which one single waveguide branches into two different waveguides with the same dimensions and characteristics of the input mode. In this case, the waveguide is gradually enlarged to twice its original width. The bridging

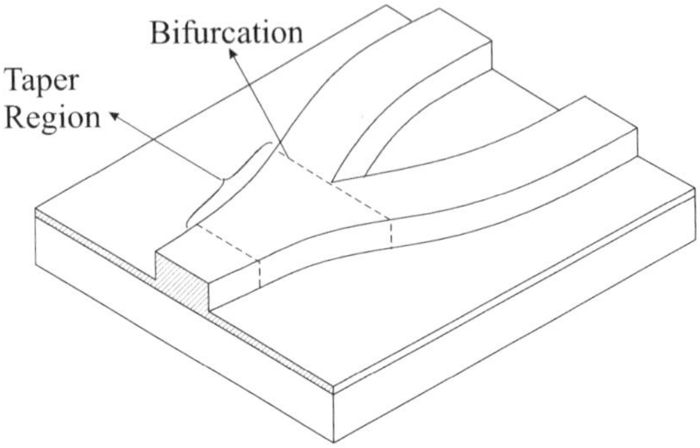

Fig. 4.2: Y-junction rib waveguide.

area of the junction is called *transition* or *taper region.* Subsequently, the waveguide is split into two separate waveguides which are progressively distanced so as to avoid coupling.

The Y-junction can mainly operate as either a divider or a combiner according to the direction of propagation of the field. Compared to the directional coupler with $L_{acc} = 1/2L_c$, that will be described later, the Y-junction has some important features which do not depend on the wavelength; furthermore, the fields from the two output ports are in phase with each other. However, one of the drawbacks of these devices is the significant power loss. For this reason, both the taper and the subsequent curved waveguides must be properly designed in order to minimise losses. Every variation in terms of waveguide features, in fact, entails the excitation of radiation modes with a consequent power loss: the slower and more gradual these changes, the lower the power loss. Typically, the

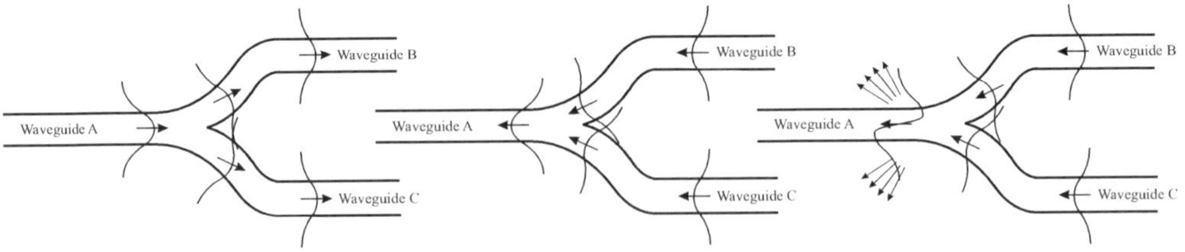

Fig. 4.3: From left to right: the excitation of the waveguide A; the excitation of the waveguides B and C in phase; the excitation of the waveguides B and C in phase opposition.

angle between the axes of the output waveguides cannot be greater than a few degrees: this results in considerably longer components – up to several hundred or a few thousand microns.

By acting on the phase relations between the modes propagating in the different waveguides, the Y-junction may also be used as an optical switch. For the sake of simplicity, let us assume that we are dealing with single-mode identical waveguides. According to figure 4.3, if the field propagates in the waveguide A on the left, all losses aside, then it will be evenly distributed between the two output waveguides B and C – and this is because the two fundamental modes are excited at the junction. Similarly, if two modes propagate in phase in the waveguides B and C, they combine at the waveguide junction to produce a field with even symmetry, with respect to the axis of the waveguide A. If the taper region is properly designed, the aforementioned field will excite the fundamental mode of the waveguide A, thus leading to a sum of powers. If, on the other hand, the two modes are in phase opposition, their combining gives rise to a field distribution with odd symmetry which will only excite the radiation modes – since the fundamental mode has even symmetry. In this case, the guided power in A is zero, hence the output power can be maximised or cancelled simply by adjusting the phase of the fields propagating in the waveguides B and C, which is very close to what happens in directional couplers.

According to what we have just explained, we can easily show that, when the waveguide A is excited, the transfer matrix is:

$$[T] = \frac{1}{\sqrt{2}} \begin{bmatrix} 1 \\ 1 \end{bmatrix};$$

on the other hand, if the field is propagated in the opposite direction, we will have:

$$[T] = \begin{bmatrix} 1 & 1 \end{bmatrix}.$$

4.2 Directional Couplers

Directional couplers are constituted by two dielectric waveguides initially not interacting with each other. As figure 4.4 clearly shows, they come close for a given amount of space – the finite length L_{coup} – and then separate once more. For this reason, the two coupled waveguides are set close enough together in the section L_{coup}: if one waveguide is excited, the field is gradually transferred to the other, and then finally reverts to the first. The transfer follows a periodic trend and comes to an end only when the waveguides are distanced enough. This aspect is taken into great consideration in photonics and it can be understood by analysing the properties of the modes of the coupler. Indeed, if we neglect the effects of the curves necessary to first bring close and then separate the two waveguides, and if we only consider the section in which the two waveguides are parallel, the structure is invariant along the propagation direction and is therefore characterised by a set of modes of its own – both guided and radiation modes – known as *supermodes*.

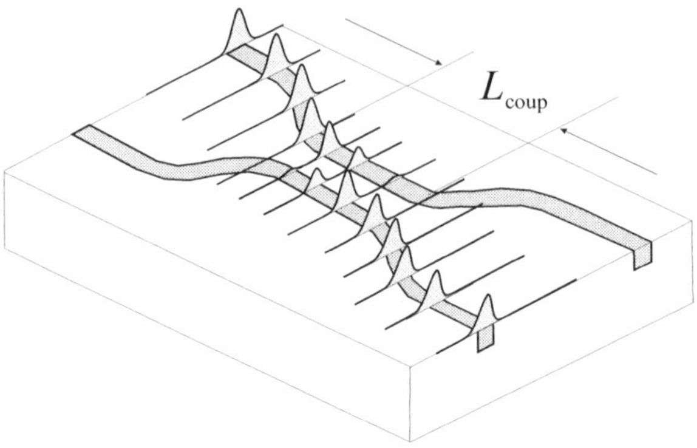

Fig. 4.4: Directional Coupler.

As far as our analysis is concerned, we will initially refer to the simplest type of coupler made of two identical and parallel slabs. As outlined in figure 4.5, such a structure can be seen as a single five-layer waveguide: two $2d$-thick high-index guiding layers centred at $x = \pm x_0$, a $2s$-thick central region of lower index, and two side claddings. By virtue of the invariance along y, a classification between TE and TM modes is still possible. If the two waveguides are identical – as is the case considered – we have a symmetrical structure with respect to the line equidistant from the two cores. As a consequence of this symmetry, we will have either *even* or *odd* supermodes. By solving the Helmholtz equation in the different layers and imposing the continuity conditions at the interfaces, we can obtain the expressions of the mode functions.

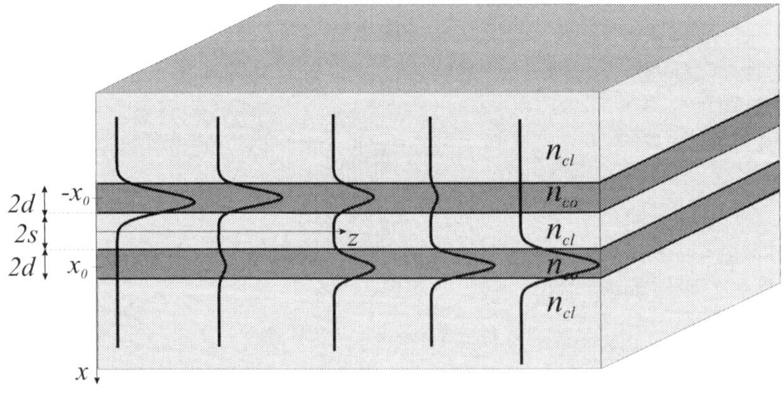

Fig. 4.5: Directional coupler constituted by two identical slabs.

In particular, for even TE modes, we have:

$$E_y(x) = \begin{cases} Ce^{-\gamma_0(x-s-2d)} & \text{if} \quad x > s+2d \\ A\cos[k_{x1}(x-s)] + B\sin[k_{x1}(x-s)] & \text{if} \quad s \leq x \leq s+2d \\ E(0)\cosh(\gamma_2 x) & \text{if} \quad |x| < s \\ A\cos[k_{x1}(x+s)] - B\sin[k_{x1}(x+s)] & \text{if} \quad -(s+2d) \leq x \leq -s \\ Ce^{\gamma_0(x+s+2d)} & \text{if} \quad x < -(s+2d) \end{cases} \tag{4.1}$$

where $E(0)$ is the value of the function in correspondence to the plane of symmetry,

$$A = E(0)\cosh(\gamma_2 s), \qquad B = E(0)Z\sinh(\gamma_2 s),$$

$$C = E(0)[\cos(2k_{x1}d)\cosh(\gamma_2 s) + Z\sin(2k_{x1}d)\sinh(\gamma_2 s)], \qquad Z = (\gamma_2/k_{x1})(n_{co}/n_{cl})^2.$$

As for odd TE modes:

$$E_y(x) = \begin{cases} -C'e^{-\gamma_0(x-s-2d)} & \text{if} \quad x > s+2d \\ A'\cos[k_{x1}(x-s)] - B'\sin[k_{x1}(x-s)] & \text{if} \quad s \leq x \leq s+2d \\ -E(0)\sinh(\gamma_2 x) & \text{if} \quad |x| < s \\ -A'\cos[k_{x1}(x+s)] - B'\sin[k_{x1}(x+s)] & \text{if} \quad -(s+2d) \leq x \leq -s \\ C'e^{\gamma_0(x+s+2d)} & \text{if} \quad x < -(s+2d) \end{cases} \tag{4.2}$$

where

$$A' = E(0)\sinh(\gamma_2 s), \quad B' = E(0)Z\cosh(\gamma_2 s),$$

$$C' = E(0)[\cosh(2k_{x1}d)\sinh(\gamma_2 s) + Z\sin(2k_{x1}d)\cosh(\gamma_2 s)].$$

Similar expressions can be found for the H_y component of TM modes. Remember that $\cosh(x) = (e^x + e^{-x})/2$ and $\sinh(x) = (e^x - e^{-x})/2$; as for the single slab, we notice that trigonometric functions describe the guided modes in high-index regions, while the same modes are described by decreasing exponentials in low-index regions. Figure 4.6 shows the profiles of the first even and the first odd supermode, that is the fundamental and the first higher-order mode, respectively. Both are constituted by a pair of lobes approximately centred on the two high-index slabs: as for the even supermode, the two lobes have the same sign; we have opposite signs, on the contrary, in the case of the odd supermode. Further analysis – here omitted for reasons of brevity – shows that,

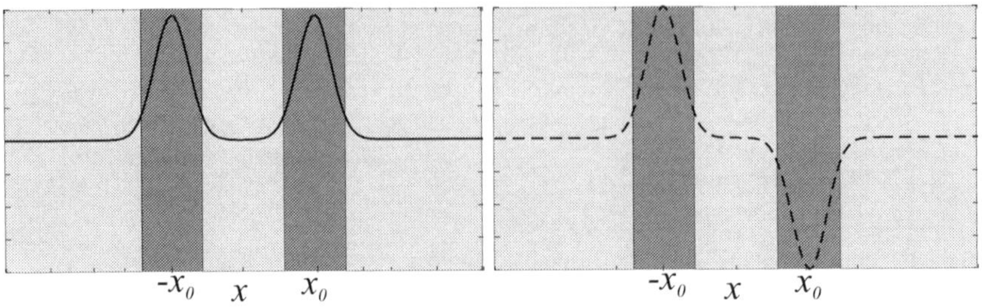

Fig. 4.6: On the left, even supermode; on the right, odd supermode.

if the two cores are distanced to such an extent that the value of $E(0)$ becomes negligible, the different lobes are then very similar to the profile of the fundamental mode of the two waveguides considered separately. In this case, the symbol $E_0(x)$ will be used to designate the mode function of the TE_0 mode of a slab centred in $x = 0$, while $E_e(x)$ and $E_o(x)$ will respectively designate the mode functions of the even and odd TE supermodes. Subsequently:

$$E_e(x) = \frac{1}{2}\left[E_0(x + x_0) + E_0(x - x_0)\right]; \tag{4.3}$$

$$E_o(x) = \frac{1}{2}\left[E_0(x + x_0) - E_0(x - x_0)\right]. \tag{4.4}$$

Figure 4.7 shows how, in this case, the in-phase or opposite-phase sum (difference) of two modes leads to a field distribution which is identical to that of the fundamental mode of the single waveguide – centred on the left waveguide (from now on, waveguide a) and on the right waveguide (waveguide b), respectively. According to (4.3) and (4.4), in fact, it follows that:

$$E_0(x + x_0) = E_e(x) + E_o(x);$$

$$E_0(x - x_0) = E_e(x) - E_o(x).$$

Despite their cutoff angular frequency being zero, the two supermodes are not degenerate; the odd supermode, in particular, is somewhat less confined in the cores, thus its effective index n_{eff_o} is slightly lower than that of the even supermode n_{eff_e}. If β_e and β_o are the phase constants of the even and odd modes respectively, then:

$$\Delta\beta = \beta_e - \beta_o = 2\pi\frac{n_{eff_e} - n_{eff_o}}{\lambda} > 0.$$

Suppose now that we only excite the waveguide a in $z = 0$:

$$\bar{e}_{in}(x) = E_0(x + x_0)\hat{\jmath}.$$

This type of field excites the in-phase sum of the two supermodes, in that:

$$\bar{e}_{in}(x) = E_0(x + x_0)\hat{\jmath} = (E_e + E_o)\hat{\jmath} = \bar{e}_e(x, 0) + \bar{e}_o(x, 0),$$

$\overline{e}_e$ and $\overline{e}_o$ indicating the fields of even and odd supermodes, respectively. Figure 4.7(a) illustrates the situation in the input region. Due to the different phase constants, during the propagation the two supermodes shift progressively out of phase. With regard to the even supermode, the profile of the field along the coupler is the following:

$$\begin{aligned}
\overline{e}(x,z) &= \overline{e}_e(x,z) + \overline{e}_o(x,z) \\
&= \left[E_e(x)e^{-j\beta_e z} + E_o(x)e^{-j\beta_o z}\right]\hat{\jmath} \\
&= \left[E_e(x) + E_o(x)e^{j\Delta\beta z}\right]e^{-j\beta_e z}\hat{\jmath} \\
&= \frac{1}{2}\left[E_0(x+x_0) + E_0(x-x_0) + E_0(x+x_0)e^{j\Delta\beta z} - E_0(x-x_0)e^{j\Delta\beta z}\right]e^{-j\beta_e z}\hat{\jmath} \\
&= \frac{1}{2}\left[E_0(x+x_0)(1+e^{j\Delta\beta z}) + E_0(x-x_0)(1-e^{j\Delta\beta z})\right]e^{-j\beta_e z}\hat{\jmath},
\end{aligned} \tag{4.5}$$

From (4.5) we can infer that the field amplitude on the waveguide a is $1/2(1+e^{j\Delta\beta z})$, while on the waveguide b is $1/2(1-e^{j\Delta\beta z})$. When the *coupling length* is equal to:

$$L_c = \frac{\pi}{\Delta\beta} = \frac{\pi}{\Delta n_{eff}}\frac{\lambda}{2\pi} = \frac{\lambda}{2\Delta n_{eff}}$$

the two modes are in phase opposition, hence:

$$\overline{e}(x, L_c) = E_0(x-x_0)e^{-j\beta_e L_c}\hat{\jmath} = \big(E_e(x) - E_o(x)\big)e^{-j\beta_e L_c}\hat{\jmath}.$$

The situation is depicted in figure 4.7(b), where the field has passed from one waveguide to the other. Figure 4.8 shows then the profile of the magnitude of $\overline{e}(x,z)$ with its periodic trend along z – with $2L_c$ period.

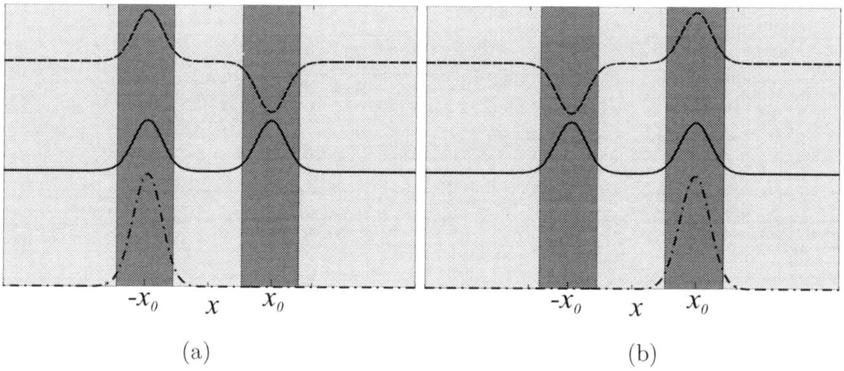

Fig. 4.7: (a) The input field and the two supermodes in phase. (b) The profile of the field and the two supermodes in case of a distance equal to L_c. Note the different sign of the odd supermode.

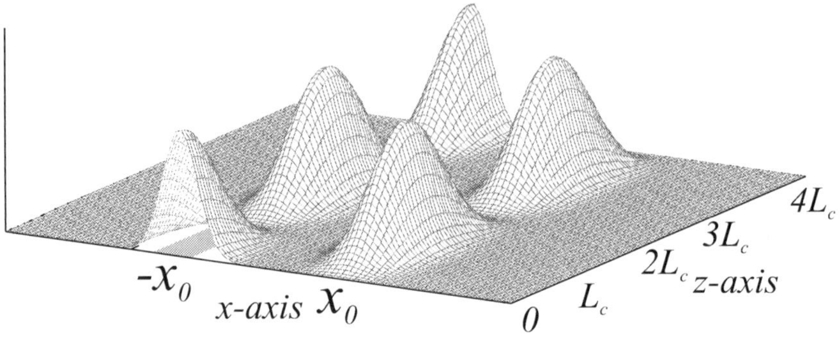

Fig. 4.8: The magnitude of $\overline{e}(x, z)$ along the coupler.

4.2.1 Rib-waveguide Couplers

Figure 4.9 gives an example of a coupler formed by two InP rib waveguides. The physical and geometrical characteristics of the two waveguides are as follows: $W = 1\mu m$, $d_1 = 0.7\mu m$, $d_2 = 0.4\mu m$, $n_c = 3.28$ and $n_s = 3.17$ – the symbol s indicating the distance which separates the two waveguides. In figure 4.10 we can see the three components of the magnetic field of the two supermodes with dominant H_x in the following conditions: $s = 2\mu m$ and $\lambda = 1550\,nm$. The effective indices of

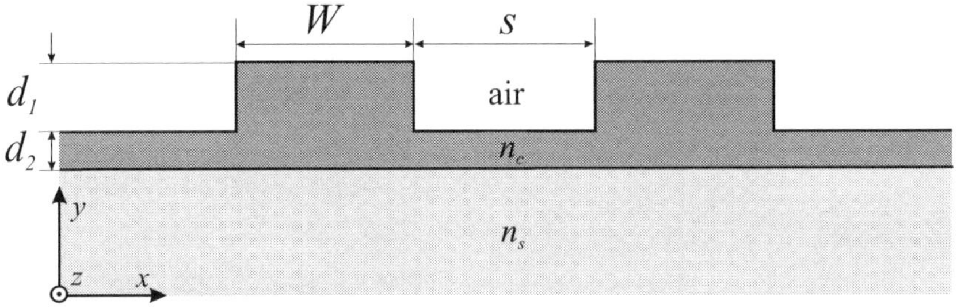

Fig. 4.9: Schematic of a rib-waveguide coupler.

even and odd supermodes are $n_{even} = 3.18507$ and $n_{odd} = 3.18453$ respectively, with a consequent coupling length equal to $L_c = 1.45mm$. Such a high value can be explained if we consider the high level of field confinement in the same waveguides, despite their being closely spaced. We can also have a pair of supermodes with dominant H_y and the following effective indices: $n_{even} = 3.18529$ and $n_{odd} = 3.18432$; in this last case, the coupling length is $L_c = 0.794mm$. The difference between the coupling lengths of the two polarisations is due to the birefringence of the two waveguides, and, more precisely, to the different degree of confinement along the x-axis. However, by adjusting the geometric dimensions of the ribs and their distance, it is possible to design a coupler which does not depend on polarisation and with the same coupling lengths. For the sake of illustration, figure 4.11

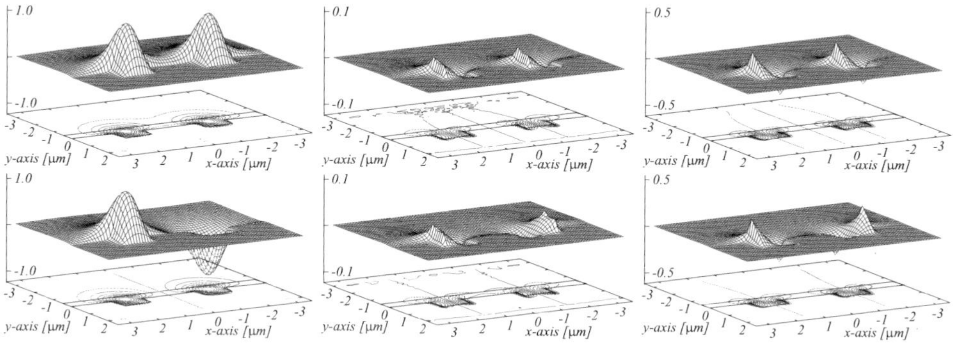

Fig. 4.10: The components of the rib-waveguide coupler from figure 4.9. From left to right: the components H_x, H_y, H_z of the magnetic field vector with respect to the even supermode H_{11}^x (top) and the odd supermode H_{11}^x (bottom).

displays the coupling lengths of the two polarisations calculated at $\lambda = 1550\,nm$, as the distance between the ribs s varies and for two different values of the width W. As the distance increases, so does L_c, but at different rates for the two polarisations. Consequently, at a certain value of s, the two lengths become identical and the coupler insensitive to the polarisation of the input field. The curves also depend on the size of the ribs: in particular, as their width W grows, the field tends to be more confined, thus increasing the coupling length. Furthermore, the dependence on s slightly changes too, thus shifting the point of intersection of the curves – *i.e.* of polarization insensitivity.

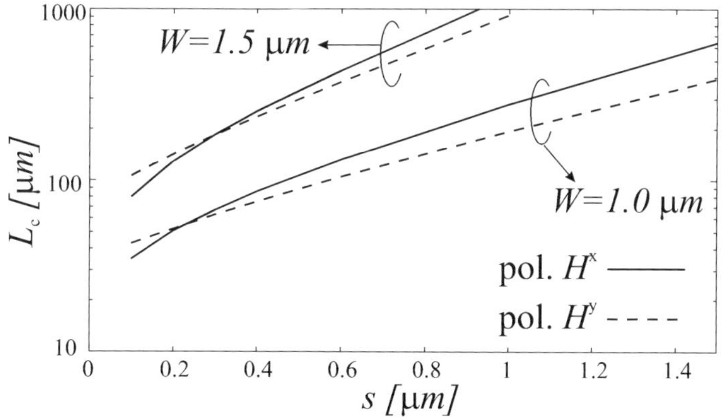

Fig. 4.11: Coupling length at 1550 nm as a function of the distance s between the two ribs, and for different values of the width W.

For this reason, it is possible to adjust the geometric parameters of the coupler so as to attain the level of polarization insensitivity for the coupling length desired.

4.2.2 Transfer Matrix

The transfer matrix is a useful method for analysing the optical components characterised by more than one input and output waveguide. Figure 4.12 shows the case of a component formed by two input and two output waveguides – assuming we are dealing with single-mode elements. $\overline{E}_{a_{in}}$, $\overline{E}_{a_{out}}$, $\overline{E}_{b_{in}}$ and $\overline{E}_{b_{out}}$ indicate the electric field mode functions of the different waveguides. The matrix relates the width of the fundamental modes A_{in} and B_{in} of the input waveguides to those of the output waveguides A_{out} and B_{out}:

$$\begin{bmatrix} A_{out} \\ B_{out} \end{bmatrix} = \begin{bmatrix} T_{11} & T_{12} \\ T_{21} & T_{22} \end{bmatrix} \cdot \begin{bmatrix} A_{in} \\ B_{in} \end{bmatrix}; \tag{4.6}$$

We call T_{11} and T_{22} "direct terms" as they relate the mode amplitudes at the input end of a

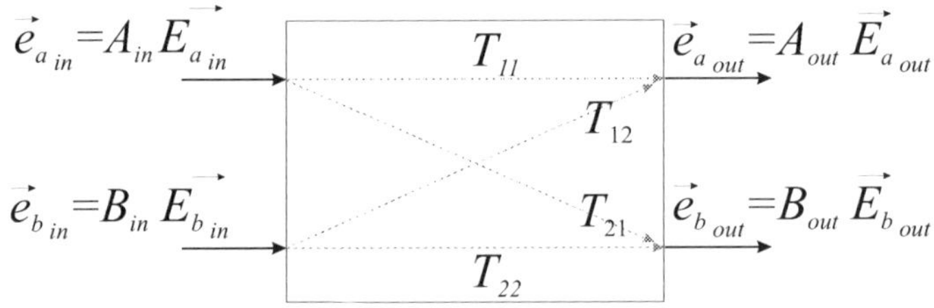

Fig. 4.12: Optical component with two input and two output waveguides.

waveguide to the output port of the same waveguide; on the other hand, when they link the mode widths of different waveguides, they are known as "crossed terms", as is the case of T_{12} and T_{21}. Cf. figure 4.13 to determine the coefficients of the transfer matrix of the coupler. From (4.6) we have:

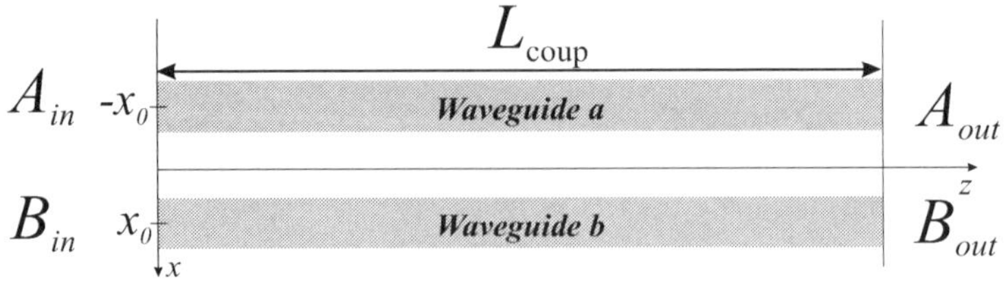

Fig. 4.13: Schematic of the coupler used to calculate the transfer matrix.

$$\begin{aligned} A_{out} &= T_{11}A_{in} + T_{12}B_{in}; \\ B_{out} &= T_{21}A_{in} + T_{22}B_{in}. \end{aligned}$$

If we only excite waveguide a with a field characterised by unit amplitude, the result is $A_{in} = 1$ and $B_{in} = 0$, therefore:

$$\begin{aligned} A_{out} &= T_{11}; \\ B_{out} &= T_{21}; \end{aligned}$$

which shall be compared to the output field expressed by (4.5), hence:

$$T_{11} = \frac{1}{2}(1 + e^{j\Delta\beta L_{coup}})e^{-j\beta_e L_{coup}};$$

$$T_{21} = \frac{1}{2}(1 - e^{j\Delta\beta L_{coup}})e^{-j\beta_e L_{coup}}.$$

By the same token, it is possible to determine the expressions of T_{12} and T_{22} by exciting waveguide b only – again with a field having uniform width. Overall, the coupler matrix has the following expression:

$$[T] = \frac{1}{2}\begin{bmatrix} 1 + e^{j\Delta\beta L_{coup}} & 1 - e^{j\Delta\beta L_{coup}} \\ 1 - e^{j\Delta\beta L_{coup}} & 1 + e^{j\Delta\beta L_{coup}} \end{bmatrix} e^{-j\beta_e L_{coup}}.$$

It is common to highlight the term $e^{-j\beta_m z}$ with

$$\beta_m = \frac{\beta_e + \beta_o}{2}$$

instead of the factor $e^{-j\beta_e z}$, thus obtaining:

$$[T] = \begin{bmatrix} \cos(\pi\frac{\Delta n_{eff}}{\lambda}L_{coup}) & -j\sin(\pi\frac{\Delta n_{eff}}{\lambda}L_{coup}) \\ -j\sin(\pi\frac{\Delta n_{eff}}{\lambda}L_{coup}) & \cos(\pi\frac{\Delta n_{eff}}{\lambda}L_{coup}) \end{bmatrix} e^{-j\beta_m L_{coup}}$$

being $\frac{\Delta\beta}{2} = \pi\frac{\Delta n_{eff}}{\lambda}$.

Before proceeding to analyse the properties of the transfer matrix, we shall make a more general statement concerning the situation considered, by taking into account the effect of the input and output waveguides of the coupler. Figure 4.14 illustrates the common case in which the two input, as well as the two output waveguides, have the same length. Let us suppose, then, we ignore the losses introduced by the curved segments of the waveguides; as a consequence, we consider that the only portion of the waveguide interested by the phenomenon of coupling is the section L_{coup}: according to what we have just explained, the waveguides simply introduce a phase shift which is identical on the two branches. Hence, the transfer matrices of the input and output waveguide pairs are:

$$[T_i] = \begin{bmatrix} 1 & 0 \\ 0 & 1 \end{bmatrix} e^{-j\phi_i}$$

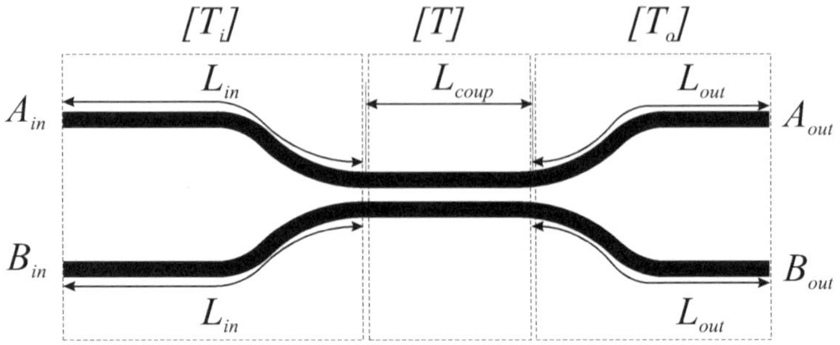

Fig. 4.14: A coupler with input and output waveguides.

and

$$[T_o] = \begin{bmatrix} 1 & 0 \\ 0 & 1 \end{bmatrix} e^{-j\phi_o}$$

where:

$$\phi_i = \int_0^{L_{in}} \beta(s)ds \qquad \phi_o = \int_{L_{in}+L_{coup}}^{L_{in}+L_{coup}+L_{out}} \beta(s)ds$$

are the phase shifts of the modes during the propagation process, and s is the curvilinear abscissa following the trend of the waveguide axis. The integrals are necessary since the changes affecting the curvature of the waveguides trigger a slight variation in terms of phase constant of the mode during the propagation process. The full expression is then:

$$[T_{coup}] = [T_o][T][T_i] = \begin{bmatrix} \cos(\pi\frac{\Delta n_{eff}}{\lambda}L_{coup}) & -j\sin(\pi\frac{\Delta n_{eff}}{\lambda}L_{coup}) \\ -j\sin(\pi\frac{\Delta n_{eff}}{\lambda}L_{coup}) & \cos(\pi\frac{\Delta n_{eff}}{\lambda}L_{coup}) \end{bmatrix} e^{-j\phi_c} \tag{4.7}$$

with $\phi_c = \phi_i + \phi_o + \beta_m L_{coup}$.

Observe that, if $A_{in} = B_{in} = a$, then according to (4.6) and (4.7):

$$A_{out} = a\left[\cos(\pi\frac{\Delta n_{eff}}{\lambda}L_{coup}) - j\sin(\pi\frac{\Delta n_{eff}}{\lambda}L_{coup})\right]e^{-j\phi_c};$$

$$B_{out} = a\left[\cos(\pi\frac{\Delta n_{eff}}{\lambda}L_{coup}) - j\sin(\pi\frac{\Delta n_{eff}}{\lambda}L_{coup})\right]e^{-j\phi_c}.$$

With reference to Euler's formulae and the definitions of Δn_{eff} and β_m, by expanding ϕ_c we obtain:

$$A_{out} = a\,e^{-j2\pi\frac{n_{eff_e}}{\lambda}L_{coup}}e^{-j(\phi_i+\phi_o)};$$

$$B_{out} = a\,e^{-j2\pi\frac{n_{eff_e}}{\lambda}L_{coup}}e^{-j(\phi_i+\phi_o)}.$$

Hence, the output field is a copy of the input field according to the fact that the symmetrical excitation of the two input waveguides corresponds to the excitation of the even supermode only, which is then propagated unaltered save for a phase shift. Similarly, the same procedure can be followed as far as the odd supermode is concerned, provided $A_{in} = -B_{in} = a$.

4.2.3 Properties and Applications

The purpose of this section is to highlight the most interesting properties of the coupler from a practical point of view.

Let us first consider the situation in which only one input waveguide is excited, *e.g.* the waveguide a; as a consequence, $A_{in} = a$ and $B_{in} = 0$. According to (4.6) and (4.7), it follows that:

$$A_{out} = a\,\cos(\pi\frac{\Delta n_{eff}}{\lambda}L_{coup})e^{-j\phi_c};$$
$$B_{out} = -ja\,\sin(\pi\frac{\Delta n_{eff}}{\lambda}L_{coup})e^{-j\phi_c}.$$

In terms of power,

$$P_a = |A_{out}|^2 = P_{in}\cos^2(\pi\frac{\Delta n_{eff}}{\lambda}L_{coup}),$$
$$P_b = |B_{out}|^2 = P_{in}\sin^2(\pi\frac{\Delta n_{eff}}{\lambda}L_{coup}),$$

being $P_{in} = |a|^2$, whose trends, according to the variations of L_{coup}, are reported in figure 4.15. If $L_{coup} = L_c/2$, the power on the two waveguides is:

$$P_a = P_b = \frac{1}{2}P_{in};$$

in this case, the device is known as *3-dB coupler* (see figure 4.15). More generally, the ratio between P_a and P_b is:

$$\frac{P_b}{P_a} = \tan^2\left(\pi\frac{\Delta n_{eff}}{\lambda}L_{coup}\right). \quad (4.8)$$

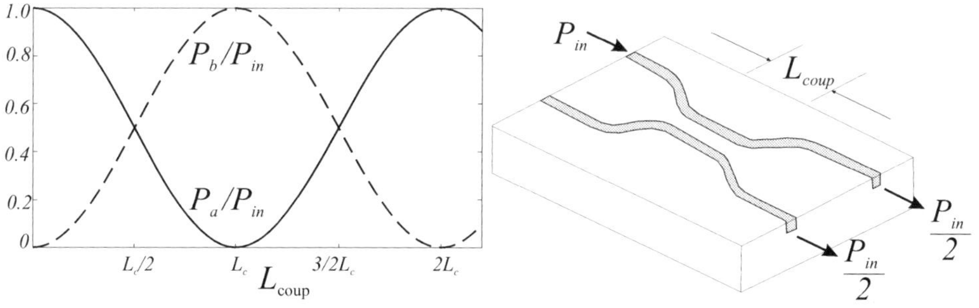

Fig. 4.15: On the left, the power trend at the output end of the coupler according to the length variation – in this case, only the waveguide a is excited. On the right, an example of a *3-dB* coupler.

As a consequence, if we want a $x : 100 - x$ coupler, the following must be true:

$$P_a = \frac{x}{100} P_{in}\,; \qquad P_b = (1 - \frac{x}{100}) P_{in},$$

hence:

$$\frac{P_b}{P_a} = \frac{100 - x}{x}. \tag{4.9}$$

By equating (4.8) and (4.9), and highlighting the coupler length, we obtain:

$$L_{coup} = \frac{\lambda}{\pi \Delta n_{eff}} \arctan\left(\sqrt{\frac{100 - x}{x}}\right).$$

As for almost all the devices with more than one waveguide, the behaviour of the coupler does not only depend on the value of the transfer matrix coefficients – that is, on the term $\Delta n_{eff} L_{coup}/\lambda$ – but it is also linked to the width and phase of the mode exciting the input waveguides.

If the coupler length is equal to the coupling length, $L_{coup} = L_c = \lambda/(2\Delta n_{eff})$, from (4.7) it follows that:

$$[T] = \begin{bmatrix} 0 & -j \\ -j & 0 \end{bmatrix} e^{-j\phi_c} = \begin{bmatrix} 0 & 1 \\ 1 & 0 \end{bmatrix} e^{-j(\phi_c + \pi/2)}.$$

If then $A_{in} = a$ and $B_{in} = b$, at the output end – apart from a common phase shift of $\phi_c + \pi/2$ – $A_{out} = b$ and $B_{out} = a$, *i.e.* the signal entering one waveguide shifts to the other. Basically, the coupler behaves as an optical switch: technically speaking, we talk about *cross-state coupling.*

If the coupler length is twice the coupling length, $L_{coup} = 2L_c = \lambda/(\Delta n_{eff})$, then:

$$[T] = \begin{bmatrix} 1 & 0 \\ 0 & 1 \end{bmatrix} e^{-j(\phi_c \pm \pi)}.$$

Since only the elements of the main diagonal are non-zero, the signal does not shift from one waveguide to the other and the final effect corresponds to that of two decoupled waveguides. In this case, the coupler is said to be in *bar state.* In all these cases, the matrices – being diagonal – cause no interaction between the signals on the two input waveguides of the coupler.

As for the 3-dB coupler, in this case the coupler length is half the coupling length, $L_{coup} = L_c/2 = \lambda/(4\Delta n_{eff})$:

$$[T] = \frac{\sqrt{2}}{2} \begin{bmatrix} 1 & -j \\ -j & 1 \end{bmatrix} e^{-j\phi_c}.$$

If we only excite one input waveguide, *e.g.* $A_{in} = a$ and $B_{in} = 0$, then:

$$A_{out} = \frac{\sqrt{2}}{2} a e^{-j\phi_c};$$
$$B_{out} = -j\frac{\sqrt{2}}{2} a e^{-j\phi_c};$$

the input field is equally divided on the two output waveguides, even if they are out of phase by $\pi/2$. On the other hand, when we excite both input waveguides, the matrix being no longer diagonal, the field distribution on the output waveguide is related not only to the field width on the input waveguides, but it also depends on the relative phase shift.

For example, supposing we excite the two input waveguides by means of two fields having the same amplitude c, yet out of phase by $-\pi/2$. Since $A_{in} = c$ and $B_{in} = -jc$, we will have on the input end:

$$A_{out} = 0;$$
$$B_{out} = -j\sqrt{2}ce^{-j\phi_c},$$

the fields are on the same output waveguide. If the field on the waveguide b is put out of phase by π with respect to the previous situation, then $A_{in} = c$ and $B_{in} = jc$; hence:

$$A_{out} = \sqrt{2}ce^{-j\phi_c};$$
$$B_{out} = 0.$$

A similar result can be achieved by putting the field on the waveguide a out of phase by π – $A_{in} = -c$ and $B_{in} = -jc$:

$$A_{out} = -\sqrt{2}ce^{-j\phi_c};$$
$$B_{out} = 0.$$

This proves that it is possible to control the waveguide from which the field emerges simply by adjusting the phase of the input field of one of the two waveguides.

If, on the other hand, the input fields are in phase, *e.g.* $A_{in} = a$ and $B_{in} = b$, then:

$$A_{out} = \frac{\sqrt{2}}{2}(a - jb)e^{-j\phi_c};$$
$$B_{out} = \frac{\sqrt{2}}{2}(b - ja)e^{-j\phi_c},$$

the input fields are both equally distributed on the output waveguides and are out of phase with each other by $\pi/2$.

The fact that the output power depends on the wavelength is another important property that shall be taken into account. Again, supposing we excite one input waveguide, *e.g.* the waveguide a. If we substitute $A_{in} = a$ and $B_{in} = 0$ in (4.6), according to (4.7) we have:

$$P_a = |A_{out}|^2 = P_{in}\cos^2\left(\pi\frac{\Delta n_{eff}}{\lambda}L_{coup}\right);$$
$$P_b = |B_{out}|^2 = P_{in}\sin^2\left(\pi\frac{\Delta n_{eff}}{\lambda}L_{coup}\right);$$

being $P_{in} = |a|^2$, whose trends – according to the wavelength variations – are reported in figure 4.16. From the chart we can infer that $P_a = P_{in}$ and $P_b = 0$ for certain wavelengths, while the opposite is true for some others. This property can be used to design multiplexers and demultiplexers for channels propagating at different wavelengths. To identify which conditions must be met in this respect, we suppose sending two signals with wavelength λ_1 and λ_2, respectively, to the input port.

Defining as L_{c1} and L_{c2} the coupling length values according to the two wavelengths considered, and setting

$$L_{coup} = 2mL_{c1} = (2n+1)L_{c2} \tag{4.10}$$

where m and n are integers, the two transfer matrices then become:

$$[T](\lambda_1) = \begin{bmatrix} \pm 1 & 0 \\ 0 & \pm 1 \end{bmatrix} e^{-j\beta_m L_{coup}}$$

$$[T](\lambda_2) = \begin{bmatrix} 0 & \pm j \\ \pm j & 0 \end{bmatrix} e^{-j\beta_m L_{coup}}.$$

As a consequence, the coupler is in bar state at λ_1 and in cross state at λ_2. Hence, if at the coupler input port the two signals propagate on different waveguides, they merge into one single waveguide at the output end: in this first case, the component performs the function of a wavelength multiplexer (cf. figure 4.16); conversely, if the signals are propagated on the same waveguide at the input port and then split on different waveguides at the output end, the device is used as a demultiplexer. These kinds of applications are found in optical amplifiers where it is necessary to multiplex on a single fiber both the signal to be amplified – with a wavelength close to $1530nm$ – and the pump beam with a $980-nm$ wavelength. However, the more similar the channel lengths, the more similar the coupling lengths and, as a consequence, the greater L_{coup} will have to be. In fact, being $n = m$ for reasons of simplicity, and indicating $\Delta\lambda = \lambda_1 - \lambda_2$, from (4.10) we can deduce that:

$$2m\frac{\lambda_1}{2\Delta n_{eff}(\lambda_1)} = (2m+1)\frac{\lambda_2}{2\Delta n_{eff}(\lambda_2)},$$

from which – if we assume to a first approximation that $\Delta n_{eff}(\lambda_1) \simeq \Delta n_{eff}(\lambda_2)$ – we obtain:

$$2m+1 = \frac{\lambda_1}{\Delta\lambda}. \tag{4.11}$$

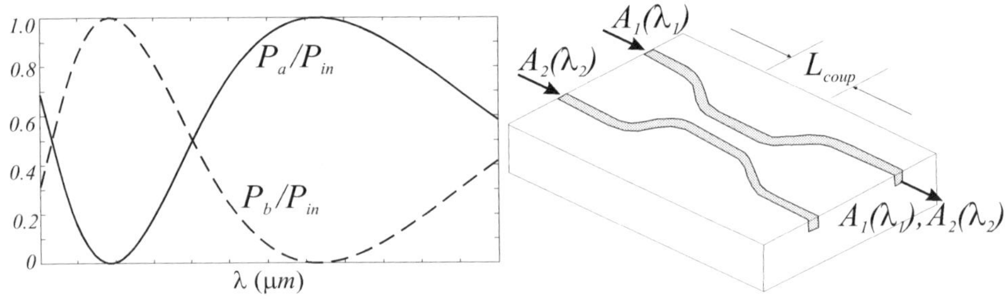

Fig. 4.16: On the left, the trend of the output power on the two waveguides according to the wavelength variation – in this case, only the waveguide a is excited. On the right, a coupler used as wavelength multiplexer.

Given that $2m + 1$ can be inferred from (4.10), the final result is:

$$L_{coup} = \frac{\lambda_1 \lambda_2}{2\Delta n_{eff} \Delta\lambda}, \tag{4.12}$$

which is valid, according to (4.11), on condition that λ_1 is an odd multiple of $\Delta\lambda$. (4.12) underlines a general characteristic of photonic components: the longer the component, the greater its ability to discriminate similar wavelengths. For example, let us determine the length required by a coupler to discriminate two different wavelengths, λ_1 and λ_2, separated by $\Delta\lambda = 10\, nm$, and assuming that $\Delta n_{eff} = 0.001$ and $\lambda_1 = 1550\, nm$. Since:

$$\frac{\lambda_1}{\Delta\lambda} = 155,$$

the condition imposed by (4.11) is fulfilled; according to (4.12) thus, the coupler length will have to be:

$$L_{coup} = \frac{1550 \cdot 1540}{2 \cdot 0.001 \cdot 10} 10^{-9} = 119.35\, mm.$$

Such dimensions effectively prevent its construction, even more so if we consider the channel spacing ($0.8 \div 0.4\ nm$) in *DWDM* systems. As we shall see later, there exist other components whose characteristics are more sharply dependent on the wavelength, thus allowing to perform multiplexing and demultiplexing operations with dimensions compatible with integrated optics.

4.2.4 Characteristic Parameters

With reference to figure 4.17, we will now delineate the main parameters characterising optical couplers.

- *Excess Loss (EL)*, defined as the ratio between the input power and the total power on the output branches.

$$EL = 10\, log_{10} \frac{P_1}{P_3 + P_4}\ (dB). \tag{4.13}$$

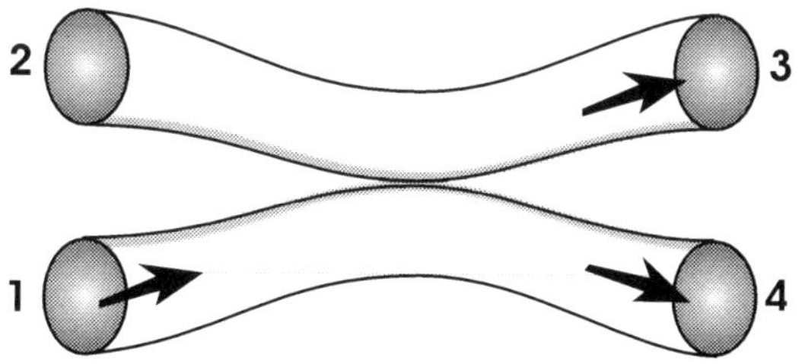

Fig. 4.17: Four-port coupler.

- *Insertion Loss (IL)*, defined as the loss of a particular optical path between the input and the output port. Depending on whether the coupler is in bar or cross state, two IL values can be identified, namely:

$$IL_{14} = 10\,log_{10}\frac{P_1}{P_4}\,(dB). \tag{4.14}$$

$$IL_{13} = 10\,log_{10}\frac{P_1}{P_3}\,(dB). \tag{4.15}$$

- *Directivity (D)*, defined as the fraction of unwanted power at port 2 with respect to the input power provided at port 1, that is to say:

$$D = -10\,log_{10}\frac{P_2}{P_1}\,(dB). \tag{4.16}$$

- *Crosstalk (XT)*, defined as the ratio between the unwanted output power at port 2 and the input power at port 1. The description is as follows:

$$XT = 10\,log_{10}\frac{P_2}{P_1}\,(dB). \tag{4.17}$$

  Crosstalk and directivity are closely related in that the two parameters have the same absolute value: directivity is usually written as a positive value, while crosstalk is negative.

- *Coupling Ratio (CR)*, defined as the ratio between the output power at port 4 and the sum of the output powers at ports 3 and 4 – *i.e.* the ratio of coupled to total output power:

$$CR = -10\,log_{10}\frac{P_4}{P_3 + P_4}\,(dB). \tag{4.18}$$

  Alternatively, we can talk about "percent coupling ratio" as the percentage distribution of optical power between the output ports:

$$CR(\%) = \left[\frac{P_4}{P_3 + P_4}\right] \times 100\% = \left[1 - \frac{P_3}{P_3 + P_4}\right] \times 100\%. \tag{4.19}$$

  As a consequence, the following equality holds:

$$IL = CR + EL. \tag{4.20}$$

- *Return Loss (RL)*, defined as the ratio between the output and input power on the input port. For obvious reasons, the RL ought to be as high as possible and is thus expressed:

$$RL = -10\,log_{10}\frac{P_1^{out}}{P_1^{in}}. \tag{4.21}$$

Case Study

In a multimode-fiber coupler the optical power launched into port 1 is $P_1 = 60\,\mu$W. The output powers at ports 2, 3 and 4 are, respectively, $P_2 = 0.004\,\mu$W, $P_3 = 26.0\,\mu$W and $P_4 = 27.5\,\mu$W.

Determine the excess loss, the insertion loss between the input and output ports 3 and 4, the directivity, or crosstalk, and the coupling ratio of the device.

Solution:

EL can be determined through the equation (4.13):

$$EL = 10\,log_{10}\frac{P_1}{P_3 + P_4} = 10\,log_{10}\frac{60}{53.5} = 0.5\,dB$$

ILs are calculated by resorting to (4.15):

$$IL_{13} = 10\,log_{10}\frac{P_1}{P_3} = 10\,log_{10}\frac{60}{26} = 3.63\,dB$$

$$IL_{14} = 10\,log_{10}\frac{P_1}{P_4} = 10\,log_{10}\frac{60}{27.5} = 3.39\,dB$$

(4.16) gives the directivity value:

$$D = -10\,log_{10}\frac{P_2}{P_1} = -10\,log_{10}\frac{0.004}{60} = 41.8\,dB$$

from which $XT = -41.8\,dB$. Finally, *CR(%)* is determined thanks to (4.19):

$$CR = \left[\frac{P_3}{P_3 + P_4}\right] \times 100 = \left[\frac{26}{53.5}\right] \times 100 = 48.6\%.$$

4.2.5 Coupled Mode Theory Approach

The directional coupler can also be studied considering the coupled mode theory (CMT), as illustrated in Appendix B. Figure 4.18 shows its transverse section, consisting of two cores with refractive indices n_a and n_b, surrounded by a cladding with refractive index n_{cl}. The CMT is based on a perturbational approach; to apply it, we should consider a coupler like the one represented in figure 4.19, *viz.* a single waveguide perturbed by a second waveguide set at a finite distance from the first one. The study of the effects produced by such perturbation will allow us to describe the performance of the device in a similar way to what was seen with respect to the even and odd supermodes approach.

Since the second waveguide only alters the index distribution on the transverse section, the perturbation $\Delta\varepsilon$ only depends on x and y, not on z. The coupling between counterpropagating modes will thus be negligible. If we also assume, for the sake of simplicity, that we are dealing with two single-mode waveguides, and that the electric field longitudinal components are zero, we will have $K^z_{\nu\mu} = 0$. According to these conditions, (B.27) will result in:

$$\frac{dA_a}{dz} = -jK^t_{aa}A_a - jK^t_{ba}A_b e^{-j(\beta_b - \beta_a)z}; \tag{4.22}$$

$$\frac{dA_b}{dz} = -jK^t_{bb}A_b - jK^t_{ab}A_a e^{-j(\beta_a - \beta_b)z}; \tag{4.23}$$

where the subscripts a and b are used in lieu of ν and μ to indicate the two waveguides considered. The effect of the perturbation causes a variation in the phase constants of the two modes due to the

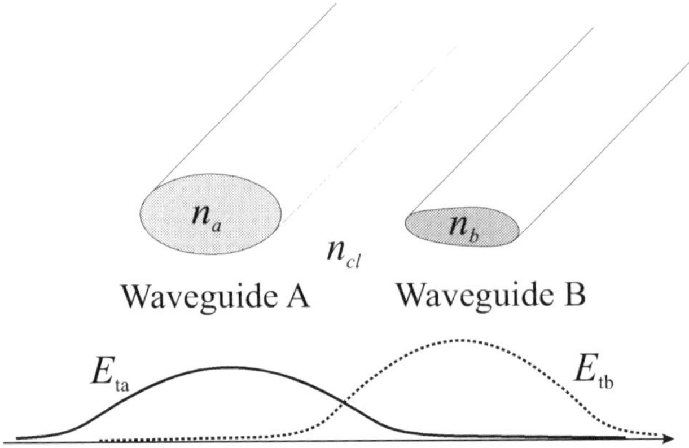

Fig. 4.18: The transverse section of a coupler and the qualitative profile of the fundamental mode of the two waveguides considered separately.

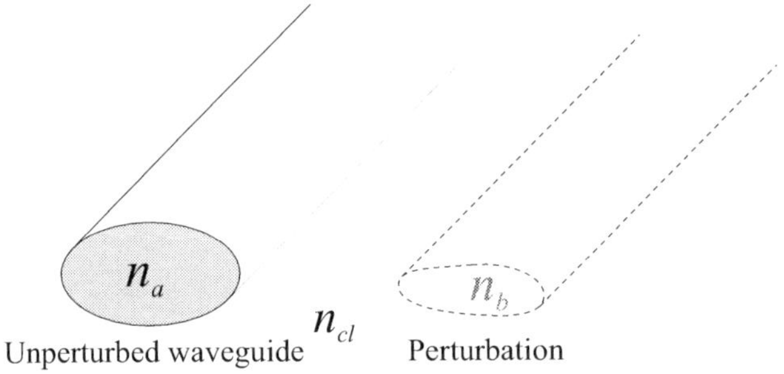

Fig. 4.19: Waveguide with perturbation.

coefficients K^t_{aa} and K^t_{bb}, as well as a coupling between the two waves due to K^t_{ab} and K^t_{ba}. Note that, without these coupling terms, the two equations would be reduced to relations such as

$$\frac{dA_\mu}{dz} = -jK^t_{\mu\mu}A_\mu,$$

with $\mu = a, b$ – that is, the simple description of the amplitude variation of the μ-th mode along z.

We shall now define the perturbation in order to calculate the coupling coefficients. If we initially assume that the waveguide a is situated *at a finite distance*, the perturbation can be attributed to the waveguide b, which modifies *the cladding of the waveguide* a by $(\epsilon_b - \epsilon_{cl})$. For this reason it must be:

$$K^t_{aa} = \frac{\omega}{4}\int\int_{-\infty}^{\infty}(\epsilon_b - \epsilon_{cl})\overline{E}_{ta}\cdot\overline{E}^*_{ta}dx\,dy; \tag{4.24}$$

$$K^t_{ba} = \frac{\omega}{4} \int\int_{-\infty}^{\infty} (\epsilon_b - \epsilon_{cl}) \overline{E}_{tb} \cdot \overline{E}^*_{ta} dx\, dy. \tag{4.25}$$

The two equations above highlight how K^t_{aa} is negligible with respect to K^t_{ba} since, as figure 4.18 shows, in the first case the integrand in the perturbed region is given by the product of the tails of the unperturbed waveguide field; in the second case, however, we have the proper solution to the unperturbed waveguide. This is why, from now on, we shall ignore the terms in K^t_{aa} and K^t_{bb} as far as (4.22) and (4.23) are concerned. To determine K^t_{ab}, we may assume that the waveguide b is situated at a finite distance and that the perturbation is produced by the waveguide a. Consequently:

$$K^t_{ab} = \frac{\omega}{4} \int\int_{-\infty}^{\infty} (\epsilon_a - \epsilon_{cl}) \overline{E}^*_{tb} \cdot \overline{E}_{ta} dx\, dy \tag{4.26}$$

and the equations regulating the directional coupler become:

$$\frac{dA_a}{dz} = -jK^t_{ba} A_b e^{-j(\beta_b - \beta_a)z} \tag{4.27}$$

$$\frac{dA_b}{dz} = -jK^t_{ab} A_a e^{-j(\beta_a - \beta_b)z}. \tag{4.28}$$

This is enough to continue our investigation. However, it is possible to determine a relation between K^t_{ab} and K^t_{ba} resulting from a condition of power conservation. Once the mode functions are normalised, the power of the μ-th mode is simply defined as:

$$P_\mu = A_\mu A^*_\mu.$$

Since the equations describing the operating principles of the directional coupler introduce two modes only, it follows that:

$$P_t = P_a + P_b = A_a A^*_a + A_b A^*_b;$$

thus, according to power conservation principles:

$$\frac{dP_t}{dz} = 0$$

from which

$$\frac{dA_a}{dz} A^*_a + A_a \frac{dA^*_a}{dz} + \frac{dA_b}{dz} A^*_b + A_b \frac{dA^*_b}{dz} = 0.$$

By substituting (4.27) and (4.28), we will have:

$$jA_a A^*_b (K^*_{ba} - K_{ab}) e^{j\delta z} + jA^*_a A_b (K^*_{ab} - K_{ba}) e^{-j\delta z} = 0$$

where we hold that:

$$\delta = \beta_b - \beta_a;$$

as the equation above has to be verified according to each value of z, it must certainly be:

$$K^*_{ab} = K_{ba} = k.$$

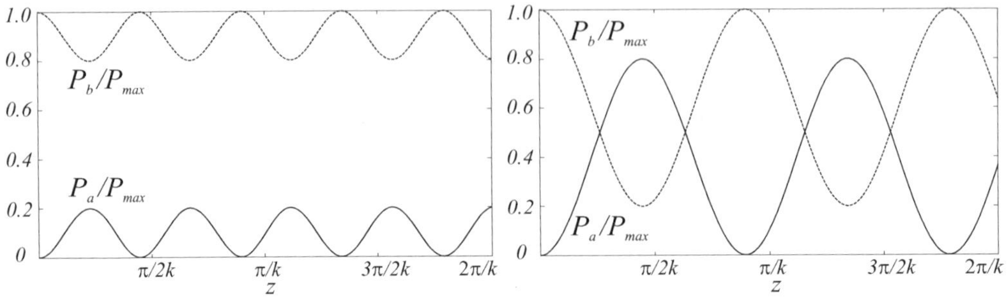

Fig. 4.20: Normalised powers of modes a and b according to the level of synchronism – *i.e.* for decreasing values of δ – with the same coupling coefficient. On the right, $\delta = 4k$; on the left, $\delta = k$.

Accordingly, the equations of the coupler become:

$$\frac{dA_a}{dz} = -jkA_b e^{-j\delta z}; \tag{4.29}$$

$$\frac{dA_b}{dz} = -jk^* A_a e^{j\delta z}. \tag{4.30}$$

For the system to be solved, (4.29) and (4.30) must necessarily be decoupled. By deriving the former and substituting the latter, the result is a quadratic equation with one variable $A_a(z)$ only:

$$\frac{d^2 A_a}{dz^2} + j\delta\frac{dA_a}{dz} + |k|^2 A_a = 0.$$

This can be solved by employing the process outlined in Appendix B.2. In particular, according to the boundary conditions at $z = 0$ – the presence of the second waveguide mode only – that is to say, by supposing that:

$$A_a(0) = 0\ ,\ A_b(0) = A_0\ ,$$

the solutions are:

$$A_a(z) = -jA_0\frac{2k}{S}e^{-j\delta z/2}\sin\frac{Sz}{2}\ , \tag{4.31}$$

$$A_b(z) = A_0 e^{j(\delta z/2)}\left(\cos\frac{Sz}{2} - j\frac{\delta}{S}\sin\frac{Sz}{2}\right), \tag{4.32}$$

where

$$S = 2\sqrt{|k|^2 + (\delta/2)^2}\ .$$

It is easy to prove that the total power is:

$$P_t = A_a A_a^* + A_b A_b^* = |A_0|^2$$

as it should be. Moreover,

$$P_a = |A_0|^2\frac{4|k|^2}{S^2}\sin^2\frac{Sz}{2}$$

$$P_b = P_t - P_a = |A_0|^2 \left(1 - \frac{4|k|^2}{S^2} \sin^2 \frac{Sz}{2}\right) .$$

These are two periodic functions. Remembering that the period of $\sin^2$ is equal to π, the period of the two powers P_a and P_b is determined by:

$$\Delta z = \frac{2\pi}{S} ;$$

in particular, the maximum value is:

$$P_{max} = |A_0|^2 \frac{4|k|^2}{S^2} .$$

The trends of P_a and P_b are different in the two cases $\delta \neq 0$ and $\delta = 0$. The exciting mode A_b

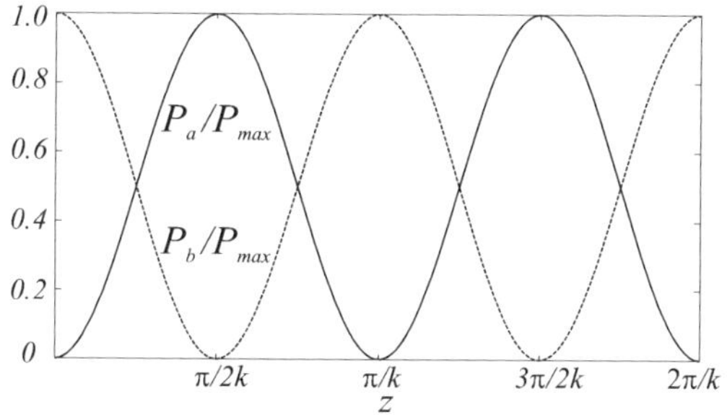

Fig. 4.21: Normalised powers of modes a and b in case of perfect synchronism ($\delta = 0$).

transmits some of its power to the mode A_a until the maximum value P_{max} is reached; the latter depends on both the coupling coefficient and the difference δ between the phase constants: for this reason, the closer we get to the condition of synchronism, the higher the maximum value will be. The situation is illustrated in figure 4.20 for two different non-zero values of δ.

On the other hand, when the condition of perfect synchronism between the two waves is reached, *i.e.* when $\delta = 0$, complete power exchange can take place – given that $P_{max} = |A_0|^2$, as illustrated in figure 4.21. Complete power exchange can occur beyond the so-called *coupling length* L_c:

$$L_c = \frac{\pi}{2|k|} .$$

This value is influenced by λ since $|k|$, too, is determined by the wavelength – because of the mode functions $\overline{E}_{ta}$ and $\overline{E}_{tb}$ depending on λ. After the excited mode has reached the maximum intensity, the exchange is reversed and power flows back to the exciting mode. High values of the coefficient k will evidently help reduce the coupling length.

In conclusion, note that the condition of synchronism – that is, the identity between the two propagation constants of the coupled modes – can also be found when it comes to modes of different order and/or modes propagating in waveguides with different geometry or index difference. For this reason, thanks to appropriate designs, it is possible to achieve mode conversion or to switch the field even in the case of highly dissimilar waveguides.

Coupler Transfer Matrix

The expressions (4.31) and (4.32) describe the trend of the solutions in the case of excitation as illustrated by the initial conditions $A_a(0) = 0$ and $A_b(0) = A_0$. If the excitation conditions of the two waveguides are exchanged, *i.e.* by imposing $A_a(0) = A_0$ and $A_b(0) = 0$, it is evident how (4.31) can now be used to describe the trend of $A_b(z)$, whereas (4.32) comes to define the trend of $A_a(z)$. Regardless of the type of excitation, these considerations allow us to describe the response of the coupler through the transfer matrix T, expressed by the following expression:

$$[T] = \begin{bmatrix} e^{j(\delta z/2)}\left(\cos\frac{Sz}{2} - j\frac{\delta}{S}\sin\frac{Sz}{2}\right) & -j\frac{2k}{S}e^{-j\delta z/2}\sin\frac{Sz}{2} \\ -j\frac{2k}{S}e^{-j\delta z/2}\sin\frac{Sz}{2} & e^{j(\delta z/2)}\left(\cos\frac{Sz}{2} - j\frac{\delta}{S}\sin\frac{Sz}{2}\right) \end{bmatrix}.$$

In the case of a device operating in conditions of perfect synchronism ($\delta = 0$), the former becomes:

$$[T] = \begin{bmatrix} \cos kz & -j\sin kz \\ -j\sin kz & \cos kz \end{bmatrix},$$

which highlights how the power distribution between the two branches not only depends on the excitation conditions, but is also closely linked to the coupling coefficient value k.

The matrix expression is formally identical to that obtained by means of the supermodes theory. According to the latter approach, the field transfer from one waveguide to another is due to the difference $\Delta\beta$ between the phase constants of the two supermodes; when it comes to the coupled mode theory, on the other hand, the transfer is caused by the coupling between the modes of the two single waveguides, calculated through the coefficient k.

Chapter 5

Filters and Interferometers

Filters are a key component in modern optical systems as they are mostly used to select one or more channels in an optical WDM system. In terms of signal transmission, they also find application as dispersion compensators or gain equalizers in optical amplifiers. In most cases, optical filters are based on interferometric structures where constructive and destructive interferences are exploited to obtain a variation of the transfer function. These two types of interference are a result of the combination of different waves with appropriate phase relationships. Since phase varies periodically at different frequencies, the interferometers are characterised by a periodic frequency response $H(\nu)$, as the example from figure 5.1 shows. The period of $H(\nu)$ is called *Free Spectral Range* (FSR_ν), whereas $FWHM_\nu$ (Full-Width at Half-Maximum) identifies the frequency range over which the power spectral density $|H(\nu)|^2$ is reduced to less than half the maximum value.
As far as photonic components are concerned, the spectral characteristics are often reported as a function of the wavelength rather than frequency. In this case, the response is no longer periodic,

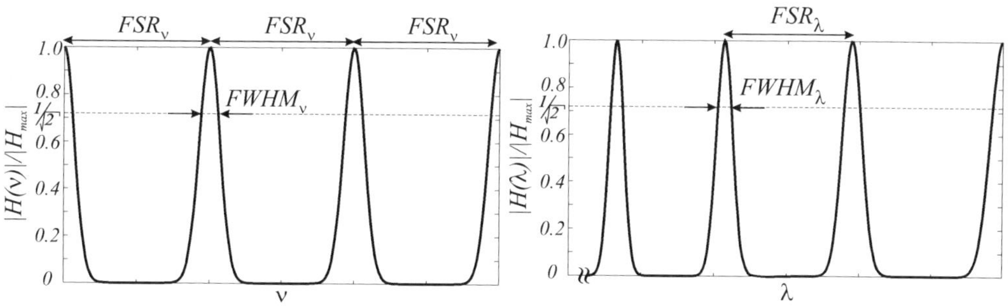

Fig. 5.1: Qualitative trend of the normalised transfer function of an optical filter according to its frequency (on the left) and wavelength (on the right).

though we still talk about FSR and $FWHM$, with implicit reference to the precise period of the harmonic response where we find the signals to be processed. By indicating with FSR_λ and $FWHM_\lambda$ the Free Spectral Range and the bandwidth in terms of wavelength respectively, we have:

$$FSR_\lambda = \frac{\lambda^2}{c} FSR_\nu; \qquad FWHM_\lambda = \frac{\lambda^2}{c} FWHM_\nu;$$

where λ is the operating wavelength.

5.1 The Fabry-Perot Interferometer

Consider a slab with refractive index equal to n_1 and thickness h, surrounded by a medium with refractive index n_2 – as figure 5.2 shows. An incident plane wave with wavelength λ is delivered onto the slab at an angle equal to θ_2. Either the electric or the magnetic field can be tangent to the interface: in the first case we consider it a TE wave, while in the second case we have a TM wave. The incident wave is partly reflected and partly transmitted: the transmitted beam subsequently strikes the second interface, and is itself partly reflected and transmitted as the process continues indefinitely. θ_1 and θ_2 represent the angles the beams form with the interface normals, and are

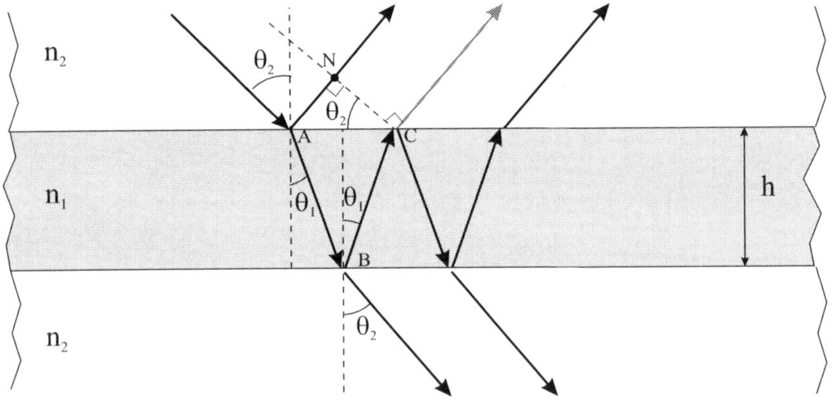

Fig. 5.2: Fabry-Pï¿½rot Interferometer.

related to the media refractive indices, n_1 and n_2, by Snell's law:

$$n_1 \sin\theta_1 = n_2 \sin\theta_2.$$

The total field in both the upper and lower half-spaces depends on the infinitely reflected and transmitted beams, each with its specific amplitude and relative phase shift – the latter, in particular, being determined by the optical path difference. In order to calculate the phase shift, consider the upper half-space, and especially the beam reflected from the first interface – represented by the black arrow in figure 5.2 – as well as the beam transmitted from the first interface, reflected from the second and again transmitted by the first – indicated by the grey arrow. The optical path of the first beam is:

$$n_2\overline{AN};$$

while that of the second is:

$$n_1\left(\overline{AB}+\overline{BC}\right).$$

Since

$$\overline{AB}=\overline{BC}=\frac{h}{\cos\theta_1};$$

$$\overline{AN}=\overline{AC}\sin\theta_2=2h\tan\theta_1\sin\theta_2;$$

it follows that the optical path difference is:

$$\begin{aligned}\Delta S &= n_1\frac{2h}{\cos\theta_1}-2hn_2\frac{\sin\theta_1}{\cos\theta_1}\sin\theta_2\\ &= \frac{2h}{\cos\theta_1}\left[n_1-n_2\sin\theta_1\sin\theta_2\right]\\ &= \frac{2h}{\cos\theta_1}n_1\left[1-\sin^2\theta_1\right]\\ &= 2hn_1\cos\theta_1;\end{aligned}$$

where Snell's law is applied. The relative phase shift is $\delta=\Delta Sk_0=2hn_1\cos\theta_1\,2\pi/\lambda$, that is:

$$\delta=4\pi h\frac{n_1\cos\theta_1}{\lambda}.$$

The amplitude $A^{(r)}$ of the total field in the upper half-space is:

$$A^{(r)}=A^{(i)}\left[r+tt'r'e^{-j\delta}+tt'r'^3e^{-j2\delta}+tt'r'^5e^{-j3\delta}+\cdots\right].$$

where r and r', respectively, indicate the reflection coefficients in medium 2 and medium 1, while t and t' represent the transmission coefficients from medium 2 to medium 1 (t) and from medium 1 to medium 2 (t'). The formula can thus be rewritten as follows:

$$A^{(r)}=A^{(i)}\left[r+tt'r'e^{-j\delta}\sum_{p=0}^{\infty}r'^{2p}e^{-jp\delta}\right].$$

The power series is of the type:

$$\sum_{p=0}^{\infty}x^p=\frac{1}{1-x}$$

with $x=r'^2e^{-j\delta}$, hence:

$$A^{(r)}=A^{(i)}\left[r+tt'r'e^{-j\delta}\frac{1}{1-r'^2e^{-j\delta}}\right].$$

According to the symmetry of the structure analysed, it follows that $r = -r'$; hence, R (reflectivity) and T (transmittance) are:

$$R = r^2 = r'^2 = -rr',$$

$$T = tt';$$

furthermore,

$$R + T = 1.$$

According to this explanation, we can rewrite the expression above in the following way:

$$\begin{aligned} A^{(r)} &= A^{(i)} \frac{r - rRe^{-j\delta} + r'Te^{-j\delta}}{1 - Re^{-j\delta}} \\ &= A^{(i)} \frac{r - r(R+T)e^{-j\delta}}{1 - Re^{-j\delta}} \\ &= A^{(i)} \frac{r(1 - e^{-j\delta})}{1 - Re^{-j\delta}} \\ &= A^{(i)} \frac{\sqrt{R}(1 - e^{-j\delta})}{1 - Re^{-j\delta}}. \end{aligned}$$

In terms of incident beam intensity $I^{(i)} = A^{(i)}A^{(i)*}$ and of reflected beam intensity $I^{(r)} = A^{(r)}A^{(r)*}$:

$$\begin{aligned} \frac{I^{(r)}}{I^{(i)}} &= \frac{R(1 - e^{-j\delta})(1 - e^{j\delta})}{(1 - Re^{-j\delta})(1 - Re^{j\delta})} \\ &= \frac{2R(1 - \cos\delta)}{1 + R^2 - 2R\cos\delta} \\ &= \frac{2R(1 - \cos\delta)}{(1 - R)^2 + 2R(1 - \cos\delta)} \\ &= \frac{F\sin^2(\delta/2)}{1 + F\sin^2(\delta/2)}, \end{aligned}$$

with

$$F = \frac{4R}{(1 - R)^2}.$$

A similar discourse can be made for the beams in the lower half-space:

$$\frac{I^{(t)}}{I^{(i)}} = \frac{1}{1 + F\sin^2(\delta/2)}. \tag{5.1}$$

The *Airy functions* describe the relations obtained for the intensity of the reflected and transmitted beam as a function of the incident beam. Note that:

$$\frac{I^{(r)}}{I^{(i)}} + \frac{I^{(t)}}{I^{(i)}} = 1;$$

abiding by the power conservation law, since the materials are sourceless and without any loss. Figure 5.3 shows the trend of $I^{(t)}/I^{(i)}$ as a function of δ for $F = 10$. The interferometer transmittance

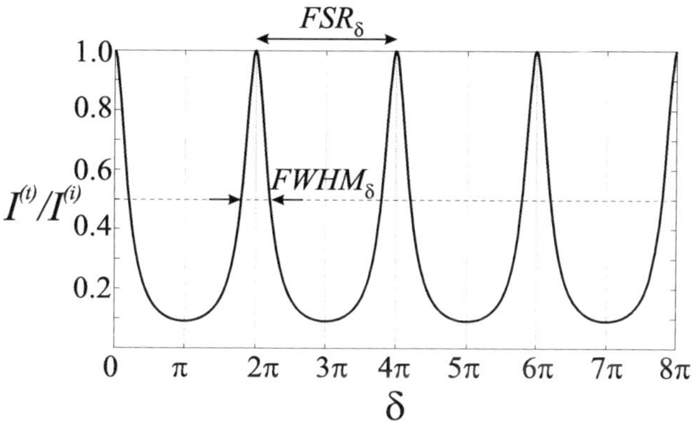

Fig. 5.3: The interferometer transmittance as a function of δ.

periodically takes the value 1 for values of δ which are multiples of 2π. According to (5.1), the $FWHM_\delta$ and FSR_δ are:

$$FWHM_\delta \simeq \frac{4}{\sqrt{F}};$$
$$FSR_\delta = 2\pi.$$

As transmittance depends on δ, its value will increase as F, or R, grows (cf. figure 5.4). The relation between the FSR_δ and the $FWHM_\delta$ is known as *finesse*:

$$\frac{FSR_\delta}{FWHM_\delta} = \frac{2\pi}{4/\sqrt{F}} = \pi\frac{\sqrt{R}}{1-R}.$$

As the reflectivity of the two interfaces increases, the interferometer becomes more selective, thus improving its quality. According to the expression of δ, the interferometer transmittance can be modified by adjusting the thickness h of the slab, the angle of incidence θ_2, or the wavelength λ. The latter dependence highlights its use as a wavelength filter. The type of dependency should therefore be made explicit in the different parameters – be it on the wavelength or, dually, on the frequency. As far as the $FWHM$ is concerned, we will have:

$$FWHM_\nu = \frac{c}{\pi h n_1 \cos\theta_1}\frac{1}{\sqrt{F}};$$

$$FWHM_\lambda = \frac{\lambda^2}{\pi h n_1 \cos\theta_1 \sqrt{F}}.$$

For the FSR, instead:

$$FSR_\nu = \frac{c}{2hn_1\cos\theta_1};$$

$$FSR_\lambda = \frac{\lambda^2}{2hn_1\cos\theta_1}.$$

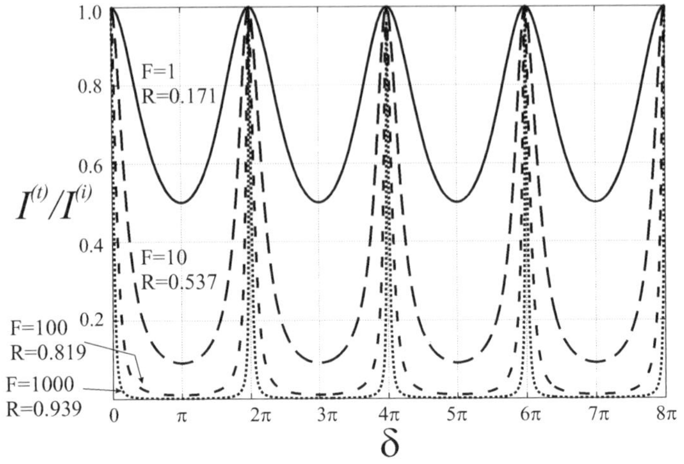

Fig. 5.4: The interferometer transmittance as a function of δ for different values of F, or R.

The higher F and h, the more selective the cavity; besides, the fact that the incidence is close to the orthogonality ($\cos\theta_1 = 1$) is another important condition which determines how selective the cavity can be. Note that, with normal incidence, $\theta_2 = \theta_1 = 0$ maximum transmittance will be:

$$\delta = 4\pi\frac{h}{\lambda}n_1 = 2m\pi$$

from which:

$$h = m\frac{\lambda}{2n_1}.$$

Maximum transmittance is reached when the interferometer thickness is equal to a multiple of half the wavelength in the material with index n_1. According to what we have seen, as the slab becomes thicker and thicker, the interferometer selectivity increases, although the number of wavelengths for which the interferometer is perfectly transparent increases as well – which can create problems in some applications. The number of such wavelengths can be reduced, but at the expense of the interferometer bandwidth. In particular, if

$$h = \frac{\overline{\lambda}}{2n_1}$$

as a consequence, only the wavelength $\overline{\lambda}$ can pass through the interferometer without being reflected; however, the bandwidth becomes:

$$\Delta\lambda = \frac{2\overline{\lambda}}{\sqrt{F}}.$$

5.2 Transmissive Diffraction Gratings

Diffraction gratings are optical components consisting of a number of slits on an opaque screen. When an electromagnetic beam strikes the screen, it is irregularly transmitted or diffracted beyond the screen, *i.e.* with varying intensity according to the phase difference of the different contributions from each of the individual slits.

Consider the case of two slits acting as point sources: through simple geometric constructions we can immediately verify that two diffracted beams are added in phase when

$$sin\theta = m\frac{\lambda}{\Lambda}, \tag{5.2}$$

with θ indicating the inclination of the rays from the normal to the plane containing the slits, λ the electromagnetic wavelength, Λ the distance between the two slits, and m an integer index that can take both positive and negative values. (5.2) indicates that the difference between the path lengths of two beams must be an integer multiple of the wavelength. Therefore, it is important to stress the fact that the condition in (5.2) remains valid also in case of a greater number of slits – as will be shown later – and is therefore taken as the simplest form of a diffraction grating equation.

We shall now illustrate the general case of a series of M aligned slits so small that the phase variations of the field within each of them can be considered to be null. The symbol Λ identifies the distance separating adjacent slits, *i.e.* the grating period; θ_i indicates the angle between the beam and the grating, while θ is used for the generic angle of diffracted rays. The situation is illustrated in figure 5.5. The ray diffracted by the first slit will be taken as the reference point for our investigation. Compared to the latter, the m-th ray, with $m = 1...M$, diffracted by the m-th

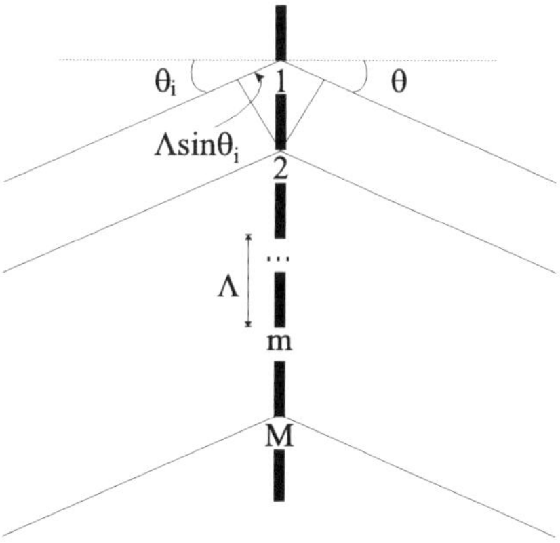

Fig. 5.5: A bright beam with an angle θ_i strikes a diffraction grating. Rays are diffracted at an angle θ.

slit, has a path difference between two equiphase wavefronts equal to:

$$m\Lambda(sin\theta_i + sin\theta).$$

The phase difference between two adjacent rays is:

$$\psi = \frac{2\pi}{\lambda}\Lambda(sin\theta_i + sin\theta),$$

while the phase difference between the m-th and the first ray is:

$$m\psi = m\frac{2\pi}{\lambda}\Lambda(sin\theta_i + sin\theta).$$

The complex amplitude $A(\theta_i, \theta)$ of the beam – obtained by adding up all the rays emerging from the grating with amplitude A_0 – can be written as follows:

$$A(\theta_i, \theta) = A_0[1 + e^{j\psi} + e^{j2\psi} + ... + e^{j(M-1)\psi}].$$

The sum of this geometric series is:

$$A(\theta_i, \theta) = A_0\,\frac{1 - e^{jM\psi}}{1 - e^{j\psi}} = A_0\,e^{j(M-1)\psi/2}\,\frac{sin(M\psi/2)}{sin(\psi/2)}. \tag{5.3}$$

The phase of $A(\theta_i, \theta)$ is indicated by the exponent – as we chose to take as reference the diffracted beam from the first slit. If, on the other hand, the reference beam to calculate the phase were chosen from the central slit, then the exponent would be:

$$A(\theta_i, \theta) = A_0\,\frac{sin(M\psi/2)}{sin(\psi/2)}. \tag{5.4}$$

We shall now use the symbol δ to denote the phase shift introduced by the inclination of the incident beam,

$$\delta = \frac{2\pi}{\lambda}\Lambda sin\theta_i, \tag{5.5}$$

which will be assumed to be fixed hereinafter. In other words, the inclination of the incident field can be described by introducing a linear phase shift throughout the grating length. Thanks to (5.5), it is possible to express (5.3) as a function of the diffraction angle θ only. With the grating centre as phase reference, the result is:

$$A(\theta, \delta) = A_0\,\frac{sin\left(M\pi\frac{\Lambda}{\lambda}sin\theta + M\frac{\delta}{2}\right)}{sin\left(\pi\frac{\Lambda}{\lambda}sin\theta + \frac{\delta}{2}\right)} \tag{5.6}$$

which is an important expression for the study of the diffraction pattern produced by this type of grating. To give an example, figure 5.6 shows the intensity distribution produced by a diffraction grating with five slits. The main lobes arise when the phase difference between adjacent beams is an integer multiple of 2π – that is to say, the path difference is equal to an integer multiple of λ. This occurs when both the numerator and denominator of (5.6) tend to zero, *i.e.* when:

$$\pi\frac{\Lambda}{\lambda}sin\theta + \frac{\delta}{2} = m\pi \tag{5.7}$$

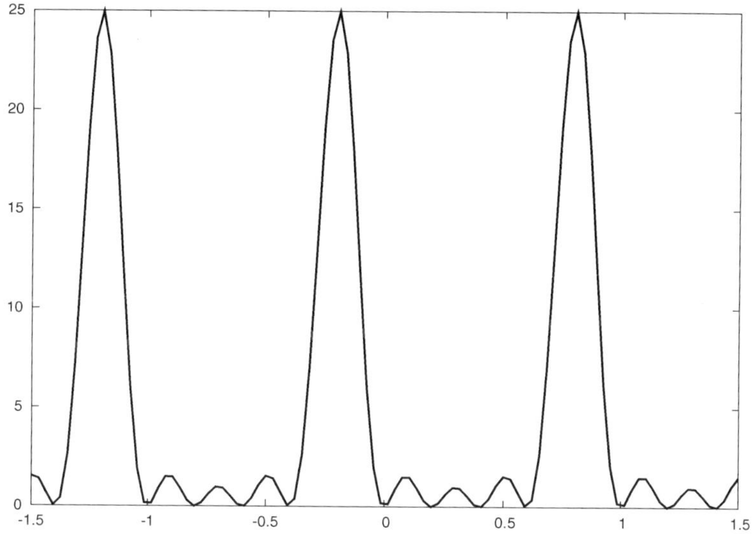

Fig. 5.6: The intensity distribution produced by a five-slit diffraction grating as a function of $sin\theta$ and under the following conditions: oblique incidence and unit amplitude A_0. In the case of normal incidence, the fringes are centred with respect to the value $sin\theta = 0$.

from which:

$$sin\theta^{(m)} = -\frac{\delta\lambda}{2\pi\Lambda} + m\frac{\lambda}{\Lambda}, \tag{5.8}$$

where the superscript (m) indicates that the transmitted angle depends on the integer m. Note that, if we reintroduce (5.5), we have:

$$sin\theta^{(m)} = -sin\theta_i + m\frac{\lambda}{\Lambda}, \tag{5.9}$$

which is formally very similar to what will be described in the next section regarding reflective diffraction gratings. In the particular case of normal incidence ($sin\theta_i = 0$), we have $\delta = 0$, with (5.8) becoming:

$$sin\theta^{(m)} = m\frac{\lambda}{\Lambda}. \tag{5.10}$$

The lobe corresponding to $sin\theta = 0$ is called zero-order lobe and represents the field propagating through the grating without diffraction. Subsequently, m-order lobes are defined by the values of $\theta^{(m)}$ as determined by (5.8) or (5.10). Note in conclusion that the maximum value of the main lobes is equal to M^2, while their amplitude $2\Delta\theta$ can be defined as the angular width centred on $\theta^{(m)}$ – its extrema being the values of theta for which the function is null. In the case of normal incidence, for the sake of simplicity, from the expression of the numerator of (5.6), we have:

$$M\pi\frac{\Lambda}{\lambda}\Delta\theta = \pm\pi$$

from which the semi-amplitude:

$$\Delta\theta = \frac{1}{M}\frac{\lambda}{\Lambda}. \tag{5.11}$$

Thus, the semi-amplitude of the grating lobes is inversely proportional to the number of slits of the grating itself. Furthermore, one should not forget that the separation between adjacent lobes does not depend on δ and is:

$$sin\theta^{(m)} - sin\theta^{(m-1)} = \frac{\lambda}{\Lambda}. \tag{5.12}$$

The separation between the lobes, in terms of $sin\theta$, is therefore constant and only depends on the ratio between the wavelength and the grating period. The phase shift δ, instead, determines the position of the lobes, which can thus be controlled by adjusting the illumination angle θ_i.

As already stated, the zeros of the function (5.6) are obtained by nullifying the numerator – the denominator being other than zero. In the more general case of oblique incidence, we have:

$$M\pi\frac{\Lambda}{\lambda}sin\theta + M\frac{\delta}{2} = p\pi, \tag{5.13}$$

provided that p is an integer while the ratio p/M is not – from which, finally:

$$sin\theta = -\frac{\delta\lambda}{2\pi\Lambda} + \frac{p\lambda}{M\Lambda}. \tag{5.14}$$

When the ratio p/M is an integer, once again we obtain the values of $sin\theta$, which give the position of the maxima of the lobes of various orders. As for small diffraction angles, we can reformulate (5.8) and (5.14) by assuming that the sine may be approximated with its argument. As a result, the position of the maxima is given by:

$$\theta^{(m)} = -\frac{\delta\lambda}{2\pi\Lambda} + m\frac{\lambda}{\Lambda}, \tag{5.15}$$

while that of the zeros is:

$$\theta = -\frac{\delta\lambda}{2\pi\Lambda} + \frac{p\lambda}{M\Lambda}. \tag{5.16}$$

5.2.1 Resolving Power

From the expression of $\theta^{(m)}$ we can verify that the direction of each lobe among the various orders depends on the wavelength. For this reason, if two beams of different wavelengths, λ_1 and λ_2, strike the grating, the result is two different distributions of the intensity fringes, similar to those shown in figure 5.6, though slightly shifted – cf. figure 5.7. Note that the position of the zero-order lobe generated at the two different wavelengths only coincides when $\delta = 0$, while the position of higher-order lobes (*i.e.* $m \geq 1$) is always different due to the different value of λ in (5.15).

If we are able to distinguish the lobes of the two fringes, then we will be able to distinguish the two different wavelengths. To do so, we shall refer to the Rayleigh Criterion, which explains that the lobes at the wavelength $\lambda \pm \Delta\lambda/2$ are distinguishable when the principal diffraction maximum

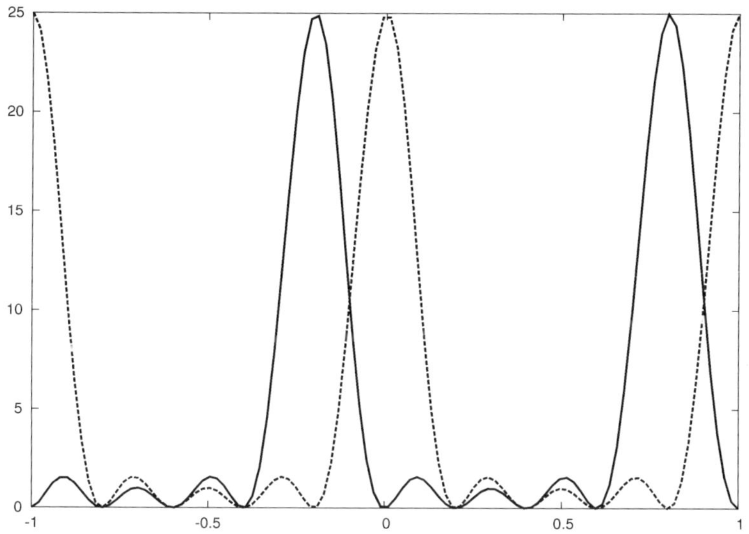

Fig. 5.7: The intensity distribution produced by a diffraction grating illuminated by two beams of different wavelength.

at one wavelength coincides with the first minimum (zero) of the other. We can now rewrite the condition (5.7) which gives the main lobes, with $\delta = 0$, for two different incident wavelengths:

$$\pi \frac{\Lambda}{\lambda_1} \theta_1 = m\pi ; \tag{5.17}$$

$$\pi \frac{\Lambda}{\lambda_2} \theta_2 = m\pi ; \tag{5.18}$$

that is to say,

$$\lambda_1 = \frac{\Lambda}{m} \theta_1 ; \tag{5.19}$$

$$\lambda_2 = \frac{\Lambda}{m} \theta_2 ; \tag{5.20}$$

from which the separation $\Delta\lambda = \lambda_2 - \lambda_1$:

$$\Delta\lambda = \frac{\Lambda}{m} \Delta\theta,$$

with $\Delta\theta = \theta_2 - \theta_1$. If we take the semi-amplitude of the lobes (5.11) – *i.e.* the distance satisfying the Rayleigh Criterion – as the angular separation $\Delta\theta$, we obtain

$$\Delta\lambda = \frac{\Lambda}{m} \frac{1}{M} \frac{\lambda}{\Lambda} = \frac{\lambda}{mM},$$

which represents the minimum spectral distance that allows two different wavelengths to be solved according to the Rayleigh Criterion. From the previous equation, we can determine the so-called *resolving power* R of the grating:

$$R = \frac{\lambda}{\Delta\lambda} = mM$$

which is the product of the fringe order multiplied by the number of slits. The order considered permits the resolving power of the grating to be increased. In the third-order spectrum, a variation $\Delta\lambda$ provides a path variation between adjacent beams which is three times that of the first order – increasing R by a factor of three.

By way of example, determine the number of slits in a grating so as to resolve two spectral lines separated by $1\,nm$ at a wavelength of $1.5\,\mu m$ by using second-order fringes:

$$M \geq \frac{1}{m}\frac{\lambda}{\Delta\lambda} = \frac{1}{2}\frac{1.5 \cdot 10^{-6}}{10^{-9}} = 750.$$

5.3 Reflective Diffraction Gratings

Reflective diffraction gratings (also known as reflection gratings) are formed by a reflecting surface into which steps or grooves have been cut. When a plane wave strikes this type of grating, beams of various orders – *i.e.* different angles – are generated by reflection. The order of the reflected beams is directly associated to the grating order, that is to say, to the number of harmonics adopted to describe its periodicity. The reflection angles of the different beams also depend on the wavelength of the incident beam, therefore the grating permits incident beams with the same angle but different wavelength to be separated spatially.

In order to explain how these gratings work, let us consider a non-uniform surface, as the one shown in figure 5.8, hit by a plane wave with an angle of incidence θ_i. For the sake of simplicity, we consider the incident beam to be completely polarised along the y-axis. The grating profile is described by the function $z = h(x) = h(x + \Lambda)$, which is periodic in x with spatial period Λ. If we assume that the grating surface is fully reflective, the incident beam $E_i\hat{\jmath}$ will only give rise to the reflected wave $E_r\hat{\jmath}$ – as no refracted beam is possible in this respect. The continuity condition on the surface $z = h(x)$ requires that:

$$\big(E_i(x,z) + E_r(x,z)\big)|_{z=h(x)} = 0 \qquad (5.21)$$

that is to say, if the amplitude E_{0i} and the wave vector $\overline{k}_i = k_{ix}\hat{\imath} + k_{iz}\hat{z}$ of the incident beam are made explicit, then:

$$E_{0i}e^{-j(k_{ix}x+k_{iz}z)}|_{z=h(x)} + E_r(x,z)|_{z=h(x)} = 0. \qquad (5.22)$$

The incident beam on the grating surface can be written as follows:

$$E_i = E_{0i}e^{-jk_{ix}x}e^{-jk_{iz}h(x)} \qquad (5.23)$$

hence, the function $e^{-jk_{iz}h(x)}$ is periodic in x with period Λ. Furthermore, since from (5.22) the reflected beam has to cancel the incident one, it must be:

$$E_r = e^{-jk_{ix}x}\, f(x,z)$$

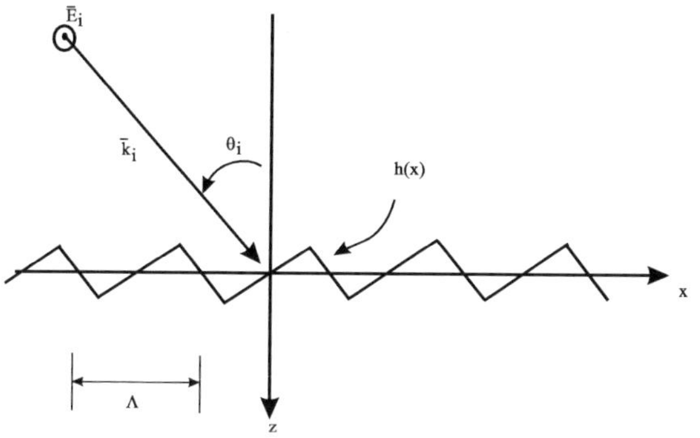

Fig. 5.8: A polarised plane wave along y strikes a reflection grating identified by the periodic function $z = h(x)$ with period Λ.

with $f(x,z)$ necessarily being a periodic function in x with period Λ. We can thus expand $f(x,z)$ in an exponential Fourier series:

$$f(x,z) = \sum_m c_m(z)\, e^{-jm\frac{2\pi}{\Lambda}x},$$

c_m being an unknown coefficient. The reflected field thus becomes:

$$E_r(x,z) = e^{-jk_{ix}x} \sum_m c_m(z)\, e^{-jm\frac{2\pi}{\Lambda}x} = \sum_m c_m(z)\, e^{-j(k_{ix}x + m\frac{2\pi}{\Lambda}x)}. \tag{5.24}$$

This field has to be a solution to Maxwell's equations, or to the Helmholtz equation:

$$\frac{\partial^2 E_r}{\partial x^2} + \frac{\partial^2 E_r}{\partial z^2} + \omega^2 \mu\varepsilon E_r = 0$$

from which, by substituting (5.24), for each m we obtain:

$$\frac{\partial^2 c_m(z)}{\partial z^2} + \left[\omega^2\mu\varepsilon - \left(k_{ix} + m\frac{2\pi}{\Lambda}\right)^2\right] c_m(z) = 0.$$

If we hold that:

$$(k_{rz}^{(m)})^2 = \omega^2\mu\varepsilon - \left(k_{ix} + m\frac{2\pi}{\Lambda}\right)^2 \tag{5.25}$$

we will have:

$$\frac{\partial^2 c_m(z)}{\partial z^2} + (k_{rz}^{(m)})^2 c_m(z) = 0$$

whose solutions can be:

$$c_m(z) = E_{0r}^{(m)} e^{jk_{rz}^{(m)} z}.$$

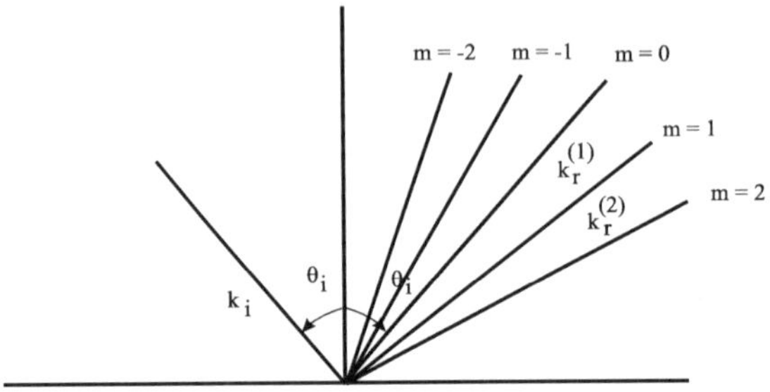

Fig. 5.9: A plane wave strikes a reflection grating and generates a set of reflected beams of m-order.

The solutions depending on $e^{-jk_{rz}^{(m)}z}$ cannot be accepted in that the completely reflective surface does not allow the propagation in the positive direction along z. The expression of the reflected electric field is thus given by:

$$E_r = e^{-jk_{ix}x} \sum_m E_{0r}^{(m)} e^{jk_{rz}^{(m)}z} e^{-jm\frac{2\pi}{\Lambda}x} \tag{5.26}$$

in which the coefficients $E_{0r}^{(m)}$ are constants which can be calculated once the boundary conditions are determined – hence depending on the form of $h(x)$. Additionally, the components of the wave vectors along z, $k_{rz}^{(m)}$, are such that the conditions in (5.25) are met. Remember that we can write:

$$(k_{rx}^{(m)})^2 + (k_{rz}^{(m)})^2 = \omega^2\mu\varepsilon; \tag{5.27}$$

hence (5.25) allows us to determine the wave vector component of the reflected beam along x, as well as the m-order angle of reflection. In fact, if we compare k_{ix} and $k_{rx}^{(m)}$, the components k_{ix} and $k_{rx}^{(m)}$ are:

$$k_{ix} = k\,sin\,\theta_i$$

$$k_{rx}^{(m)} = k\,sin\,\theta_r^{(m)}$$

from which we can deduce that:

$$k_{rx}^{(m)} = k_{ix} + m\frac{2\pi}{\Lambda} = k(sin\,\theta_i + m\frac{2\pi}{k\Lambda}) = k(sin\,\theta_i + m\frac{\lambda}{\Lambda})$$

and finally:

$$sin\,\theta_r^{(m)} = sin\,\theta_i + m\frac{\lambda}{\Lambda} \tag{5.28}$$

which constitutes the specific law of reflection of the grating. The important aspect of the expression in (5.28) is represented by the fact that the angle of reflection depends on both m and the wavelength. If we consider a flat surface, we have $\Lambda = \infty$, thus (5.28) can be reduced to the usual law of reflection, with the angle of reflection equal to the angle of incidence. In the case of finite Λ, whether it is due to mechanically-engraved grooves, or piezoelectric-induced or acousto-optical deformations, we have reflected beams of various orders, as described in figure 5.9; in this particular situation, the angle of reflection (the zero-order one excluded) depends on the wavelength. It follows that two incident beams with the same angle, yet different wavelength, have different higher-order reflection angles. Note also that it must always be $|sin\,\theta_r^{(m)}| \leq 1$ in order to have uniform plane waves in lieu of reflected beams: therefore, the smaller the value of λ/Λ, the greater the value of m – that is, of higher-order reflected beams. In the case of a large wavelength with respect to the perturbation Λ, the only possible value can thus be $m = 0$, and the incident beam basically behaves as if the surface were flat. Conversely, if we consider very large values of m, to such an extent that the condition $|sin\,\theta_r^{(m)}| \leq 1$, cannot be satisfied, as a result the reflected beams will be evanescent waves which do not contribute to the propagation at a great distance. For this reason too, we deem the lower-order reflected beams $m = \pm 1, \pm 2$ to be of interest.

The reflection gratings are particularly useful in case one requires a spectral splitting of the incident field or a wavelength selection, such as in monochromators, spectrum analysers or tunable lasers. Consider the latter, for example – analysed in chapter 12. The reflection grating usually replaces one of the mirrors of the cavity in order to reflect the beam of order -1 in the same direction of the incident beam, that is to say:

$$sin\,\theta_r^{(-1)} = -sin\,\theta_i. \tag{5.29}$$

Note that, in order to intervene on the wavelength, the zero-order reflected beam must be excluded; thus, all angles corresponding to positive orders – $+1, +2, ...$ – have angles greater than $\theta_r^{(0)}$ (cf. figure 5.10), therefore they cannot but be on the same side of the zero-order reflected beam, with respect to the normal. By imposing (5.29) in (5.28), it follows that:

$$-2sin\,\theta_i = -\frac{\lambda}{\Lambda}$$

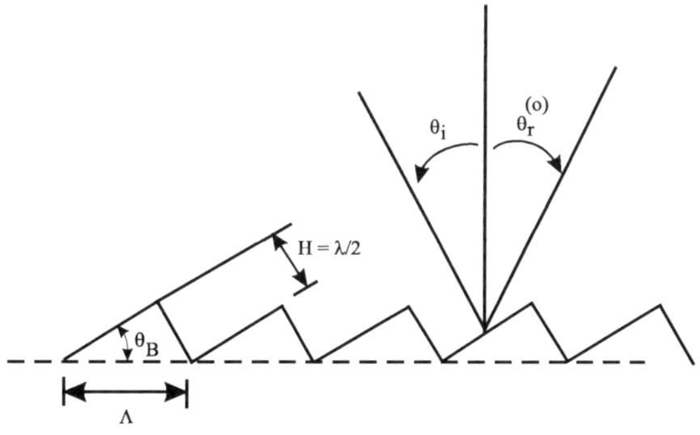

Fig. 5.10: Tunable laser grating employing a reflected beam of $m = -1$ order.

that is,

$$\Lambda \, sin\, \theta_i = \frac{\lambda}{2} \tag{5.30}$$

which, once the grating is set, determines the wavelength to be reflected in the direction of the incident beam. It is evident that, as θ_i varies – in other words, as the grating inclination varies with respect to the axis of the laser cavity – the equation (5.30) can be satisfied for different values of λ, which permits the tunability of laser radiation. On the other hand, once the wavelength is set, from the equation (5.30) it is possible to infer not only the value of the grating period Λ, but also the groove height ($\Lambda \, sin\, \theta_i$ or $\Lambda \, sin\, \theta_B$ in figure 5.10), which guarantees the in-phase sum between all beams reflected by the subsequent corrugations, *viz.* the value $\lambda/2$.
Before concluding, we would like to discuss briefly a possible method of determining the amplitude of the different reflected beams. This is possible if the grating profile is known; for this reason, and for the sake of simplicity, we shall consider a cosine-type profile:

$$z = h(x) = h_0 \, cos \frac{2\pi}{\Lambda} x. \tag{5.31}$$

If we resort to the expressions of the incident (5.23) and reflected electric field (5.26), the condition imposed by (5.21) becomes:

$$E_{0i} e^{-jk_{iz}z} + \sum_m E_{0r}^{(m)} e^{jk_{rz}^{(m)}z} e^{-jm\frac{2\pi}{\Lambda}x} = 0. \tag{5.32}$$

By assuming small deformations, we can develop the exponential $e^{-jk_{iz}z}$ with the first terms of a Taylor expansion; subsequently, by replacing the equation (5.31), we obtain:

$$e^{-jk_{iz}z} \approx 1 - jk_{iz}z = 1 - jk_{iz}h_0 \, cos \frac{2\pi}{\Lambda}x = 1 - jk_{iz}h_0 \frac{e^{j\frac{2\pi}{\Lambda}x} + e^{-j\frac{2\pi}{\Lambda}x}}{2}.$$

If the term $e^{jk_{rz}^{(m)}z}$ is thus ignored, the previous formula (5.32) becomes:

$$E_{0i}\left(1 - jk_{iz}h_0 \frac{e^{j\frac{2\pi}{\Lambda}x} + e^{-j\frac{2\pi}{\Lambda}x}}{2}\right) = -\sum_m E_{0r}^m e^{-jm\frac{2\pi}{\Lambda}x}$$

from which:

$$\begin{aligned} E_{0i}\left[1 - j\frac{k_{iz}h_0}{2}\left(e^{j\frac{2\pi}{\Lambda}x} + e^{-j\frac{2\pi}{\Lambda}x}\right)\right] = \ldots & - E_{0r}^{(-2)} e^{j2\frac{2\pi}{\Lambda}x} - E_{0r}^{(-1)} e^{j\frac{2\pi}{\Lambda}x} - E_{0r}^{(0)} \\ & - E_{0r}^{(1)} e^{-j\frac{2\pi}{\Lambda}x} - E_{0r}^{(2)} e^{-j2\frac{2\pi}{\Lambda}x} - \ldots \end{aligned}$$

Once the coefficients that multiply the same exponential are equated, the result is as follows:

$$\begin{aligned} E_{0r}^{(0)} &= -E_{0i} \\ E_{0r}^{(1)} &= E_{0r}^{(-1)} = j\frac{k_{iz}h_0}{2} E_{0i} \\ E_{0r}^{(m)} &= E_{0r}^{(-m)} = 0 \quad \text{per m} \geq 2 \end{aligned}$$

The last condition is a result of the expansion of the exponential $e^{-jk_{iz}z}$ being truncated after the first terms. In practice, non-zero amplitudes can obviously be observed even for $m \geq 2$. The reflected beams of ± 1 order have thus an amplitude which is a function of E_{0i} and which depends on the following relation:

$$\frac{2\pi}{\lambda_z}\frac{h_0}{2} = \pi\frac{h_0}{\lambda_z}$$

with λ_z indicating the wavelength in the z-direction.

Note that, in terms of reflection, the same effect would be reproduced by substituting each grating period with an isotropic point source so as to have an array of antennas that – thanks to well-known techniques in this domain – can be studied in order to analyse the forms and amplitudes of the lobes in the different directions.

5.4 The Multimode Interferometer (MMI)

The study of the directional coupler has pointed out how, in the case of two non-degenerate modes, the progressive phase shift occurring between them produces a total field whose features change in the course of the propagation progress. A generalisation of this phenomenon is exploited by MMI (multimode interference) components, basically constituted by a waveguide called *multimode region* with a large core enough to support a great number of guided modes. The interference between the different modes within the region produces a number of field distributions that gradually change during the propagation process and which are periodically repeated: this is known as the *self-imaging principle* (cf. figure 5.11). It is possible to control the characteristics of the field on each waveguide by selecting the proper dimensions of the multimode region in terms of length, and by carefully positioning the output waveguides. Figure 5.12 shows the simple case of an MMI in which the multimode region is excited by a single input waveguide and the output field emerges through two waveguides; more generally, it is possible to build $N \times M$ MMIs, *i.e.* optical devices with N input waveguides and M output waveguides.

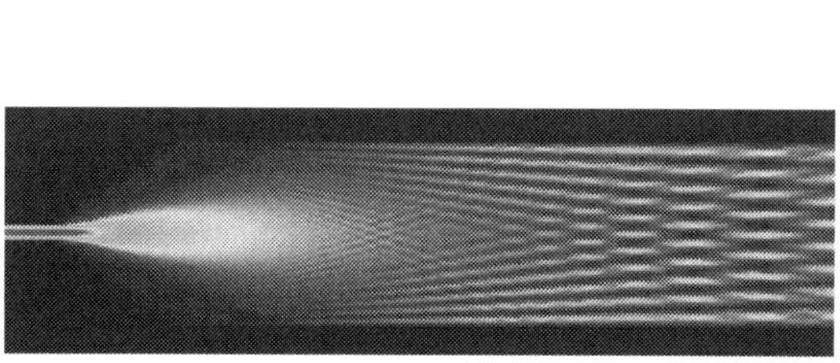

Fig. 5.11: The distribution of the total electric field inside an MMI.

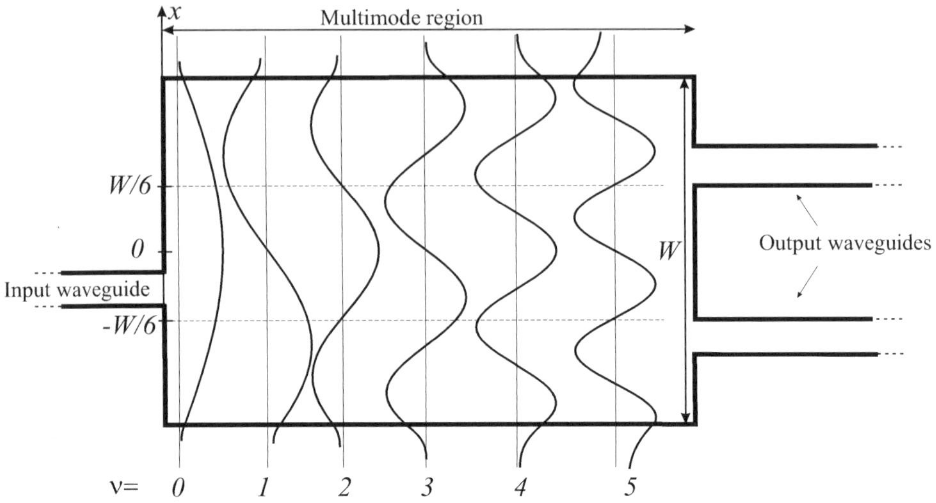

Fig. 5.12: The multimode region of an MMI with width W and index n_{co}.

5.4.1 The Self-imaging Principle

To summarise, the self-imaging principle is a property of multimode waveguides by which the profile of the input field is reproduced in single or multiple images at periodic intervals along the propagation direction of the waveguide. For our analysis we shall concentrate on planar structures and specifically consider the TE modes only – as the same process can be applied for the TM modes, too. Figure 5.12 presents the case of a multimode slab of width W and index n_{co} which can support N guided modes. The phase shift between the different modes in the propagation process is at the basis of its operating principles: for this reason, the first step of our study will be to calculate the analytical expression of the different phase constants. Hence, the paraxial approximation (cf. paragraph 2.6.1) is used by virtue of the large number of guided modes; the following expression can be assumed to be the phase constant of the νth mode:

$$\beta_\nu \simeq \frac{2\pi}{\lambda} n_{co} - (\nu + 1)^2 \frac{\pi\lambda}{4n_{co}W^2}, \tag{5.33}$$

with $\nu = 0, 1, 2, ..., N-1$. According to (5.33) we will have:

$$\beta_0 - \beta_\nu \simeq \frac{\nu(\nu+2)\pi}{3L_\pi} \tag{5.34}$$

where L_π is the beat length of the fundamental mode with the first high-order mode:

$$L_\pi = \frac{\pi}{\beta_0 - \beta_1} \simeq \frac{4n_{co}W^2}{3\lambda}. \tag{5.35}$$

If we ignore the contribution of the radiation modes – which would be utterly reasonable given the large number of guided modes – and considering the waveguide input field $\overline{e}_{ing} = e_{ing}(x)\hat{\jmath}$, its profile $e_{ing}(x)$ can be decomposed into the contributions of all N guided modes:

$$e_{ing}(x) = \sum_{\nu=0}^{N-1} A_\nu E_\nu(x)$$

where

$$A_\nu = \frac{\displaystyle\int_{-\infty}^{\infty} e_{ing}(x) E_\nu(x) dx}{\displaystyle\int_{-\infty}^{\infty} E_\nu^2(x) dx}. \tag{5.36}$$

As a result of the propagation of the N modes, the field at a distance $z = L$ from the input section is:

$$e(x, L) = \sum_{\nu=0}^{N-1} A_\nu E_\nu(x) e^{-j\beta_\nu L};$$

by factorising the phase term of the fundamental mode and resorting to (5.34), we obtain:

$$\begin{aligned} e(x, L) &= \left(\sum_{\nu=0}^{N-1} A_\nu E_\nu(x) e^{-j(\beta_\nu - \beta_0)L} \right) e^{-j\beta_0 L} \\ &= \left(\sum_{\nu=0}^{N-1} A_\nu E_\nu(x) e^{j\frac{\nu(\nu+2)\pi}{3L_\pi} L} \right) e^{-j\beta_0 L}. \end{aligned} \tag{5.37}$$

Hence, the total field profile is determined by the input field through the coefficients A_ν and by the properties of the *modal phase factor*:

$$e^{j\frac{\nu(\nu+2)\pi}{3L_\pi} L}.$$

Then, it is possible to prove that:

$$\nu(\nu+2) \quad \text{is} \quad \begin{cases} \text{even} & \text{if} \quad \nu \quad \text{is even} \\ \text{odd} & \text{if} \quad \nu \quad \text{is odd} \end{cases} \tag{5.38}$$

and

$$E_\nu(-x) = \begin{cases} E_\nu(x) & \text{if} \quad \nu \quad \text{is even} \\ -E_\nu(x) & \text{if} \quad \nu \quad \text{is odd} \end{cases} \tag{5.39}$$

that is to say, the function $E_\nu(x)$ is even if ν is even, otherwise it is odd in the case where ν is odd. By analysing (5.37) we can infer that if:

$$e^{j\frac{\nu(\nu+2)\pi}{3L_\pi} L} = 1 \quad \text{with} \quad \nu = 0, \cdots N-1 \tag{5.40}$$

then the total field

$$e(x, L) = \left(\sum_{\nu=0}^{N-1} A_\nu E_\nu(x) \right) e^{-j\beta_0 L} = e_{in}(x) e^{-j\beta_0 L}$$

is – up to a phase shift – a copy of the input field. We can consider (5.40) to be satisfied if $L/(3L_\pi)$ is an even integer, that is:

$$L = p\,3L_\pi \quad \text{with} \quad p = 2, 4, 6, \cdots$$

since the product of an integer by an even integer is always even. This condition applies to the situation in which all the relative phase shifts of the different modes are integer multiples of 2π; in this case, the modes interfere with each other with the same phase relationships of the input field: as a consequence, the field generated will be the exact copy of the input field.

If, on the other hand:

$$e^{j\frac{\nu(\nu+2)\pi}{3L_\pi}L} = (-1)^\nu \quad \text{with} \quad \nu = 0, \cdots N-1 \tag{5.41}$$

then

$$e(x, L) = \left(\sum_{\nu \text{ even}} A_\nu E_\nu(x) - \sum_{\nu \text{ odd}} A_\nu E_\nu(x) \right) e^{-j\beta_0 L}.$$

In this case, according to (5.39):

$$e(-x, L) = \left(\sum_{\nu \text{ even}} A_\nu E_\nu(x) + \sum_{\nu \text{ odd}} A_\nu E_\nu(x) \right) e^{-j\beta_0 L} = e_{in}(x) e^{-j\beta_0 L}.$$

i.e. the field profile is then a mirror image of the input field with respect to the centre of the multimode region. This second condition can be applied when the relative phase shifts are alternately odd and even multiples of π.. Therefore, the even modes will all be in phase with each other, as well as the odd modes, though the latter are in phase opposition with respect to the even modes. To satisfy the conditions from (5.41), it is necessary that:

$$\nu(\nu+2)L/(3L_\pi) = \begin{cases} \text{even} & \text{if} \quad \nu \quad \text{is even} \\ \text{odd} & \text{if} \quad \nu \quad \text{is odd;} \end{cases}$$

according to the properties of the product of natural numbers, this means that $L/(3L_\pi)$ should be odd, which is true when:

$$L = p\,3L_\pi \quad \text{with} \quad p = 1, 3, 5, \cdots$$

Identical copies or mirror images of the input field are alternately reproduced along the multimode region at a distance of $3L_\pi$ from each other, as shown in figure 5.13.

Besides individual images, there are specific sections in which it is possible to observe multiple images of the input field. Halfway between the identical and the mirror images, *i.e.* when:

$$L = \frac{p}{2} 3L_\pi \quad \text{with} \quad p = 1, 3, 5, 7, \cdots$$

the total field is:

$$e(x, \frac{p}{2}3L_\pi) = \left(\sum_{\nu=0}^{N-1} A_\nu E_\nu(x) e^{j\nu(\nu+2)p\frac{\pi}{2}} \right) e^{-j\phi_0}$$

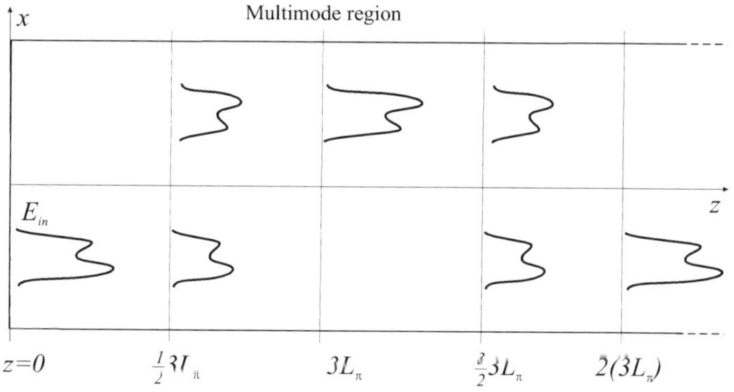

Fig. 5.13: The profile of the total field within the multimode region.

with $\phi_0 = \beta_0(p/2)3L_\pi$. As p is an even integer, and according to (5.38), (5.39) and the expressions below:

$$\sum_{\nu \text{ odd}} A_\nu E_\nu(x) = \frac{1}{2}\left[e_{in}(x) - e_{in}(-x)\right], \qquad \sum_{\nu \text{ even}} A_\nu E_\nu(x) = \frac{1}{2}\left[e_{in}(x) + e_{in}(-x)\right],$$

the total field can be reformulated as follows:

$$\begin{aligned} e(x, \frac{p}{2}3L_\pi) &= \left(\sum_{\nu \text{ even}} A_\nu E_\nu(x) + (-j)^p \sum_{\nu \text{ odd}} A_\nu E_\nu(x)\right) e^{-j\phi_0} \\ &= \left(\frac{1+(-j)^p}{2} e_{in}(x) + \frac{1-(-j)^p}{2} e_{in}(-x)\right) e^{-j\phi_0}. \end{aligned}$$

The field is given by the sum of two pairs of amplitude $1/\sqrt{2}$ and in quadrature with each other. Each pair consists of two mirror images of the input field facing each other with respect to the waveguide axis. If $p = 1$, *i.e.* at $L = 3/2L_\pi$, we have:

$$\begin{aligned} e(x, \frac{1}{2}3L_\pi) &= \left(\frac{1-j}{2} e_{in}(x) + \frac{1+j}{2} e_{in}(-x)\right) e^{-j\phi_0} \\ &= \frac{1}{\sqrt{2}}\left(e_{in}(x) + j e_{in}(-x)\right) e^{-j(\phi_0+\pi/4)}, \end{aligned}$$

which corresponds exactly to the output field of a coupler with $L_{coup} = 1/2L_c$.

More generally, it is possible to prove that, at the following distance:

$$L = \frac{p}{N}3L_\pi$$

with p and N being integers with no common divisors, the total field is formed by N generally non-equidistant copies of the input field with amplitude $1/\sqrt{N}$.

With respect to a generic distribution of the input field, the relations found so far do not postulate any type of relation between the coefficients A_ν, which is a case of *general interference*. If the multimode region is properly excited, it becomes possible to control the values of the coefficients, which permits copies of the input field at a lower distance compared to what has been found so far. This brings an obvious advantage when it comes to reducing the dimensions of the components: in this second situation, we have an example of *restricted interference*. Note that, if all the integers $\nu(\nu+2)$ were divisible by an integer q, *i.e.* $\nu(\nu+2) = mq$ with m integer, then the modal phase factor would become:

$$e^{j\nu(\nu+2)\pi L/(3L_\pi)} = e^{jmq\pi L/(3L_\pi)}$$

with its periodicity passing from $6L_\pi$ to $3L_\pi 2/q$. Consequently, all the other properties outlined above would be fulfilled over distances scaled by a factor of $1/q$. Unfortunately, only specific subsets of ν and not all its possible values can be divided by a common integer. For example [1]

$$\mathrm{mod}_3[\nu(\nu+2)] = 0 \quad \text{for} \quad \nu \neq 2, 5, 8, \cdots$$

However, in order to extend this property to all the terms of the summation (5.37), it is sufficient that:

$$A_\nu = 0 \quad \text{for} \quad \nu = 2, 5, 8, \cdots \tag{5.42}$$

That is to say, if the input field of the multimode region is such that it does not excite the modes with ν = 2, 5, 8, $\cdots$, the period of the modal phase factor is reduced by a factor of three. As a consequence, for

$$L = pL_\pi$$

we alternately have identical copies or mirror images of the input field, while for

$$L = \frac{p}{2}L_\pi$$

we will have pairs of images facing each other; finally, for

$$L = \frac{p}{N}L_\pi$$

the result will be N copies. One way to selectively excite the modes as in (5.42) is to employ a pair of waveguides centred in $x \pm W/6$. As figure 5.12 reveals, the modes with $\nu = 2, 5, 8, \cdots$ all present a zero with odd symmetry in this position, while the input waveguide modes have even symmetry. As a consequence, the corresponding overlap integrals (5.36) are all zero. When the condition (5.42) is satisfied, the pairs of modes $0-1$, $3-4$, $6-7$, etc. all have the same phase relationships, since:

$$\beta_0 - \beta_1 = \frac{\pi}{L_\pi}; \quad \beta_3 - \beta_4 = \frac{3\pi}{L_\pi}; \quad \beta_6 - \beta_7 = \frac{5\pi}{L_\pi}; \quad \cdots$$

Thus, in $z = L_\pi/2$ the odd mode within each pair is phase-shifted by $\pi/2$ with respect to the even mode, while in $z = L_\pi$ the phase difference is equal to π. Therefore, if at the end of the multimode

[1]The result of the function $\mathrm{mod}_y(x)$ is the remainder of the division x/y.

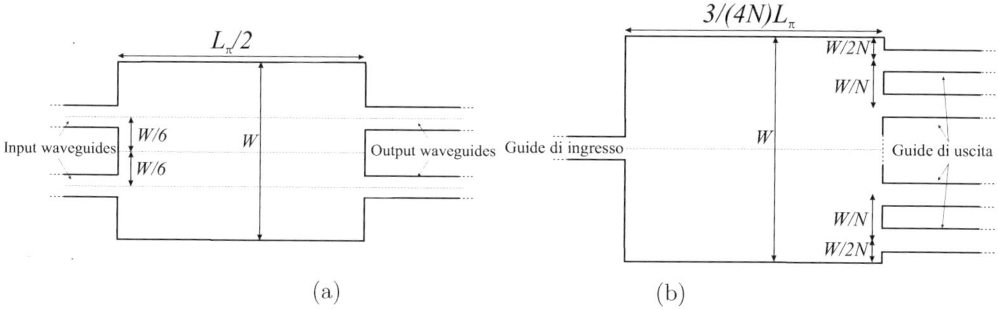

Fig. 5.14: On the left, a coupler. On the right, a $1/N$ power divider.

region, in position $\pm W/6$, we have two output waveguides which are identical to the relative input waveguides, the result is a component with the same transfer matrix of the directional coupler, with $L_{coup} = 1/2L_c$, yet less space-consuming and less sensitive to imperfections caused by the production processes – cf. figure 5.14(a).

Finally, since:

$$\text{mod}_4[\nu(\nu+2)] = 0 \quad \text{for} \quad \nu \quad \text{even},$$

the periodicity of the modal phase factor decreases by a factor of four simply by placing the input waveguide in the middle of the multimode region. By doing so, in fact, no mode with odd symmetry will be excited. As a result, we will have a certain number of distributions with N copies of the input field; these will be organised symmetrically with respect to the multimode region axis, as well as equally spaced (by W/N) and with amplitude $1/\sqrt{N}$; finally, their distance from the input section will be:

$$L = \frac{p}{N}\frac{3L_\pi}{4}.$$

For the number of guided modes to be at least N, it is necessary that:

$$v = \pi\frac{W}{\lambda}\sqrt{n_{co}^2 - n_{cl}^2} > (N-1)\frac{\pi}{2};$$

by highlighting W:

$$W > \frac{(N-1)\lambda}{2\sqrt{n_{co}^2 - n_{cl}^2}}$$

which is substituted in (5.35):

$$L_\pi > \frac{\lambda}{3}(N-1)^2\frac{n_{co}}{n_{co}^2 - n_{cl}^2}.$$

The right side of the inequality sets a limit which is lower than the length of the interferometer. As figure 5.15 proves, this limit is reduced as the core-cladding step index increases.

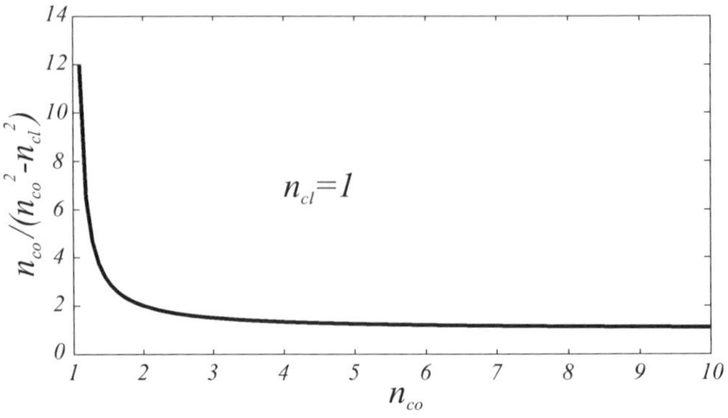

Fig. 5.15: The trend of the lower limit to the length of the MMI as the core index varies when the cladding is made of air.

5.5 The Mach-Zehnder Interferometer

The Mach-Zehnder interferometer comprises two decoupled waveguides having different length or physical characteristics. The waveguides are connected at the two ends by power splitters or combiners – typically, Y-junctions, MMI or 3-dB couplers (fig. 5.16). In the course of our study, we shall refer to the different coupling design mentioned.

The transfer matrix can be determined according to some simplifying assumptions: the fundamental mode only is propagated in the waveguides; the waveguides are decoupled apart from the sections relative to power dividers; all curves are adiabatic; the phase constants of the mode do not change along the two arms of the interferometer. We can divide the interferometer into three blocks: two couplers and a phase shifter. The transfer matrix is therefore:

$$[T]_{mach} = [T]_{coup2} \cdot [T]_{shif} \cdot [T]_{coup1}. \tag{5.43}$$

The phase shifter consists of two decoupled waveguides of different length and electromagnetic characteristics. By indicating with L_a and L_b the lengths of the two branches, respectively, and with β_a and β_b the phase constants of the mode in the two branches, the matrix becomes:

$$\begin{aligned}[T]_{shif} &= \begin{bmatrix} e^{-j\beta_a L_a} & 0 \\ 0 & e^{-j\beta_b L_b} \end{bmatrix} = \\ &= \begin{bmatrix} e^{-j\left(\frac{\beta_a L_a}{2}-\frac{\beta_b L_b}{2}\right)} & 0 \\ 0 & e^{-j\left(\frac{\beta_b L_b}{2}-\frac{\beta_a L_a}{2}\right)} \end{bmatrix} e^{-j\frac{\beta_a L_a+\beta_b L_b}{2}} = \\ &= \begin{bmatrix} e^{-j\frac{\pi}{\lambda}\Delta L_{eff}} & 0 \\ 0 & e^{j\frac{\pi}{\lambda}\Delta L_{eff}} \end{bmatrix} e^{-j\frac{2\pi}{\lambda}L_{eff}} \end{aligned} \tag{5.44}$$

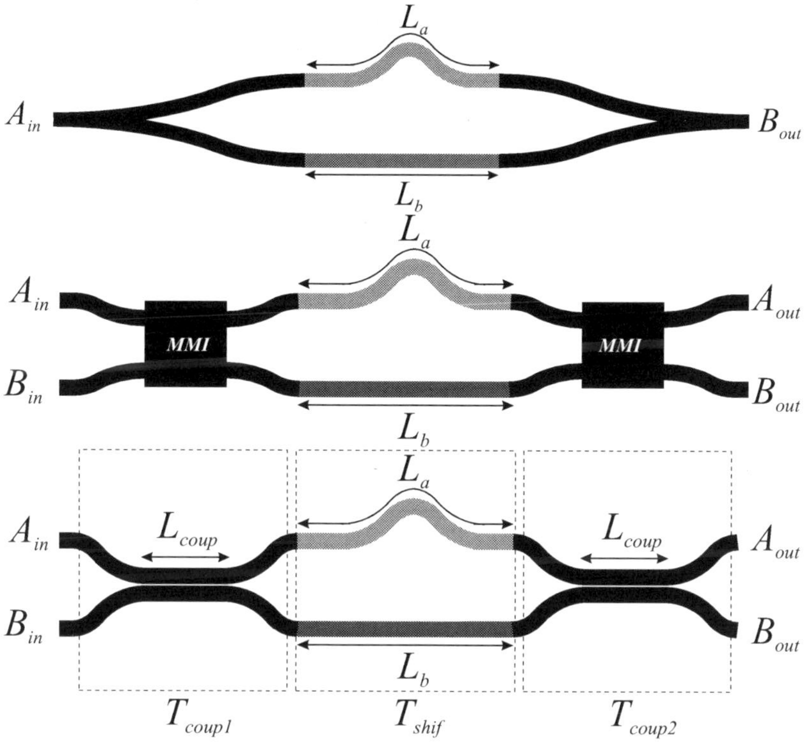

Fig. 5.16: Three examples of a Mach-Zehnder interferometer: Y-junction, MMI and 3-db coupler.

being

$$\Delta L_{eff} = (n_{effa}L_a - n_{effb}L_b), \qquad (5.45)$$

$$\frac{2\pi}{\lambda}L_{eff} = \frac{\beta_a L_a + \beta_b L_b}{2}$$

with

$$L_{eff} = \frac{n_{effa}L_a + n_{effb}L_b}{2}.$$

ΔL_{eff} quantifies the interferometer imbalance; from (5.45), this can be determined by adjusting both the effective index and the distance travelled by the mode in the two branches. As far as the two couplers are concerned – as discussed in Chapter 4, and since $L_{coup1} = L_{coup2} = L_{coup} = \frac{1}{2}L_c = \frac{1}{2}\frac{\lambda}{2\Delta n_{eff}}$ – the transfer matrices are:

$$[T]_{coup1} = [T]_{coup2} = \frac{1}{\sqrt{2}}\begin{bmatrix} 1 & -j \\ -j & 1 \end{bmatrix} e^{-j\phi_c}. \tag{5.46}$$

By substituting (5.44) and (5.46) in (5.43), and computing the matrix products, we can obtain the interferometer transfer matrix:

$$[T]_{mach} = \begin{bmatrix} \sin(\pi\frac{\Delta L_{eff}}{\lambda}) & \cos(\pi\frac{\Delta L_{eff}}{\lambda}) \\ \cos(\pi\frac{\Delta L_{eff}}{\lambda}) & -\sin(\pi\frac{\Delta L_{eff}}{\lambda}) \end{bmatrix} e^{-j\phi_t}$$

with $\phi_t = 2\phi_c + \frac{2\pi}{\lambda}L_{eff} + \pi/2$.

Let us now analyse the main conditions of excitation. If the input signal is propagated in one waveguide only, *e.g.* the waveguide a, then $B_{in} = 0$; consequently, the output signal is:

$$A_{out} = A_{in}\sin(\pi\frac{\Delta L_{eff}}{\lambda})e^{-j\phi_t};$$

$$B_{out} = A_{in}\cos(\pi\frac{\Delta L_{eff}}{\lambda})e^{-j\phi_t}.$$

In terms of power:

$$\begin{aligned} P_a &= |A_{out}|^2 = P_{a\,in}\sin^2(\pi\frac{\Delta L_{eff}}{\lambda}) = P_{a\,in}\sin^2(\pi\frac{\Delta L_{eff}}{c}\nu) \\ P_b &= |B_{out}|^2 = P_{a\,in}\cos^2(\pi\frac{\Delta L_{eff}}{\lambda}) = P_{a\,in}\cos^2(\pi\frac{\Delta L_{eff}}{c}\nu) \end{aligned} \tag{5.47}$$

since $P_{a\,in} = |A_{in}|^2$. On the two output waveguides we clearly have a periodic power trend, with the period being:

$$FSR_\nu = \frac{c}{\Delta L_{eff}};$$

in terms of wavelength:

$$FSR_\lambda = \frac{\lambda^2}{\Delta L_{eff}}.$$

By imposing $P_a/P_{a\,in} = 1/2$, we can calculate the $FWHM$:

$$FWHM_\nu = \frac{1}{2}FSR_\nu = \frac{c}{2\Delta L_{eff}};$$

$$FWHM_\lambda = \frac{1}{2}FSR_\lambda = \frac{\lambda^2}{2\Delta L_{eff}}.$$

The percentage of power allocated on the two waveguides thus depends on the ratio $\Delta L_{eff}/\lambda$. From (5.45), it is clearly possible to control the characteristics of the two output waveguides by modifying the effective index of the mode in the two branches and/or by adjusting the length of the branches. The first method is mainly used to design voltage-controlled switches and high-performance optical modulators, in which cases an interferometer with two branches of equal length $L_a = L_b = L$ and different effective indices is generally chosen. The refractive index of the medium, and therefore the effective index of the mode, is altered by means of an electric field applied through appropriate metallisation on one or both waveguides; hence:

$$\Delta n_{eff} = AV$$

V being the potential difference between the electrodes, and A a constant which depends on the material and geometry of the device. As a consequence:

$$\Delta L_{eff} = \Delta n_{eff} L = A\,V\,L$$

and is therefore controlled by the voltage V. For the signal to come out from the waveguide a it is necessary to apply a voltage V_a such that $AV_aL/\lambda = (2m+1)/2$ with integer m. On the other hand, a voltage V_b such that $AV_bL/\lambda = m$ is required to make sure that the signal comes out from the waveguide b. For a detailed discussion on modulators, cf. Chapter 10.

The second technique is primarily employed for designing filters in DWDM systems. In this case, the two branches are taken to be identical, $n_{eff1} = n_{eff2} = n_{eff}$, yet with different lengths,

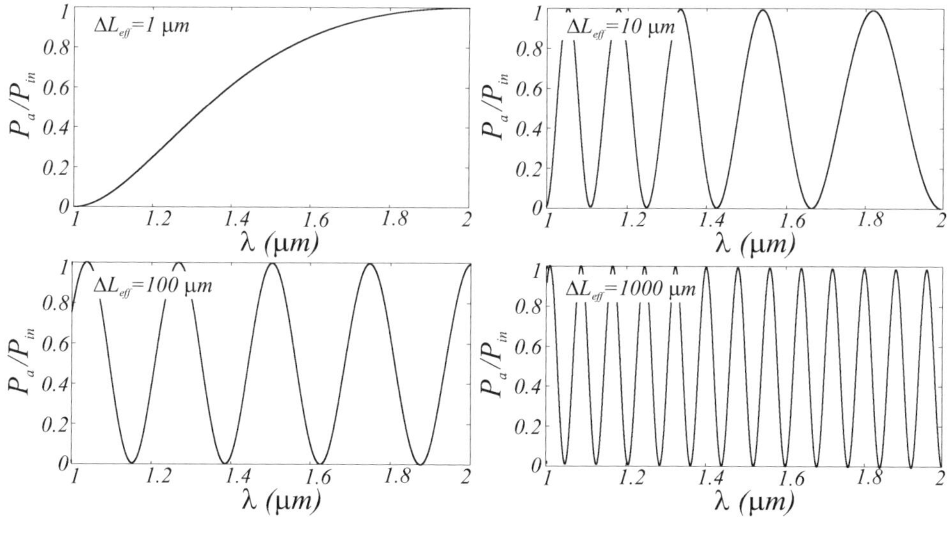

Fig. 5.17: The trend of $P_a(\lambda)$ for different values of ΔL_{eff}.

$L_a = L_b + \Delta L$; as a consequence:

$$\Delta L_{eff} = n_{eff} \Delta L.$$

According to (5.47), the channels whose wavelengths satisfy the relation

$$\lambda = \frac{2n_{eff}\Delta L}{(2m+1)}$$

will be on the output waveguide a, while the channels with

$$\lambda = \frac{n_{eff}\Delta L}{m}$$

will be on the waveguide b. With the Free Spectral Range (FSR) being the distance between two consecutive maxima, to ensure that two adjacent channels separated by $\Delta\lambda$ are routed on two different waveguides, the following must be true:

$$\Delta\lambda = \frac{FSR_\lambda}{2} = \frac{\lambda^2}{2n_{eff}\Delta L}.$$

In terms of frequency:

$$\Delta\nu = \frac{FSR_\nu}{2} = \frac{c}{2n_{eff}\Delta L}.$$

The trends of P_a/P_{in} for some values of ΔL_{eff} are reported in figure 5.17. In this case too, as we have seen for the directional coupler, it is necessary to increase the distance travelled, to the detriment of size, so as to discriminate better the different wavelengths. With reference to the ITU standard, which sets the channel spacing to $50\,GHz$ (about $0.4\,nm$ to $1550\,nm$), it follows that

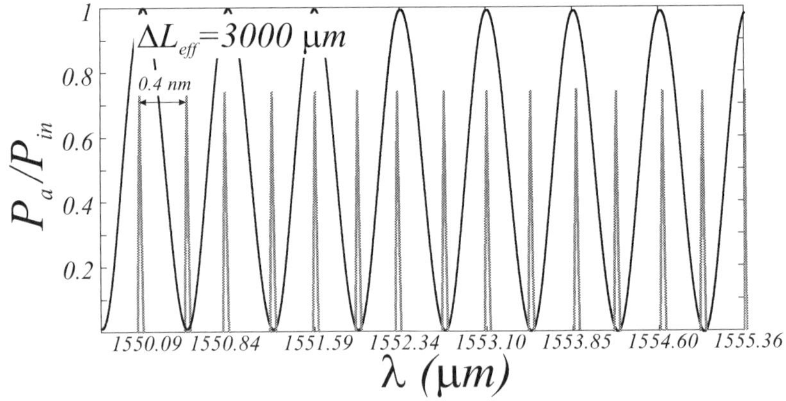

Fig. 5.18: The trend of $P_a(\lambda)$ for $\Delta L_{eff} = 3000\,\mu m$ (*i.e.* $\Delta\lambda = 0.4\,nm$). Superimposed are the spectra of the channels from a WDM system according to the ITU standard.

$n_{eff}\Delta L = 3000\,\mu m$ – cf. figure 5.18. If the waveguide is made of silica, we can estimate $n_{eff} \simeq 1.45$, from which:

$$\Delta L = \frac{\Delta L_{eff}}{n_{eff}} = 2084\,\mu m;$$

while if the material employed is indium phosphide, $n_{eff} \simeq 3.2$, from which:

$$\Delta L = \frac{\Delta L_{eff}}{n_{eff}} = 945\,\mu m.$$

It is clear that there is a strong link between the dimensions of the component and the value of the refractive index of the material used to create the same component, in the sense that the former can be reduced as the latter increases. Cf. figure 5.19 for an example of a demultiplexer obtained using three Mach-Zehnder interferometers. Each Mach-Zehnder interferometer separates the input channels on the two output ports; consequently, each output port contains only half of the input channels, which are now double-spaced. The total number of Mach-Zehnder interferometers necessary to demultiplex N channels is equal to $N-1$. The first interferometer must meet the strictest specifications as it will have to demultiplex channels with lower spacing, while the subsequent devices may even have more flexible features.

Before concluding, we shall highlight the fact that, in our analysis, the characteristics of the couplers are held to be constant at different frequencies. Yet, this cannot be valid when the interferometer works over a broad bandwidth. Figure 5.20 illustrates the characteristics of an interferometer in which the couplers work as power dividers at a wavelength of 1550 nm. As we move away from this value, the coupler behaviour is no longer symmetric, which obviously affect the performance of the interferometer.

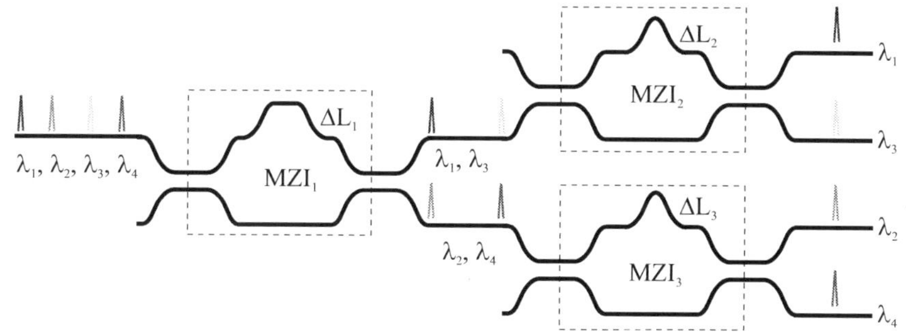

Fig. 5.19: Demultiplexer for WDM systems based on Mach-Zehnder interferometers.

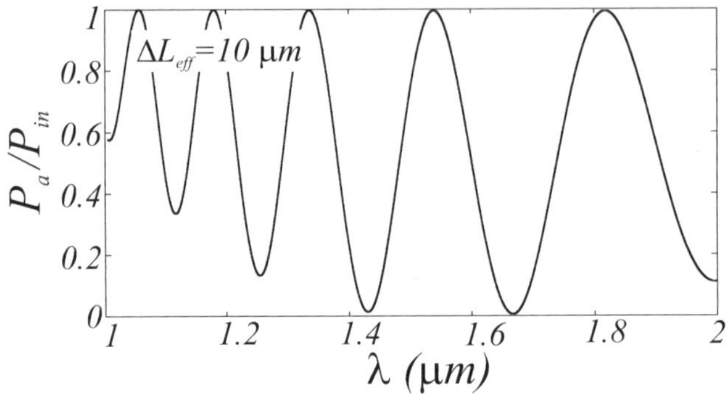

Fig. 5.20: The trend of $P_a(\lambda)$ according to two coupling lengths L_{c1} and L_{c2} depending on the wavelength.

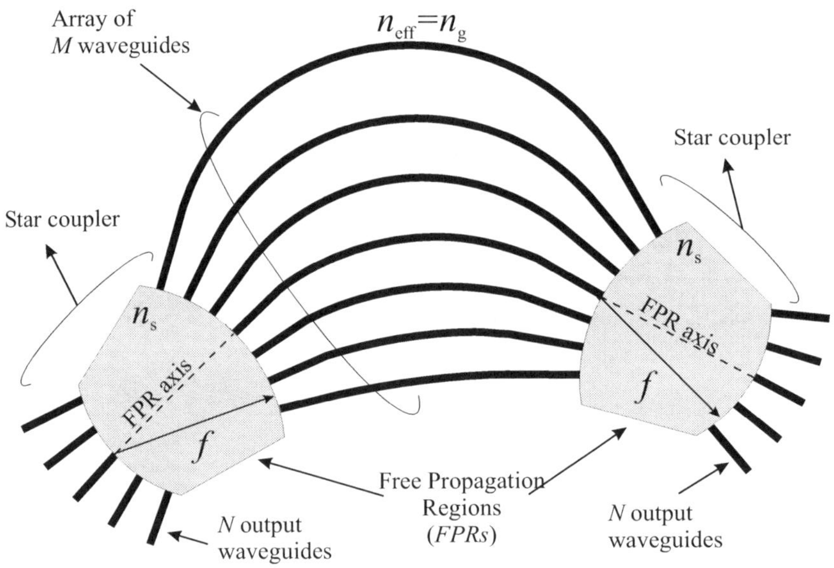

Fig. 5.21: Diagram of an AWG.

5.6 Arrayed-Waveguide Gratings (AWG)

An Arrayed-Waveguide Grating (AWG) can be seen as an extension of the Mach-Zehnder interferometer. The latter is constituted by a 2×2 coupler, used as a power divider, followed by a pair of waveguides of different length and finally a second 2×2 coupler, used as a power combiner. On the other hand, the AWG is constituted by an $N \times M$ power divider, typically known as *star coupler*, followed by an array of M waveguides of different length and an $M \times N$ combiner. With its characteristics largely depending on the wavelength, the AWG has become a key component in current WDM optical systems, *e.g.* as a wavelength multiplexer or demultiplexer, or again to realise the so-called *Wavelength Routing.* Multiplexers and demultiplexers in integrated optics based on 256-channel AWGs with $0.2\,nm$ spacing have been commercially available for some years.

Figure 5.21 schematises the basic structure of an AWG: N input waveguides traverse a multimode region which is so wide that, on a first approximation, we can consider a case of free propagation in it – from which the name Free Propagation Region (FPR). Both sides of the FPR are shaped like an arc of circumference with radius f, whose centre of curvature corresponds to the midpoint of the opposite side. The segment joining the two centres of curvature is the FPR axis. The incoming light then enters a bundle of M identical single-mode waveguides (typically, $M = 2N$) whose fundamental mode has an effective index equal to n_g. The waveguides are of different length, with a constant length increment ΔL as we pass from one waveguide to the adjacent one. Finally, the waveguides traverse a second FPR with the same characteristics as the first one and the N waveguides are propagated through the output star coupler.

Let us consider now the situation shown in figure 5.22, where a signal of wavelength λ enters the first FPR through the waveguide corresponding to the FPR axis. Once inside the FPR, the wavefronts tend to become curved due to the effect of free propagation. As their centre of curvature coincides with that of the opposite side, the waveguides are all excited in phase. Subsequently, the different contributions run through the bundle of waveguides and reach the second FPR as they undergo a constant change of phase due to the different length of the waveguides. The phase shift between two adjacent waveguides is:

$$\Delta\phi_g = 2\pi n_g \frac{\Delta L}{\lambda}.$$

The different contributions coming from the array propagate to the second FPR and interfere with each other, generating an interference figure with maxima and minima on the opposite side. The maxima correspond to the point in which the different contributions gather with a relative phase shift multiple of 2π:

$$\Delta\phi = Q2\pi, \tag{5.48}$$

with Q integer. These phase shifts are due in part to $\Delta\phi_g$, in part to the different distances travelled inside the FPR. The latter contribution can be better considered by referring to figure 5.23 in which we take into account two adjacent waveguides whose axes are spaced by Δx_i – with θ_o being the angle formed by the direction of propagation with the FPR axis. The two contributions follow two different distances, l_1 and l_2. Since $f >> \Delta x_i$, it is possible to apply the paraxial approximation and consider the two directions to be parallel to each other, as shown in figure 5.23. The path difference Δl is:

$$\Delta l = l_2 - l_1 = \Delta x_i \sin\theta_o,$$

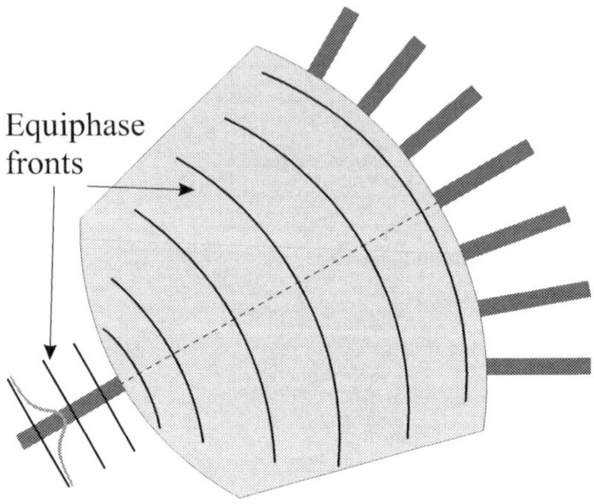

Fig. 5.22: Wavefront evolution in the first FPR.

which causes a phase shift equal to:

$$\Delta\phi_{FPR} = 2\pi n_s \frac{\Delta l}{\lambda} = 2\pi n_s \sin\theta_o \frac{\Delta x_i}{\lambda},$$

with n_s being the FPR refractive index. The total phase shift is thus:

$$\Delta\phi = \Delta\phi_g + \Delta\phi_{FPR} = \frac{2\pi}{\lambda}\left(n_g \Delta L + n_s \sin\theta_o \Delta x_i\right).$$

By substituting the expression of $\Delta\phi$ in (5.48), we obtain the condition determining the maxima of the output field:

$$n_g(\lambda)\Delta L + n_s \sin\theta_o \Delta x_i = Q\lambda. \tag{5.49}$$

As the output waveguides are typically located near the FPR axis, we can assume that $\theta_o << 1$, hence $\sin\theta_o \simeq \theta_o$; this allows us to rewrite the expression from (5.49) as follows:

$$n_g(\lambda)\Delta L + n_s \theta_o \Delta x_i = Q\lambda. \tag{5.50}$$

We have a case of constructive interference for $\theta_o = 0$, which occurs along a particular direction depending on the wavelength – commonly indicated with λ_0:

$$n_{g_0}\Delta L = Q\lambda_0, \tag{5.51}$$

with $n_{g_0} = n_g(\lambda_0)$, from which we can obtain:

$$\Delta L = Q\frac{\lambda_0}{n_{g_0}}. \tag{5.52}$$

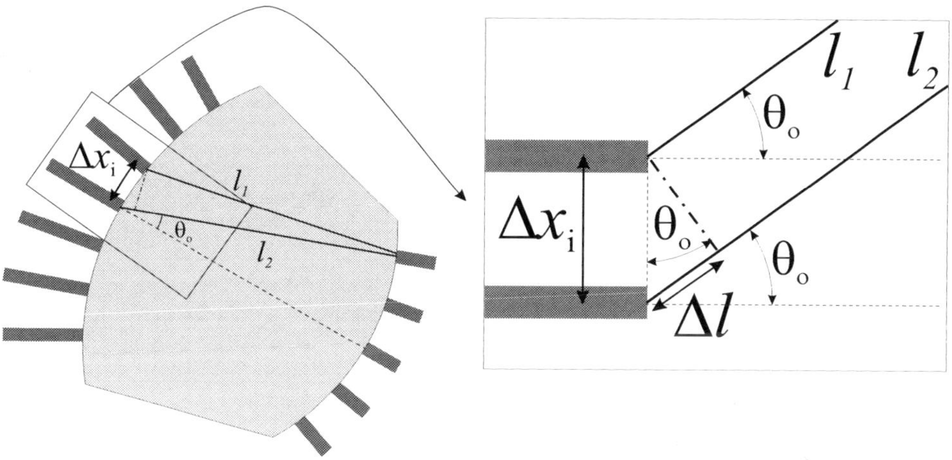

Fig. 5.23: On the left, the second FPR; on the right, path difference calculation.

With reference to figure 5.24, by laying out the output waveguides according to the directions along which, at different wavelengths, we have constructive interference, it is possible to separate the different WDM channels on the different waveguides. If two adjacent output waveguides are separated by Δx_o, the variation $\Delta\theta_o$ necessary in order to pass from the first one to the other is:

$$\Delta\theta_o = \frac{\Delta x_o}{f}, \tag{5.53}$$

which must correspond, in the range of λ_0, to a variation $\Delta\lambda$ such that:

$$\Delta\theta_o = \left.\frac{d\theta_o}{d\lambda}\right|_{\lambda_0} \Delta\lambda. \tag{5.54}$$

By equating (5.53) and (5.54), two adjacent output waveguides are spaced by (5.54) by two wavelengths separated by $\Delta\lambda$. Hence, we differentiate (5.50) with respect to λ and highlighting $d\theta_o/d\lambda$; by resorting to (5.51), we have:

$$\frac{d\theta_o}{d\lambda} = \frac{1}{n_s \Delta x_i}\left(Q - n_{g_0}^{'}\Delta L\right) = \frac{Q}{n_{g_0} n_s \Delta x_i}\left(n_{g_0} - \lambda_0 n_{g_0}^{'}\right) = \frac{Q}{n_{g_0} n_s \Delta x_i} N_g \tag{5.55}$$

with $n_{g_0}^{'} = \frac{dn_g}{d\lambda}|_{\lambda_0}$ and $N_g = n_{g_0} - \lambda_0 n_{g_0}^{'}$ called *group index*. Note that the derivative of n_g with respect to the wavelength describes the signal dispersion occurring in the propagation process within the waveguides. If we ignore this effect, the effective index coincides with the group index introduced above. By substituting the expression found in $\Delta\theta_o$, the result is:

$$\Delta\theta_o = \frac{Q}{n_{g_0} n_s \Delta x_i} N_g \Delta\lambda. \tag{5.56}$$

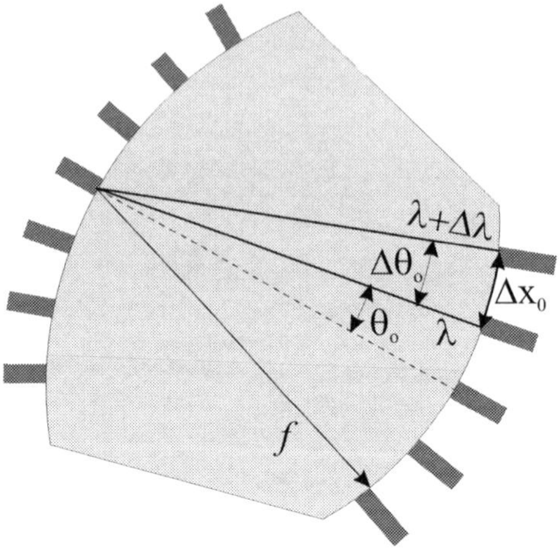

Fig. 5.24: Detail of the output FPR.

By equating (5.53) and (5.56), and highlighting $\Delta\lambda$, it follows that:

$$\Delta\lambda = \frac{n_{g_0} n_s}{N_g} \frac{\Delta x_i \Delta x_o}{Q f}.$$

Above is the variation $\Delta\lambda$ which allows two output maxima, for two different wavelengths, on two waveguides spaced by Δx_o. With θ_o fixed, there actually exist infinite wavelengths that are a solution to (5.50) and can be obtained by varying Q. This means that the AWG frequency response is periodic. To calculate the AWG Free Spectral Range, both the wavelengths λ_0 and $\overline{\lambda} = \lambda_0 + FSR_\lambda$ need to come out of the same waveguide. With reference to (5.51), and imposing

$$\overline{n}_g \Delta L = (Q-1)\overline{\lambda},$$

with $\overline{n}_g = n_g(\overline{\lambda})$, we develop the formula and obtain:

$$\overline{n}_g \Delta L = (Q-1)(\lambda_0 + FSR_\lambda) = n_{g_0}\Delta L - \lambda_0 + QFSR_\lambda - FSR_\lambda.$$

Next, we reorder the expression, use the approximation $\overline{n}_g - n_{g_0} \simeq n'_{g_0} FSR_\lambda$ and resort to (5.52): (5.52):

$$-\lambda_0 + QFSR_\lambda - FSR_\lambda = \Delta L \left(\overline{n}_g - n_{g_0}\right) = \Delta L n'_{g_0} FSR_\lambda = \frac{n'_{g_0}}{n_{g_0}} Q\lambda_0 FSR_\lambda$$

which produces:

$$FSR_\lambda = \frac{\lambda_0}{\dfrac{Q}{n_{g_0}}\left(n_{g_0} - \dfrac{n_{g_0}}{Q} - \lambda_0 n'_{g_0}\right)}.$$

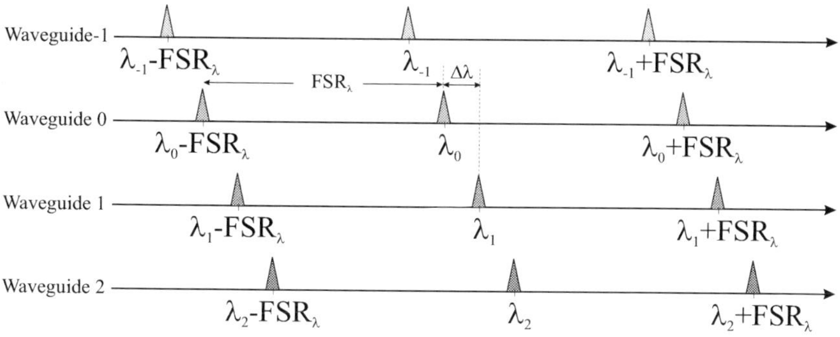

Fig. 5.25: Spectral response of the different output waveguides.

As $Q >> 1$, then $\frac{n_{g0}}{Q} << n_{g0}$, hence:

$$FSR_\lambda = \frac{\lambda_0}{Q}\frac{n_{g0}}{N_g} \tag{5.57}$$

which constitutes the FRS expression we were looking for.

The spectral response of the different output waveguides is thus illustrated in figure 5.25. In fact, the FSR is the wavelength range that can be employed without risking two different wavelengths coming out of the same waveguide. This limits the maximum number of wavelengths that can be demultiplexed to:

$$INT\left[\frac{FSR_\lambda}{\Delta\lambda}\right] = INT\left[\frac{f}{n_s \Delta x_i \Delta x_o}\lambda_0\right],$$

where $INT[x/y]$ is the integer part of the x/y ratio. Let us now consider the case shown in figure 5.26, where a signal of wavelength λ enters the first FPR through an input waveguide which forms an angle θ_i with the axis. Compared to the previous case, the different contributions combined in the second FPR present an additional phase shift $\Delta\phi_{FPR1}$ due to the fact that the equiphase fronts are no longer in phase when they reach the array of waveguides. According to the arguments presented above, and assuming that the spacing of the array of waveguides in correspondence to the first FPR is still Δx_i, we can prove that the additional phase shift is:

$$\Delta\phi_{FPR1} = \frac{2\pi}{\lambda} n_s \Delta x_i \sin\theta_i \simeq \frac{2\pi}{\lambda} n_s \Delta x_i \theta_i$$

hence the condition of constructive interference becomes:

$$n_s \Delta x_i \left(\theta_i + \theta_o\right) + n_g \Delta L = Q\lambda. \tag{5.58}$$

The expression (5.58) shows how θ_o – and thus the waveguide from which a signal of wavelength λ comes – depends on θ_i, *i.e.* the waveguide from which the signal entered. This is a fundamental characteristic as far as *wavelength routing* is concerned, *i.e.* the ability to route channels in a WDM system. We shall focus on this last aspect in the next paragraph.

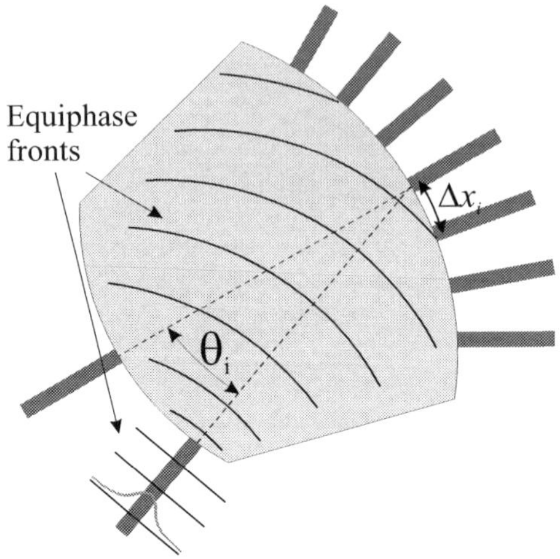

Fig. 5.26: Diagram of an input FPR when the field is propagated from a non-central waveguide.

Chapter 6

Non-reciprocal Components

Many of the materials used to propagate optical signals have anisotropic characteristics, which means that the dielectric permittivity depends on the direction. Crystals such as quartz (SiO_2), rutile (TiO_2) or lithium niobate ($LiNbO_3$) belong to this category. By exploiting their anisotropic characteristics, it is possible to design non-reciprocal components and devices, which modify their behaviour according to specific port chosen – be it the input or the output one. Among these, we shall concentrate on interesting devices for optical signal processing such as polarisers, isolators and circulators. Further investigation on the electromagnetic field behaviour propagating in anisotropic crystals is required to understand how these components work in detail.

6.1 Uniaxial Crystals Propagation

Let us consider a specific anisotropic material with *uniaxial crystals*, *i.e.* crystal structures having only one axis of symmetry, known as *optical axis*. In this case, one of the axes of the main coordinate system – for example, the z-axis – is considered to be parallel to the optical axis: as a consequence, the behaviour of the material is equivalent along the directions belonging to the plane perpendicular to this axis. Therefore, two of the three main indices of the dielectric permittivity tensor will be equal; hence, it is possible to write:

$$\hat{\varepsilon} = \varepsilon_0 \begin{pmatrix} n_t^2 & 0 & 0 \\ 0 & n_t^2 & 0 \\ 0 & 0 & n_z^2 \end{pmatrix}$$

where n_t indicates the common value of n_x and n_y. In the course of our study it will become clear why the two different values of the refractive indices, n_t and n_z, are commonly referred to as n_o and n_e – respectively, the *ordinary* and *extraordinary refractive index*; thus, a uniaxial crystal is described by the following dielectric tensor:

$$\hat{\varepsilon} = \varepsilon_0 \begin{pmatrix} n_o^2 & 0 & 0 \\ 0 & n_o^2 & 0 \\ 0 & 0 & n_e^2 \end{pmatrix}.$$

We shall now verify that plane waves solutions exist in a sourceless anisotropic crystal with no losses, considered uniaxial for reasons of simplicity. The electromagnetic field must be a solution to the homogeneous Maxwell's equations:

$$\overline{\nabla} \times \overline{E} = -j\omega\mu_0 \overline{H} \tag{6.1}$$

$$\overline{\nabla} \times \overline{H} = j\omega\varepsilon_0 \begin{pmatrix} n_o^2 & 0 & 0 \\ 0 & n_o^2 & 0 \\ 0 & 0 & n_e^2 \end{pmatrix} \begin{pmatrix} E_x \\ E_y \\ E_z \end{pmatrix} \tag{6.2}$$

where we held that $\mu = \mu_0$. The second equation can be thus condensed as we highlight the transverse component of the electric field:

$$\overline{\nabla} \times \overline{H} = j\omega\varepsilon_0 (n_o^2 \overline{E}_t + n_e^2 E_z \hat{z}) \tag{6.3}$$

in which the transverse component of the electric field $\overline{E}_t = E_x\hat{x} + E_y\hat{y}$ has been separated from the longitudinal component E_z, while $\hat{\imath}$, $\hat{\jmath}$ and $\hat{z}$ are the unit vectors of the orthogonal Cartesian coordinate system x, y, z. The solutions we are looking for are of the following type:

$$\overline{E} = \overline{E}_0 e^{-j\overline{k}\cdot\overline{r}} = \overline{E}_0 e^{-jk\hat{k}\cdot\overline{r}};$$

$$\overline{H} = \overline{H}_0 e^{-j\overline{k}\cdot\overline{r}} = \overline{H}_0 e^{-jk\hat{k}\cdot\overline{r}};$$

where $\overline{E}_0$ and $\overline{H}_0$ are constants complex vectors that do not depend on the position, $k\hat{k}$ is the wave vector in which the propagation direction $\hat{k}$ is highlighted, and $\overline{r}$ is the generic position vector. The wave vector module k thus represents the unknown eigenvalue of the problem. By making the wavenumber in vacuum $k_0 = \omega\sqrt{\mu_0\varepsilon_0}$ explicit, we can write:

$$k = \tilde{n}k_0 \tag{6.4}$$

with $\tilde{n}$ as the unknown index. For simplicity of notation, from here onward we shall simply use the symbol k, and only at the end of our work will we resort to $\tilde{n}$.

By substituting the expression of $\overline{E}$ in the first Maxwell's equation, and reminding the reader that the vector $\overline{E}_0$ is constant, we obtain:

$$\overline{\nabla} \times \overline{E} = \overline{\nabla} \times (\overline{E}_0 e^{-jk\hat{k}\cdot\overline{r}}) = \overline{\nabla} e^{-jk\hat{k}\cdot\overline{r}} \times \overline{E}_0 = -jk\hat{k} \times \overline{E}.$$

Thus, from the first Maxwell's equation, the magnetic field can be written as a function of the electric field:

$$\overline{H} = \frac{jk}{j\omega\mu_0}\hat{k} \times \overline{E}. \tag{6.5}$$

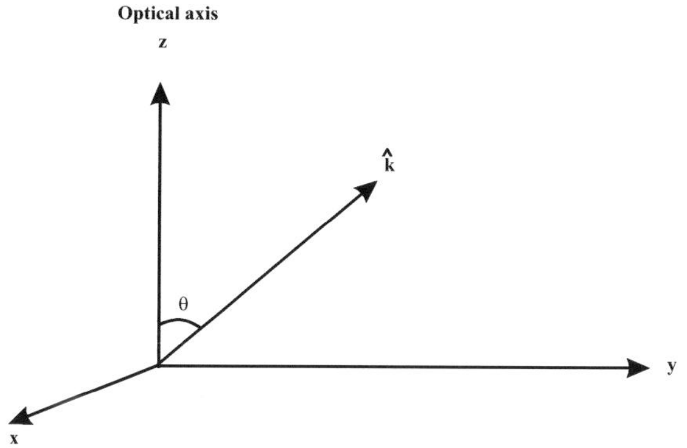

Fig. 6.1: The main coordinate system x, y, z with the z-axis aligned with the optical axis. The unit vector $\hat{k}$ determines the direction of propagation.

Similarly, the left side of (6.3) can be reformulated as:

$$\overline{\nabla} \times \overline{H} = -jk\hat{k} \times \overline{H}$$

hence:

$$jk\overline{H} \times \hat{k} = j\omega\varepsilon_0(n_o^2\overline{E}_t + n_e^2 E_z\hat{z}).$$

The following equation can be determined by resorting to the expression of $\overline{H}$ given by (6.5):

$$-(jk)^2[(\hat{k} \times \overline{E}) \times \hat{k}] = \omega^2\mu_0\varepsilon_0(n_o^2\overline{E_t} + n_e^2 E_z\hat{z}). \tag{6.6}$$

The equation (6.6) above is the eigenvalue equation to be solved: to begin with, the eigenvalue jk has to be calculated first, subsequently we will be able to obtain the corresponding eigenvectors $\overline{E}$ and $\overline{H}$. Before we move on, however, we shall reformulate our coordinate system and attribute particular importance, on the one hand, to the plane formed by the optical axis $\hat{z}$ and the direction of propagation $\hat{k}$, and, on the other, to the direction orthogonal to the same plane. Defining as θ the angle formed by the optical axis and the direction of propagation, the unit vectors $\hat{u}^{'}$ and $\hat{u}^{''}$ are defined as follows:

$$\hat{u}^{'} = \frac{\hat{k} \times \hat{z}}{sin\theta};$$

$$\hat{u}^{''} = \hat{z} \times \hat{u}^{'}.$$

With these definitions, $\hat{u}^{'}$, $\hat{u}^{''}$, $\hat{z}$ form a coordinate system obtained by counterclockwise rotation around the z-axis – with respect to the original system $\hat{x}$, $\hat{y}$, $\hat{z}$ – as a function of the direction of propagation of the field. In particular, $\hat{u}^{''}$ always belongs to the plane formed by $\hat{k}$ and $\hat{z}$, therefore it follows that:

$$\hat{k} = \hat{u}^{''} sin\theta + \hat{z} cos\theta.$$

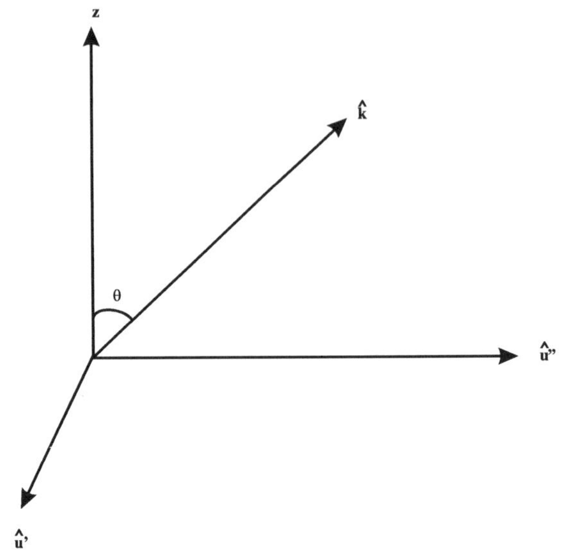

Fig. 6.2: The coordinate system $\hat{u}^{'}, \hat{u}^{''}, \hat{z}$.

The direction of $\hat{u}^{'}$ is then perpendicular to the plane formed by the direction of propagation and the optical axis. We can thus represent the electric field according to the new coordinate system with $(E^{'}, E^{''}, E_z)$ as its components. With respect to the vector identity $\overline{A} \times \overline{B} \times \overline{C} = (\overline{A} \cdot \overline{C})\overline{B} - (\overline{A} \cdot \overline{B})\overline{C}$, the equation (6.6) becomes:

$$-(jk)^2[\overline{E} - (\hat{k} \cdot \overline{E})\hat{k}] = \omega^2 \mu_0 \varepsilon_0 (n_o^2 \overline{E_t} + n_e^2 E_z \hat{z}).$$

By substituting the expression of $\hat{k}$ previously found, we can obtain:

$$\begin{aligned} -(jk)^2[\overline{E} - ((\hat{u}^{''} sin\theta + \hat{z} cos\theta) \cdot \overline{E})(\hat{u}^{''} sin\theta + \hat{z} cos\theta)] &= \omega^2 \mu_0 \varepsilon_0 (n_o^2 \overline{E}_t + n_e^2 E_z \hat{z}), \\ -(jk)^2[\overline{E} - (E^{''} sin\theta + E_z cos\theta)(\hat{u}^{''} sin\theta + \hat{z} cos\theta)] &= \omega^2 \mu_0 \varepsilon_0 (n_o^2 \overline{E}_t + n_e^2 E_z \hat{z}), \end{aligned}$$

which is projected onto the three axes $\hat{u}^{'}$, $\hat{u}^{''}$ and $\hat{z}$ so as to produce the following system of equations:

$$-(jk)^2 E^{'} = n_o^2 k_0^2 E^{'}; \tag{6.7}$$

$$(jk)^2[(sin\theta\, E^{''} + cos\theta\, E_z) sin\theta - E^{''}] = n_o^2 k_0^2 E^{''}; \tag{6.8}$$

$$(jk)^2[(sin\theta\, E^{''} + cos\theta\, E_z) cos\theta - E_z] = n_e^2 k_0^2 E_z. \tag{6.9}$$

The system (6.7) - (6.9) consists of three equations in three unknowns $E^{'}$, $E^{''}$, E_z with eigenvalue jk. However, the first equation only involves the component $E^{'}$ and is thus decoupled from the other two, which are, instead, coupled through $E^{''}$ and E_z. This permits to identify two types of solutions: the first produces *ordinary waves*, with $E^{'}$ obtained from (6.7) and $E^{''} = E_z = 0$; the second type, on the other hand, leads to *extraordinary waves*, with $E^{'} = 0$ and the components $E^{''}$ and E_z being solutions to (6.8) and (6.9). Considering the linearity of the problem, the total field can be represented as a combination of the two solutions according to the following scheme:

$$\begin{cases} \textit{Ordinary wave} \\ E^{'} \leftarrow \ (6.7) \\ E^{''} = E_z = 0 \end{cases} \qquad \begin{cases} \textit{Extraordinary wave} \\ E^{'} = 0 \\ E^{''},\ E_z \leftarrow \ (6.8), (6.9) \end{cases}$$

6.1.1 Ordinary Waves

In order to find this solution, we shall only consider (6.7):

$$-(jk)^2 E^{'} = n_o^2 k_0^2 E^{'} .$$

A non-zero solution is obtained provided the following holds true:

$$(jk)^2 = -n_o^2 k_0^2, \tag{6.10}$$

from which:

$$k = \tilde{n} k_0 = n_o k_0 .$$

The index $\tilde{n}$ takes thus the constant value of the ordinary refractive index n_o. The electric field has the direction of $\hat{u}'$ since $E' \neq 0$ and $E'' = E_z = 0$, hence it can be written as:

$$\overline{E} = A e^{-j n_o k_0 \hat{k} \cdot \overline{r}} \ \hat{u}' , \tag{6.11}$$

where A is a complex constant; on the other hand, the expression of the magnetic field will be:

$$\overline{H} = \frac{A}{\eta'} \ e^{-j n_o k_0 \hat{k} \cdot \overline{r}} \ \hat{k} \times \hat{u}' \tag{6.12}$$

with $\eta' = \frac{\eta_0}{n_o}$ and $\eta_0 = \sqrt{\frac{\mu_0}{\varepsilon_0}}$. It is interesting to calculate the Poynting vector $\overline{S}$ to determine the direction in which the energy is propagated:

$$\overline{S} = \frac{\overline{E} \times \overline{H}^*}{2} = \frac{|A|^2}{2\eta'} \hat{u}' \times (\hat{k} \times \hat{u}') = \frac{|A|^2}{2\eta'} \hat{k} .$$

The following three points summarise what has been explained so far: the electromagnetic field is linearly polarised, as (6.11) and (6.12) demonstrate; the expressions of the electromagnetic field seem to indicate that the ordinary wave is a TEM wave since both the electric and the magnetic field are perpendicular to the direction $\hat{k}$; finally, the energy propagates along the direction of propagation.

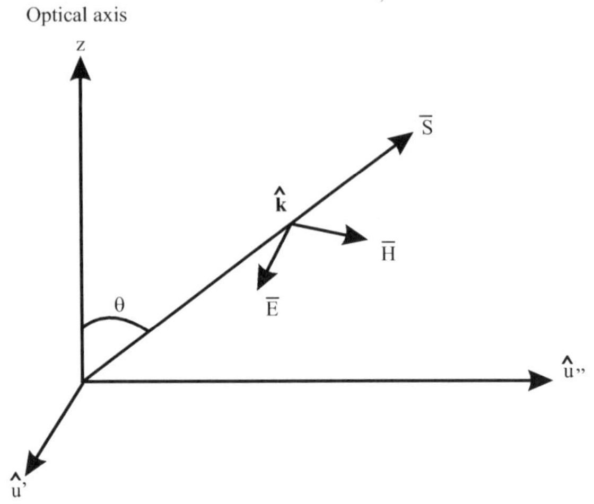

Fig. 6.3: The electromagnetic field $(\overline{E}, \overline{H})$ and the Poynting vector $\overline{S}$ in case of an ordinary wave.

6.1.2 Extraordinary Waves

The extraordinary wave can be described by solving the system (6.8) - (6.9), which can be rewritten as follows:

$$\begin{aligned} ((jk)^2 cos^2\theta + n_o^2 k_0^2)\, E'' + (jk)^2 cos\theta sin\theta\, E_z &= 0; \\ (jk)^2 cos\theta sin\theta\, E'' + ((jk)^2 sin^2\theta + n_e^2 k_0^2)\, E_z &= 0. \end{aligned}$$

The determinant of the system must be zero for the solution to be non-trivial:

$$((jk)^2 cos^2\theta + n_o^2 k_0^2)((jk)^2 sin^2\theta + n_e^2 k_0^2) - (jk)^4 cos^2\theta sin^2\theta = 0,$$

from which:

$$(jk)^2 (n_e^2 cos^2\theta + n_o^2 sin^2\theta) + k_0^2 n_o^2 n_e^2 = 0,$$

hence, according to (6.4):

$$-k^2 = -\tilde{n}^2 k_0^2 = -\frac{n_o^2 n_e^2}{n_e^2 cos^2\theta + n_o^2 sin^2\theta}\, k_0^2. \tag{6.13}$$

It follows that the unknown index is:

$$\tilde{n} = \frac{n_o n_e}{\sqrt{n_e^2 \cos^2\theta + n_o^2 \sin^2\theta}}\,. \tag{6.14}$$

The expression of $\tilde{n}$ illustrates that the characteristics of the extraordinary wave depend on θ, which is the angle formed by the direction of propagation and the optical axis. The electric field $\overline{E}$ has a

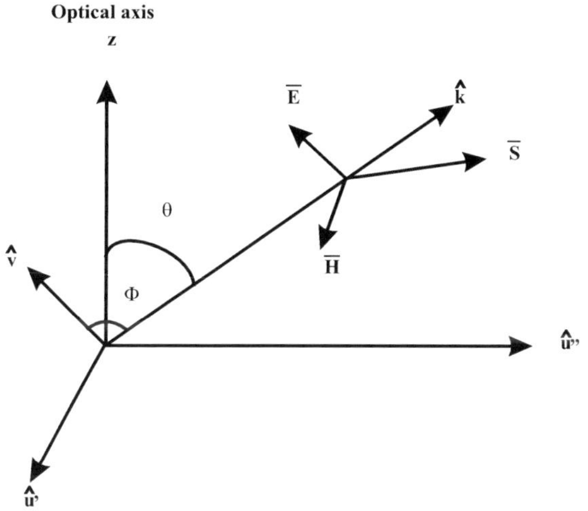

Fig. 6.4: The electromagnetic field $(\overline{E}, \overline{H})$ and the Poynting vector $\overline{S}$ in case of an extraordinary wave.

general direction $\hat{v}$ – determined by the two non-zero components E'' and E_z – and forms an angle Φ with the direction of propagation; the direction of the magnetic field, instead, is perpendicular to the plane formed by the directions $\hat{k}$ and $\hat{v}$, with $\hat{u}'$ as the only component in this respect (cf. figure 6.4). Below are the expressions of the electromagnetic field ($\overline{E}$, $\overline{H}$), the ratio between the electric field components and the expression of the angle Φ.

$$\begin{aligned} \overline{E} &= B e^{-j\tilde{n}k_0\hat{k}\cdot\overline{r}}\hat{v} \\ \overline{H} &= \frac{B}{\eta''} e^{-j\tilde{n}k_0\hat{k}\cdot\overline{r}}\hat{u}' \\ \frac{E_z}{E''} &= -\frac{n_o^2}{n_e^2}\tan\theta \\ \sin\Phi &= \frac{n_e^2\cos^2\theta + n_o^2\sin^2\theta}{\sqrt{n_e^4\cos^2\theta + n_o^4\sin^2\theta}} \end{aligned}$$

where $\eta'' = \frac{\eta_0}{\tilde{n}\sin\Phi}$, considering that $\hat{v}\times\hat{k} = -\hat{u}'\sin\Phi$. The Poynting vector of the extraordinary wave can be expressed as follows:

$$\overline{S} = \frac{\overline{E}\times\overline{H}^*}{2} = \frac{|B|^2}{2\eta''}(\hat{v}\times\hat{u}')\,.$$

Hence, the energy does not propagate along $\hat{k}$ as this direction generally does not coincide with the direction $\hat{v}\times\hat{u}'$. The electromagnetic field is linearly polarised, though the electric field $\overline{E}$, unlike the magnetic field, is not perpendicular to $\hat{k}$ – with the exception of special cases: the extraordinary

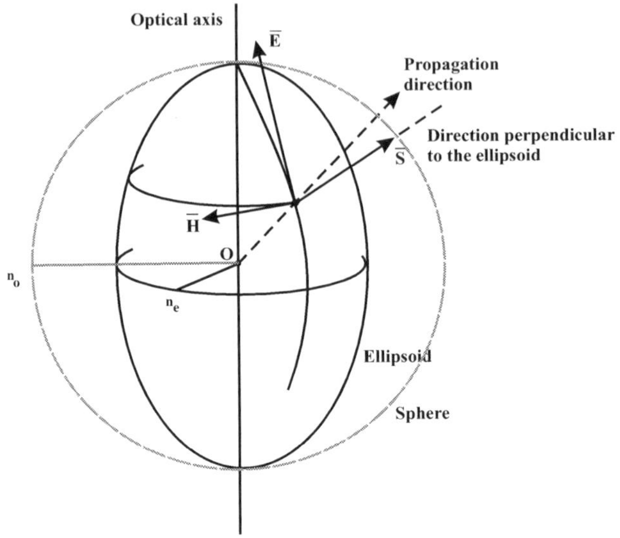

Fig. 6.5: Ellipsoid of the wave vectors for the extraordinary wave indicating the propagation direction of the equiphase planes, of the Poynting vector $\overline{S}$ and of the electromagnetic field ($\overline{E}$, $\overline{H}$). The spherical ordinary wave surface is represented here in grey.

wave is thus a TM wave. Furthermore, its characteristics depend on the direction of propagation with respect to the optical axis: as the former varies, so does $\tilde{n}$. To make an example, considering (6.14), for a value of θ equal to 0, we obtain $\tilde{n} = n_o$, which is the same value of the ordinary wave: the electric field is perpendicular to the direction $\hat{k}$, along which the energy propagates. With $\theta = \pi/2$, we have $\tilde{n} = n_e$, which means that the electric field is parallel to the optical axis and perpendicular to the direction of propagation $\hat{k}$ – which is also the direction of the Poynting vector.

6.1.3 Wave Vector Surface & Index Ellipsoid

The *wave vector surface* – or *normal surface* – can be directly derived from (6.13) rewritten as follows:

$$k^2(n_e^2 cos^2\theta + n_o^2 sin^2\theta) = n_o^2 n_e^2 k_0^2. \tag{6.15}$$

Indeed, as the components of the wave vector in a Cartesian system are

$$\begin{aligned} k_x &= k\, sin\theta\, cos\phi = \tilde{n}k_0\, sin\theta\, cos\phi, \\ k_y &= k\, sin\theta\, sin\phi = \tilde{n}k_0\, sin\theta\, sin\phi, \\ k_z &= k\, cos\theta = \tilde{n}k_0\, cos\theta, \end{aligned}$$

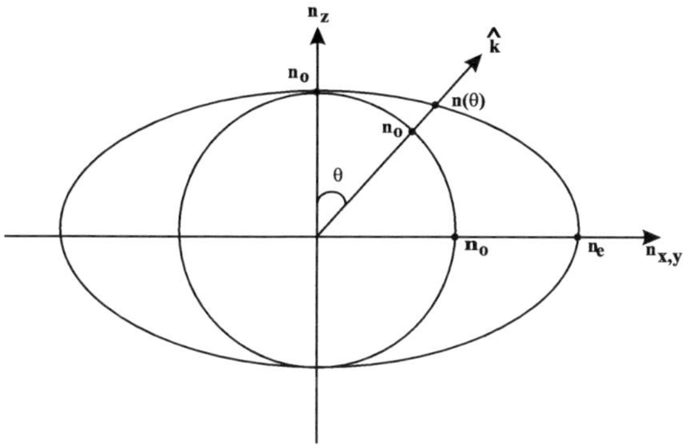

Fig. 6.6: The circle and ellipse of the wave vectors, respectively, for the ordinary and extraordinary wave, for a uniaxial crystal with positive birefringence.

the equation from (6.15) becomes:

$$k_x^2 n_o^2 + k_y^2 n_o^2 + k_z^2 n_e^2 = n_o^2 n_e^2 k_0^2$$

and finally, after both members are divided by $n_o^2 n_e^2 k_0^2$, we will have:

$$\frac{(k_x/k_0)^2}{n_e^2} + \frac{(k_y/k_0)^2}{n_e^2} + \frac{(k_z/k_0)^2}{n_o^2} = 1. \tag{6.16}$$

As far as wave vectors are concerned, this equation represents an ellipsoid of revolution – with the n_e semi-axis along x and y, and n_o along z – and permits to calculate the components of the wave vector as well as the refractive index of the extraordinary wave. In a similar vein, thanks to (6.10), it is possible to determine the ordinary wave surface. As the latter solely depends on the ordinary index, in this case we have a spherical surface of radius n_o. Hence, the surfaces of the wave vectors in uniaxial crystals are symmetrical with respect to the optical axis, as illustrated in figure 6.5.

Important information can be retrieved from these surfaces, for example about the value of the refractive index as seen by the extraordinary wave, or about the direction of the electromagnetic field $\overline{E}$ and $\overline{H}$, or again about the orientation of the Poynting vector with respect to both the direction of propagation and the crystal optical axis. In particular, the length of the radius from the origin to the surface of the ellipsoid or sphere, along the direction of propagation, is the refractive index seen by the wave considered; the electric and the magnetic field are parallel to the meridian and the parallel passing through the point of intersection between the surface itself and the direction of propagation; finally, the direction of the Poynting vector is perpendicular to the surface in the point of intersection indicated above. Thus, the direction in which the energy is transmitted is different if we consider an ordinary or an extraordinary wave: as for the former, it is always parallel to the propagation direction; when it comes to extraordinary waves, instead, the energy is transmitted along a direction which is parallel to the direction of propagation only if the propagation occurs in the direction of the optical axis, or is perpendicular to it.

As a result of the rotational symmetry of the wave vectors surfaces, we often make reference to the curves obtained from their intersection with a plane containing the optical axis. As for ordinary waves, we have a circle, while for extraordinary waves, an ellipse. Figure 6.6 shows the case of a uniaxial crystal with *positive birefringence*, $n_e > n_o$. The opposite situation of *negative birefringence* is instead characterised by $n_o > n_e$, with a dual situation – *i.e.* the ellipse being contained within the circle and the two curves always touching in correspondence of the two points defining the optical axis.

For the sake of completeness, thanks to their symmetry, we can represent the surfaces of the wave vectors only in the first octant of the three-dimensional space, as illustrated in figure 6.7 (respectively, for an isotropic, uniaxial and biaxial medium), where we consider the birefringence of the uniaxial crystal to be positive.

As an alternative to the normal surface, we often use the *index ellipsoid*, also known as *optical indicatrix* or *Fresnel ellipsoid*, which is a quadratic representation in the Cartesian space of the dielectric impermeability tensor $\hat{\eta} = \varepsilon_0 \hat{\varepsilon}^{-1}$,

$$\sum_{ij} \eta_{ij} x_i x_j = 1.$$

By taking the principal axes of the crystal as our coordinate system, the index ellipsoid is given by:

$$\frac{x^2}{n_x^2} + \frac{y^2}{n_y^2} + \frac{z^2}{n_z^2} = 1,$$

where $1/n_x^2$, $1/n_y^2$ and $1/n_z^2$ are the main values of the tensor $\hat{\eta}$. In the case of a uniaxial crystal, the index ellipsoid is an ellipsoid of revolution around the optical axis, while it degenerates into a sphere for an isotropic medium. In order to determine the indices value and the polarisation of the two waves generated *e.g.* in a uniaxial crystal, we can resort to the following geometrical method. Once the wave propagation direction has been identified, the plane passing through the origin and perpendicular to the propagation direction itself defines an ellipse whose semi-axes identify the values of the refractive indices seen by the ordinary and extraordinary wave; besides, the same semi-axes also describe the directions – orthogonal to each other – of the electric displacement vector of the two waves; the latter gives us the components of the electric field through the dielectric permittivity tensor.

Before concluding, let us remember that what has been said so far can also be valid for the more general case of a biaxial medium, with different values with respect to the three main coefficients of dielectric permittivity. Both the waves which are solutions to Maxwell's equations depend, in these materials, on the propagation direction. The indices surfaces provide the same information as seen previously. In particular, the points where they touch define the directions of the optical axes along which the waves naturally see the same refractive index.

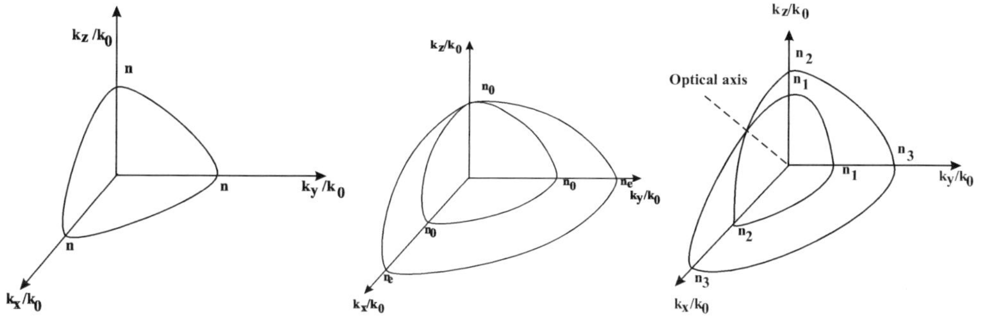

Fig. 6.7: The wave vectors surfaces in case of isotropic (left), uniaxial (middle) and biaxial (right) media.

6.1.4 Birefringence

As we have seen, in uniaxial anisotropic crystals, two solutions – the ordinary and the extraordinary wave – propagate through the material in different ways: this generates the so-called phenomenon of birefringence, as illustrated in figure 6.8, where a wave from an isotropic medium, such as air, strikes a uniaxial crystal. For simplicity, the optical axis is in the plane of incidence. In this case,

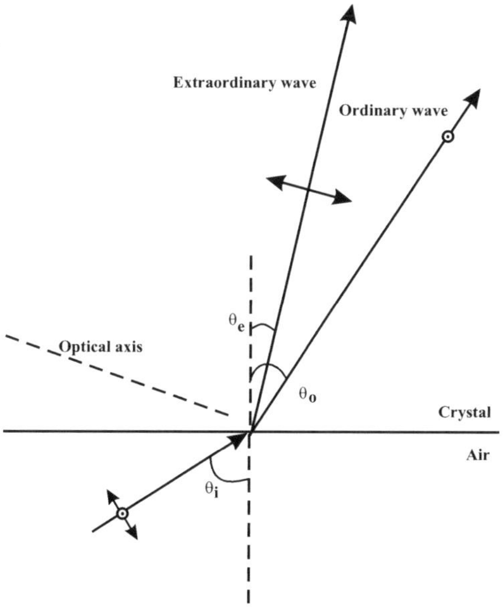

Fig. 6.8: The birefringence of an electromagnetic field which is incident on a uniaxial crystal.

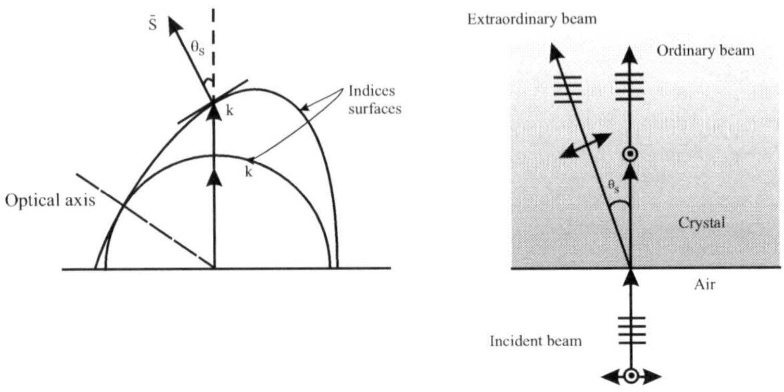

Fig. 6.9: The birefringence of an electromagnetic field which is incident on a uniaxial crystal, in case of normal incidence, described using the index surface.

Snell's law can be verified with two different conditions:

$$n_{air}\, sin\, \theta_i = n_o\, sin\, \theta_o = \tilde{n}\, sin\, \theta_e\,, \tag{6.17}$$

θ_i being the angle of incidence. In fact, since the wave in the anisotropic crystal is separated into its ordinary and extraordinary components, with two different indices n_o and $\tilde{n}$ respectively, it follows that two different angles of refraction will have to result from (6.17) – their value θ_o and θ_e clearly depending on the value of n_o and $\tilde{n}$, as well as on the orientation of the optical axis.

The transmitted ordinary wave has an electric field which is always orthogonal to the optical axis and the direction of propagation – here, to the plane of incidence – while the electric field of the extraordinary wave is parallel to the plane of incidence. It is possible to describe this phenomenon in graphical form by resorting to the index surfaces, as in figure 6.9 in the particular case of normal incidence. If the optical axis is neither parallel nor perpendicular to the surface of separation of the two media, and it has a generic inclination with respect to the same surface of separation, the extraordinary wave will not be able to propagate its energy in a direction that is orthogonal to the surface of separation – which is possible, instead, as far as ordinary waves are concerned. According to the rules described above, the Poynting vector of the extraordinary wave will thus be orthogonal to the ellipse, with the consequent separation of the two components.

In conclusion, note that, by considering the optical axis perpendicular to the interface between the two media, the two waves in the crystal see the same index and propagate in the same direction without separating. On the other hand, if the optical axis is parallel to the interface, the directions of propagation of the wavefronts and of the Poynting vector coincide not only with each other, but also for the ordinary and extraordinary wave – though the latter see two different indices, n_o and n_e respectively. Hence, the two waves carry the energy in the same direction, but travel at different phase speeds. The reader is free to investigate the behaviour of the field in the crystal for generic angles of incidence and orientations of the optical axis.

6.2 Polarisation Converters

Polarisation converters are devices used to change the wave polarisation by retarding a particular field component; for this reason, they are also known as *phase retarders* or *waveplates*. These components exploit the anisotropic properties of the medium they are made of. In order to analyse this phenomenon, consider the general solution in an anisotropic medium made of a uniaxial crystal, represented by the linear combination of the ordinary and extraordinary wave. The form of the electric and the magnetic field will thus be as follows:

$$\overline{E} = Ae^{-jn_o k_0 \hat{k}\cdot\overline{r}}\,\hat{u}' + Be^{-j\tilde{n}k_0\hat{k}\cdot\overline{r}}\hat{v}\,; \tag{6.18}$$

$$\overline{H} = \frac{A}{\eta'}e^{-jn_o k_0 \hat{k}\cdot\overline{r}}\,(\hat{k}\times\hat{u}') + \frac{B}{\eta''}e^{-j\tilde{n}k_0\hat{k}\cdot\overline{r}}\hat{u}'\,. \tag{6.19}$$

As the expressions above demonstrate, the electromagnetic field depends on the angle θ, hence the characteristics of the wave are linked to the propagation direction $\hat{k}$. While for the ordinary and extraordinary wave the polarisation is linear, for the generic wave – which is a combination of the two – the polarisation is not constant during the propagation process. This is due to the fact that the ordinary and extraordinary waves have different phase velocities, c/n_o and $c/\tilde{n}$ respectively, and add different contributions to the phase of the total wave; as a consequence, this phase shift causes a polarisation variation. This concept will be further clarified in the course of our discussion.

We shall now take into account the value of $\theta = \pi/2$. From (6.14), we know that $\tilde{n} = n_e$ and the electric field of the extraordinary wave is parallel to the optical axis z and perpendicular to the propagation direction $\hat{k}$. Consider $s = \hat{k}\cdot\overline{r}$ and, to make calculations easier, $A = B$. In this case, the electric field $\overline{E}$ can be rewritten as:

$$\overline{E} = A[e^{-jn_o k_0 s}\,\hat{u}' + e^{-jn_e k_0 s}\,\hat{z}]$$

The electric field $\overline{E}$ is in the plane determined by the direction of $\hat{u}'$ and $\hat{z}$. By expressing the complex constant A as $|A|e^{j\phi}$, the electric field in the time and space domain can be formulated as:

$$\begin{aligned}
\overline{\mathcal{E}}(t,s) &= \mathcal{R}e\{\overline{E}(s)e^{j\omega t}\}\\
&= \mathcal{R}e\{A[e^{-jn_o k_0 s}\,\hat{u}' + e^{-jn_e k_0 s}\,\hat{z}]e^{j\omega t}\}\\
&= |A|(\mathcal{R}e\{e^{-jn_o k_0 s}e^{j\phi}e^{j\omega t}\,\hat{u}'\} + \mathcal{R}e\{e^{-jn_e k_0 s}e^{j\phi}e^{j\omega t}\,\hat{z}\})\\
&= |A|(\cos(\omega t - n_o k_0 s + \phi)\,\hat{u}' + \cos(\omega t - n_e k_0 s + \phi)\,\hat{z}\,).
\end{aligned} \tag{6.20}$$

For a generic value of s, as said above, the contributions of the ordinary – the first term – and extraordinary wave – the second term – to the phase of the generic wave are not equal since $n_o \neq n_e$.

Hence, there is a phase difference $\Delta\phi$ equal to:

$$\Delta\phi = (n_o - n_e)k_0 s\,.$$

Given that $\Phi = \omega t - n_o k_0 s + \phi$, it is possible to rewrite $\overline{\mathcal{E}}(t,s)$ in a more condensed form:

$$\overline{\mathcal{E}}(t,s) = |A|\{\cos\Phi\,\hat{u}' + \cos\left(\Phi + \Delta\phi\right)\hat{z}\}\,,$$

which highlights how the second wave is retarded – or advanced, depending on the sign of $\Delta\phi$ – with respect to the first. For this reason, we also talk about *fast* and *slow axes*, or *advanced* and *retarded waves*. As s varies, so does $\Delta\phi$ and, consequently, the polarisation. It is thus possible to rotate the polarisation of a wave propagating in an anisotropic crystal by selecting the appropriate thickness, in relation to its ordinary and extraordinary indices.

We shall start by considering a wave in $s = 0$ ($\overline{r} = 0$). From (6.20) we will have:

$$\overline{\mathcal{E}}(t, s=0) = |A|\cos(\omega t + \phi)[\hat{u}' + \hat{z}]$$

with linear polarisation. During the propagation process, it is possible to identify some particular cases as s varies:

- $\Delta\phi = m2\pi \quad \rightarrow \quad \cos\Phi = \cos\left(\Phi + \Delta\phi\right)$

which corresponds to a value of s equal to:

$$s = m\frac{\lambda_0}{|n_o - n_e|}$$

with λ_0 being the wavelength in vacuum. In this case, the field polarisation coincides with the initial one in $s = 0$.

- $\Delta\phi = (2m-1)\pi \quad \rightarrow \quad \cos\Phi = -\cos\left(\Phi + \Delta\phi\right)$

i.e. for values of s determined by

$$s = (2m-1)\frac{\lambda_0}{2|n_o - n_e|}\,.$$

Subsequently, we will have *half-wave plates*, and the polarisation is still linear, though perpendicular (rotated by 90°) with respect to the original polarisation in $s = 0$. Moreover, it can be easily verified how these retarders can turn a clockwise circular polarisation into counterclockwise one.

- $\Delta\phi = (2m-1)\pi/2 \quad \rightarrow \quad \cos\left(\Phi + \Delta\phi\right) = \pm\sin\Phi$

that is to say, for values of s given by

$$s = (2m-1)\frac{\lambda_0}{4|n_o - n_e|}\,,$$

we obtain *quarter-wave plates*, and the polarisation is circular – be it clockwise or counterclockwise, according to the sign $\pm$. Conversely, if the polarisation of the input field is circular, this will become linear.

Different lengths will result in arbitrary phase shifts $\Delta\phi$ with consequent elliptic output polarisations.

6.3 Optical Isolators

An optical isolator is an important component used in different devices which allows the transmission of light in only one specific direction. For example, semiconductor lasers are usually followed by an isolator, which is often integrated on the same chip, to prevent unwanted feedback into the laser cavity and avoid damage or unstable emissions; in optical amplifiers, isolators are used to block the counterpropagating amplified spontaneous emission (ASE), which would excessively degrade the signal-to-noise ratio.

Optical isolators can be obtained by exploiting the so-called *magneto-optic effects*, which permit to control the anisotropy of certain materials by means of a proper magnetostatic field. The Faraday effect is the best-known magneto-optic effect: it causes the variation of some elements of the dielectric permittivity tensor and permits to rotate the polarisation direction of the field by an angle θ – known as *optical rotatory power*. Its value is given by:

$$\theta = VBL\,, \tag{6.21}$$

where V is the Verdet constant, in $[rad/m \cdot T]$, B the static magnetic flux density applied to the medium and L its length. The rotation can be controlled by acting on the field B and the thickness of the material employed.

In order to study this phenomenon from an analytical point of view, we shall follow a similar approach to that adopted for uniaxial media – namely, we need to determine the eigenvalues and eigenvectors of Maxwell's equations considering an anisotropic medium described by a proper dielectric permittivity tensor. Note that the media exhibiting the Faraday effect are non-reciprocal – that is to say, $\hat{\varepsilon} \neq \hat{\varepsilon}^T$ – and that the off-diagonal elements are responsible not only for the coupling between the transverse components of the field, but also for the polarisation rotation.

In particular, as the medium is non-reciprocal, the polarisation rotation stays in the same relative direction even if the direction of propagation is reversed. As figure 6.10 (left) illustrates then, when

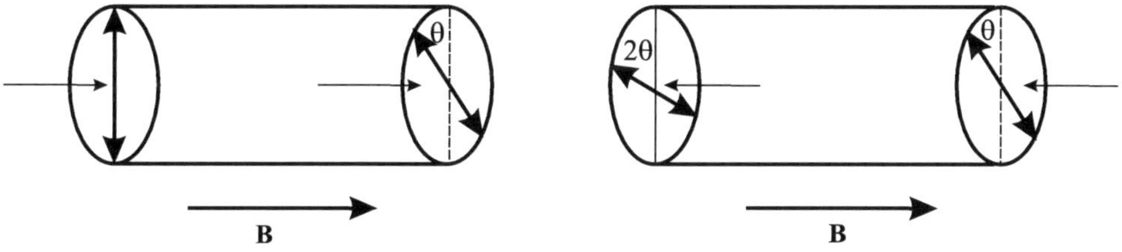

Fig. 6.10: The polarisation rotation in a medium exhibiting the Faraday effect. The direction of rotation is invariant with respect to the propagation direction and is only determined by the direction of the applied magnetic field – on the left, light travels in the forward direction; on the right, light travels backwards.

a wave travels through a Faraday rotator, its polarisation undergoes a rotation equal to θ. If, subsequently, the wave is reflected and travels backwards through the rotator (figure 6.10, right), its polarisation on the output section is rotated by $2\,\theta$ with respect to the input section value. Finally, note that the direction of rotation, *i.e.* the sign of the angle θ, is only determined by the direction of the magnetic flux density B applied to the medium.

It is important to point out the fact that isolators exploit the characteristic of non-reciprocity of Faraday rotators, which are dimensioned so as to ensure a 45° field rotation. Figure 6.11 schematises the components of an optical isolator: an input polariser, a Faraday rotator, and an output polariser. When a light beam travels through the device in the forward direction, only the vertical component is propagated through the input polariser; next, the Faraday rotator will rotate the polarisation by 45°, and the output polariser, which is oriented at 45° too, enables the light to be transmitted through the isolator. Conversely, light traveling in the backward direction becomes polarized at 45° by the second polariser; as the field travels back, the Faraday rotator will again rotate the polarization by 45°. Therefore, the light is polarised horizontally and, since the first polariser is still vertically aligned, the light will be extinguished. These isolators are sensitive to the polarisation of

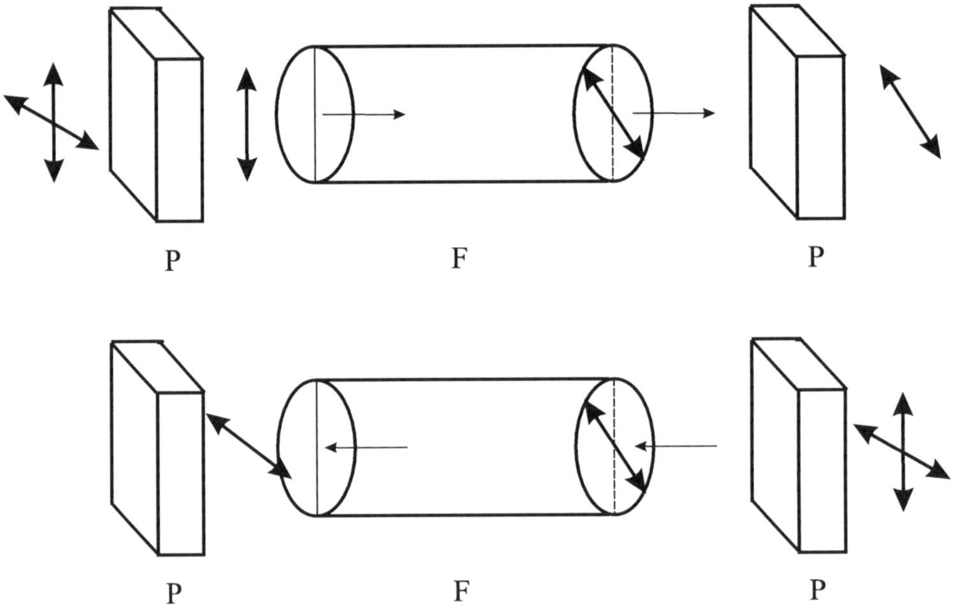

Fig. 6.11: An isolator formed by a vertical polariser (P), a Faraday rotator (F) and a second polariser at 45° (P). The light beam is transmitted when it propagates from left to right (top), while it is blocked when it travels in the opposite backward direction (bottom).

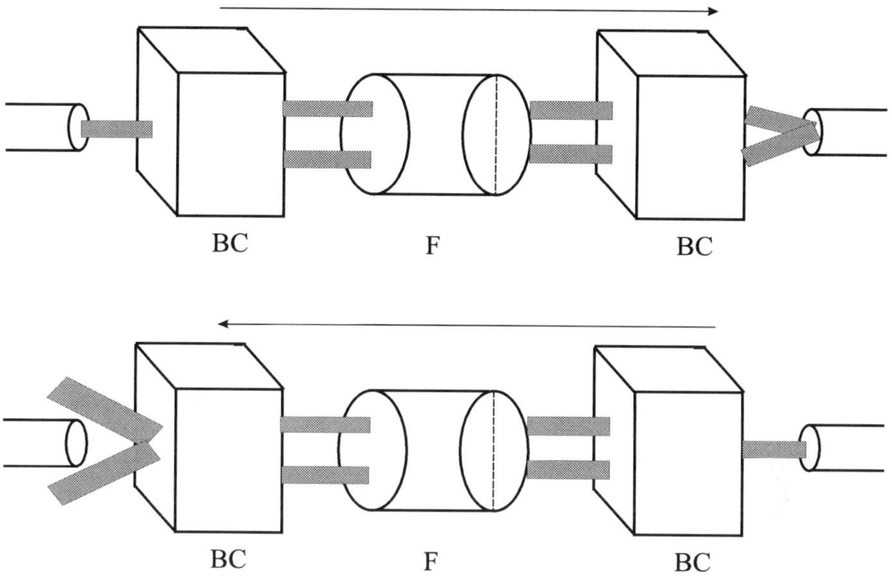

Fig. 6.12: A fiber-coupled isolator insensitive to polarisation constituted by a birefringent crystal (BC), a Faraday rotator (F) and a second birefringent crystal (BC). The light beam is transmitted when it propagates from left to right (top); conversely, the two orthogonal beams diverge (bottom).

the input beam, hence the components of the incident field on the first polariser that are not in the vertical direction are evidently lost.

For this reason, other types of isolators have been developed for telecommunication systems: these are directly fiber-coupled and insensitive to input polarisation – a very important aspect given the continuous variation of polarisation in single-mode fibers. An example of a fiber-coupled isolator is represented in figure 6.12. At the input end, light is projected onto a birefringent anisotropic crystal, typically made of titanium oxide, where it is split into the two orthogonal waves. Then the two waves pass through the Faraday rotator, only to be recombined by the second birefringent crystal and, finally, transmitted to the output fiber. In the backward direction, the second birefringent crystal still divides the light beam into two orthogonal waves; however, due to the non-reciprocal Faraday rotator, the two waves will not have the same polarisation obtained from the process of direct propagation. Hence, the first birefringent crystal further deflects the two waves rather than focus them on the input fiber.

6.4 Optical Circulators

An optical circulator is a non-reciprocal device in micro-optics or fiber used for routing light between three different ports. In particular, a signal entering port one is emitted from port two, but if light travels back through port two, it will exit from port three. Besides, four-port circulators with a similar routing mechanism are also on the market. A typical three-port circulator for micro-optics is constituted by six optical elements, as outlined in figure 6.13: a prism reflector, a beam splitter, two birefringent crystals commonly referred to as a *Birefringent Walk-off Crystals* (BWC), a Faraday

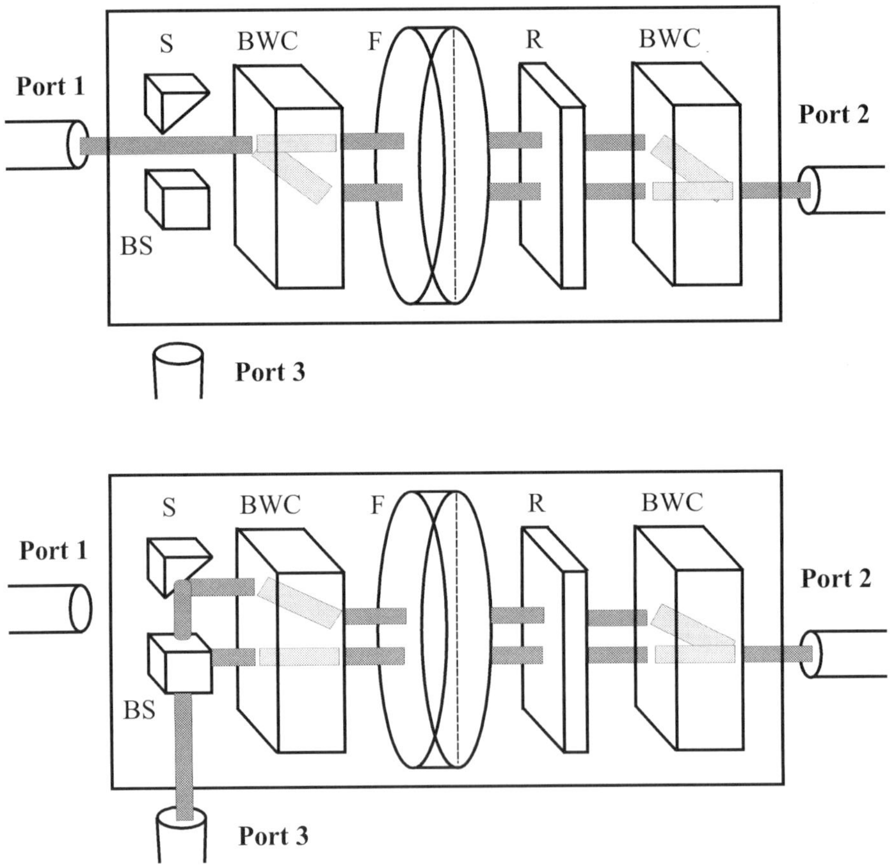

Fig. 6.13: A circulator for micro-optics constituted by a prism reflector (S), a polarising beam splitter (BS), two birefringent crystals (BWC), a Faraday rotator (F) and a phase retarder (R). In the forward direction (top), a beam enters the input port 1 and is projected through port 2. In the opposite direction (bottom) the beam enters port 2 and exits from port 3.

rotator and a phase retarder. The properties of the anisotropic crystals are the key components in these circulators. Consider *e.g.* a light beam with any polarisation entering port 1 and striking the first BWC, where its two orthogonal components are split: the ordinary wave travels in a straight line, while the extraordinary wave is deflected. These undergo two 45° rotations as they go first through the Faraday rotator and then through the retarder. Subsequently, they enter the second BWC with their polarisations rotated by 90° each (which is given by the sum of the two previous rotations); as a result, the extraordinary wave is seen as ordinary and vice versa: the former can continue undisturbed, while the latter (*i.e.* the original extraordinary wave, now ordinary wave) is deflected to meet the other wave at the output port.

In the backward direction, light enters port 2 and is still separated into two waves by the first BWC, which deflects the extraordinary component. The retarder rotates the polarisations of both waves by 45°, but the non-reciprocal Faraday rotator brings the polarisations back to their original state; hence, the extraordinary wave travels through the second BWC and is further diverted to strike the reflector, which deflects it towards port 3. On the other hand, the ordinary wave travels once again in a straight line to reach the splitter and join the extraordinary component. Hence, the two waves travel as a single beam, which is diverted towards port 3. To summarise, from port 1 to port 2, the Faraday rotator and the phase retarder produce an equal and concordant rotation on the two waves, while from port 2 to port 3 their effect becomes null due to the non-reciprocity of the Faraday effect.

Faraday rotators used in micro-optical circulators are very small, *e.g.* $1\,mm \times 1\,mm \times 0.5\,mm$; in this respect, the insertion losses are less than $0.1\,dB$ and the reflections on the faces less than 0.2%. Circulators generally exhibit insertion losses less than $1\,dB$ and high isolation (more than $60\,dB$) over a wide range, which allows to cover all the wavelengths of interest in telecommunications. Unfortunately, the greatest disadvantage with these devices is constituted by their dimen-

Fig. 6.14: A connectorised circulator with three output ports of different colours.

sions. However, smaller three- and four-port connectorised circulators (with dimensions of the order of $50\,mm \times 5\,mm$) are available on the market and can operate at $1310\,nm$ or $1550\,nm$ – cf. figure 6.14.

The employment of circulators is fundamental for those systems where bidirectional communications are required; these devices are therefore essential to perform add-drop functions or for reasons of dispersion compensation, as we have already seen in the chapter about gratings.

Chapter 7

Optical Routing

Modern telecommunication networks have been focused on developing all-optical WDM systems for years now. Optical devices, such as the so-called *optical cross connects* or *optical add-drop multiplexers*, are necessary to achieve this purpose, as they can dynamically route the channels transmitted at different wavelengths, thus ensuring high performance in terms of quality, versatility and reliability. WDM channel management is a fundamental aspect of all optical systems, be it spectrally – *i.e.* by adding and subtracting signals at a given wavelength from appropriate sections of the network – or spatially – that is to say, by switching channels at any frequency on different waveguides or fibers. As stated above, optical systems have to be dynamic in order to reconfigure signal routing according to the network specifications; besides, it is important to guarantee the property of scalability, which ensures simple and cheap system updates as both traffic and users increase.

In order to meet these requirements, these devices are based on particular components: wavelength multiplexers and demultiplexers or *arrayed waveguide gratings* are in fact often integrated together to achieve ever more complex functionalities, such as those typical of the so-called *Wavelength Blockers* (WB), *Wavelength Selective Switches* (WSS) or *Planar Lightwave Circuits* (PLC).

Without going into too much detail, this chapter will describe these devices on a qualitative point of view, which will be presented as exhaustively as possible nonetheless: the reader will be able to understand how these components work and how they can be used, both individually and as elements constituting more complex all-optical signal processing circuits.

7.1 AWG

Let us start by considering an AWG as described in chapter 5 operating in a WDM system where $N = 2p + 1$ channels are associated with the wavelengths $\lambda_{-p}, \lambda_{-(p-1)}, ..., \lambda_0, ..., \lambda_m, ...\lambda_p$, with:

$$\lambda_m = \lambda_0 + m\Delta\lambda,$$

$\Delta\lambda$ indicating the channel spacing. The input and output waveguides of the AWG are identified by two indices, h and k respectively, which take the value zero in correspondence of the central waveguides, as shown in figure 7.1. As a consequence, we can write:

$$\theta_o = k\Delta\theta \quad \text{and} \quad \theta_i = h\Delta\theta$$

supposing $\Delta x_i = \Delta x_o = \Delta x$ and $\Delta\theta = \Delta x/f$. To facilitate the understanding of the mechanism regulating wavelength routing, we deem it necessary to simplify our discussion and assume we can ignore waveguide dispersion. Therefore, since $\frac{dn_g}{d\lambda} = 0$, then $n_{g_0} = N_g$, and (5.56) becomes:

$$\Delta\theta_o = \frac{Q}{n_s\Delta x}\Delta\lambda. \tag{7.1}$$

To begin with, let us consider the case in which all the different wavelength signals pass through the central waveguide. With $\theta_i = 0$, and with reference to (5.51), the formula from (5.58) can be rewritten as:

$$n_s\Delta x k\Delta\theta = Q\,m\Delta\lambda.$$

Finally, by substituting (7.1) we can obtain:

$$k = m.$$

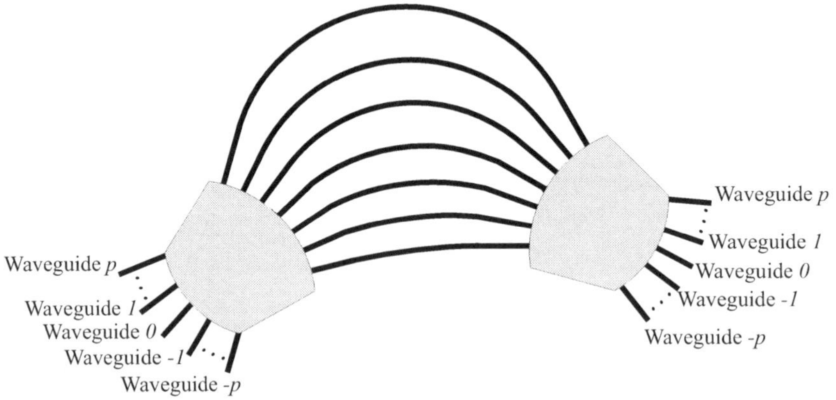

Fig. 7.1: Diagram of a wavelength-routing AWG.

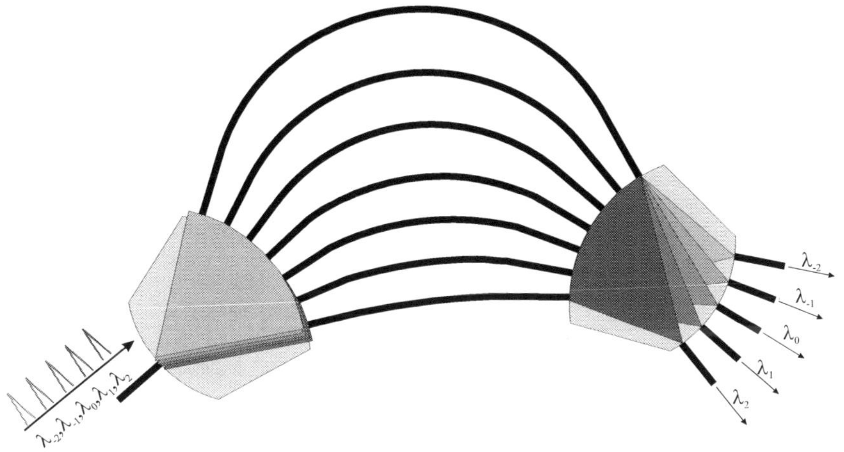

Fig. 7.2: AWG demultiplexer.

Each channel comes out from a different waveguide, as figure 7.2 shows. With respect to other components, the AWG can either operate as an optical multiplexer or demultiplexer of a great number of closely spaced channels through a single component, with a significant reduction in terms of costs and dimensions. For the sake of example, figure 7.3 displays the spectra of 80 output waveguides of an AWG with spacing of $0.4\,nm$ between adjacent channels. We shall now extend this analysis and assume that the signal with wavelength λ_m enters the generic h-th input waveguide. In this case, (5.58) becomes:

$$n_s \Delta x \Delta\theta(h+k) = Q\, m\Delta\lambda.$$

Similarly to what we have seen above, we will have:

$$h + k = m.$$

For each wavelength there is a different relation between input and output waveguide. For example, if we consider a signal with λ_0, since $m = 0$, it follows that $k = -h$, *i.e.* if the signal enters waveguide 4, it will come out from waveguide -4. For a signal with λ_1, being $m = 1$, then $k = 1 - h$, hence if it enters waveguide 4, it will exit from waveguide -3.

The link between wavelengths and input and output waveguides is synthesized by the *connection matrix*, reported in the table in figure 7.4 in the case $N = 5$. Each row represents an input waveguide, each column an output waveguide and each V_{ij} element describes the value of the wavelength which enters the waveguide associated with the i-th row and comes out from the waveguide associated with the j-th column. A row, or a column, can be obtained from the adjacent one through a cyclic shifting of the elements. Since the spectral response is periodic, this property can be explained by the fact that two wavelengths which differ by FSR_λ have the same mapping between the input and

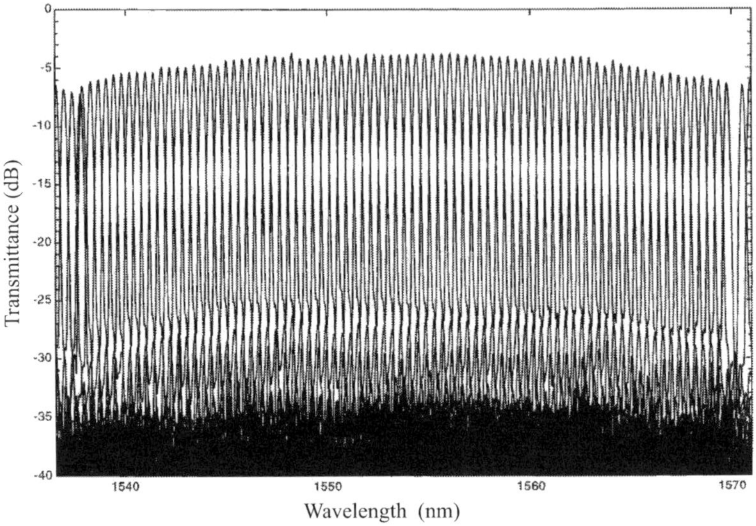

Fig. 7.3: Output spectrum of an 80 × 80 AWG with a 0.4-nm spacing.

$[V]=$

Output waveguide -2	Output waveguide -1	Output waveguide 0	Output waveguide 1	Output waveguide 2	
λ_1	λ_2	λ_{-2}	λ_{-1}	λ_0	Input waveguide -2
λ_2	λ_{-2}	λ_{-1}	λ_0	λ_1	Input waveguide -1
λ_{-2}	λ_{-1}	λ_0	λ_1	λ_2	Input waveguide 0
λ_{-1}	λ_0	λ_1	λ_2	λ_{-2}	Input waveguide 1
λ_0	λ_1	λ_2	λ_{-2}	λ_{-1}	Input waveguide 2

Fig. 7.4: AWG connection matrix.

output waveguides, thus becoming indistinguishable. Note that each wavelength appears only once in each row or column, which permits to simultaneously apply N wavelengths to all the N input waveguides, for a total of N^2 different signals. As figure 7.5 demonstrates, the AWG separates the signals applied to the same input port, by distributing them on the N outputs, with no risk of collision with the signals applied to the other input ports and having the same wavelengths. Hence, the device has the important characteristic of providing a complete connection between the N input and N output waveguides, ensuring N^2 connections in total without any risk of blocking and by only using N wavelengths.

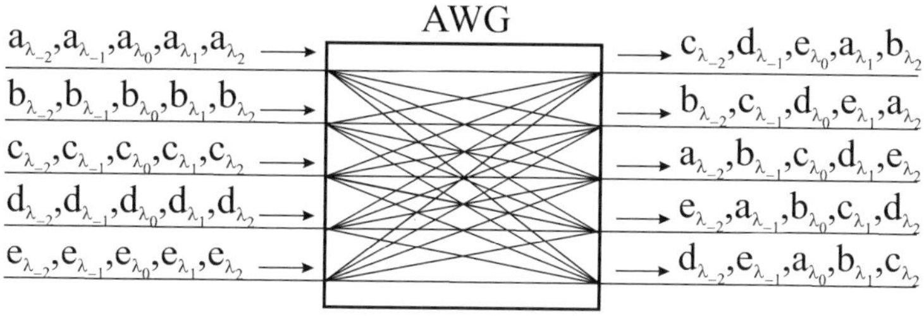

Fig. 7.5: Channel routing in a WDM system by means of an AWG.

7.2 Wavelength Blockers

The wavelength blocker (WB) is a two-port device that receives an input signal with several wavelengths and sends them to the output port once they are properly attenuated; their power is thus equalised and they are completely attenuated, as figure 7.6 clearly illustrates. The top image shows a WB employed as an equaliser, in which each channel can be individually attenuated; underneath, the device is utilised to completely block some of the channels, *viz.* the ones indicated by the dashed line.

The WB can selectively act on any wavelength with attenuations up to 20 dB and can be remotely reconfigured in a few milliseconds. Furthermore, it can operate on the entire C-band and L-band, has a high extinction ratio (typically 40 dB), low polarization losses, low PMD (polarization mode dispersion), and has a high passband at 3 dB, with transfer function with a very flat profile and

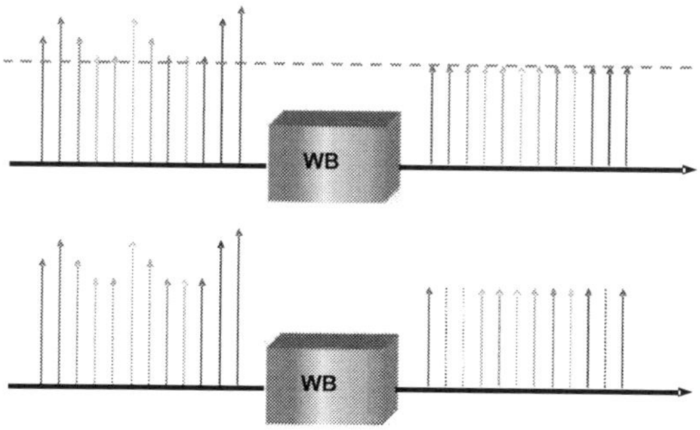

Fig. 7.6: A WB used as an equalizer (top) or to block the three channels indicated by the dashed line (bottom).

ripple of the order of 0.3 dB for channel spacings of 50 and 100 GHz.

According to figure 7.7, the elements composing a WB are the following: an input port receiving the multi-channel beam, a dispersive element separating the beam spatially according to the wavelength, an array of attenuator elements which can be controlled individually to selectively operate on some portions of the dispersed beam, a second dispersive element receiving the different components of the original beam and multiplexing them into a single multi-channel beam, and, finally, the output port that transmits the modified beam. As previously stated, the array of attenuators is designed so as to attenuate or block a variable and non-consecutive number of channels, without risking any kind of interference.

This process typically takes place with discrete elements and through free-space propagation, as represented in figure 7.8. The multi-channel signal input goes through the circulator and is collimated on a transmission diffraction grating which spatially separates the different wavelength channels. Subsequently, the lens forwards each channel on its own attenuator; the channel is then reflected again on the grating which multiplexes all the channels into a single beam. Hence, a single physical element performs the two functions of multiplexing and demultiplexing – which are instead logically separated in figure 7.7. Finally, the multiplexed beam reaches the output port through the circulator.

In addition to that, another widespread type of structure is the one based on $1 \times N$ AWGs which separate the different input channels on N output waveguides. This is followed by the controlled attenuation of the various signals. The free-space solution is nonetheless to be preferred as it is less distorting with respect to AWGs.

The most common technologies employed to fabricate attenuators are essentially two: one is based on *microelectromechanical systems* (MEMS, cf. figure 7.8) and is used for developing movable micromirrors; the second exploits *digital liquid crystals* (LCs), and their attenuation can be controlled by an external voltage. By electrically manoeuvring the MEMS mirrors orientation, the reflected beam – depicted in blue in figure 7.8 – is moved slightly off-axis as it travels back to the collimator; as a consequence, the power of the signal at the output port is suitably, if not completely, reduced. Hence, the attenuation is obtained mechanically by reducing the amount of light that reaches the collimator.

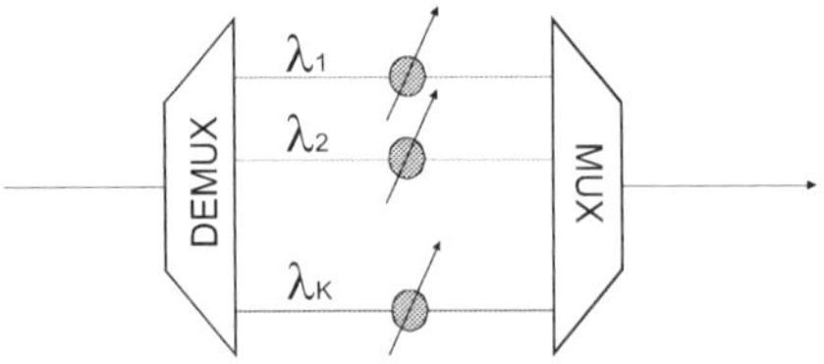

Fig. 7.7: Schematic of a WB formed by an input multiplexer, the attenuators on the different channels and a multiplexer situated before the output port.

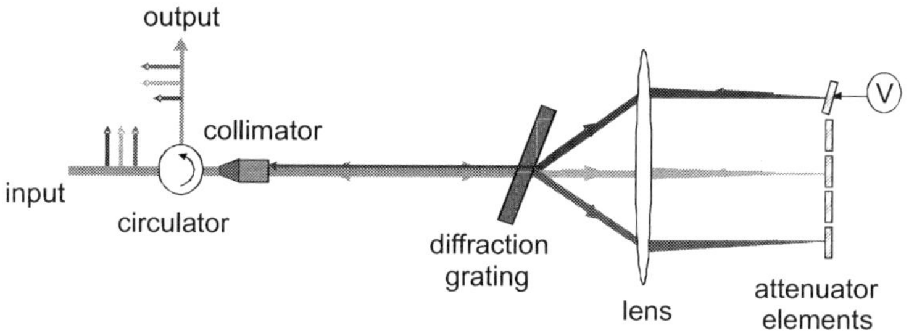

Fig. 7.8: WB operating principle. The blue signal is properly attenuated by way of a mirror rotation.

What has been described can obviously be applied to every channel in the time interval desired and independently from the other channels. Hence, the WB has obviously vast potentials in an all-optical processing system: rather than constantly operating with the same attenuation and at a fixed wavelength, the WB is extremely dynamic and can work on different channels; besides, the attenuation can be controlled by simply sending electrical signals to the mirrors.

For a thorough explanation, figure 7.9 is a schematic of the WB based on digital LCs.
Since the attenuation of LCs is sensitive to polarization, light is polarized immediately after the collimator by means of a birefringent crystal – which separates the extraordinary and the ordinary waves – and a half-wave plate that rotates the extraordinary wave by 90°. The polarized light strikes the digital LC, which introduces an attenuation controlled by a voltage signal, and is reflected back to the output port by the mirrors.
Compared to the MEMS, the WBs based on LCs are more common because the attenuator element is cheaper, simpler and more reliable in terms of production, since there are no mechanical components in motion. Additionally, WBs can be operated at less than 10 V.

7.3 Wavelength Selective Switching

Wavelength selective switching (WSS) components are used to meet the increasing demand for flexible nodes as they incorporate all functions necessary for a reconfigurable optical network: multiplexing, demultiplexing, spatial switching, power-level control and channel blocking. WSSs are constituted by one input and N output ports that essentially allow to control and route any WDM channel to any output port of an optical network. In particular, a WSS can be used in the following cases: to equalise different input channels and bring them all to the same output end with the same power level; to selectively block certain channels and bring the others to the same output port; to route groups of signals to different output ports, *i.e.* any wavelength can be routed to any port; finally, to demultiplex a multi-channel signal and send each wavelength to a different port. All the different functions are illustrated in figure 7.10).

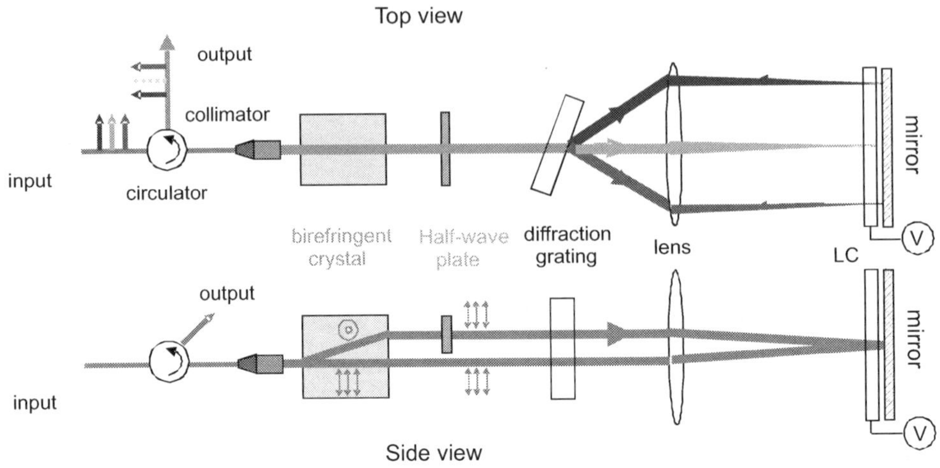

Fig. 7.9: WB based on liquid crystals.

The multiplexing and demultiplexing operations require a particularly flat passband filter, low chromatic and polarization dispersion, as well as low polarization-dependent losses.

Ideally, cascading several WSS modules in subsequent nodes in an optical network should not degrade signal quality. The control of the power level of each channel in the WSS must ensure that the system is kept equalised; hence, the attenuations introduced for this reason must guarantee the quality of the signal transmitted, notwithstanding the level of attenuation applied. The routing process must be fast – within a few milliseconds – and should never interfere with the uninvolved channels.

Consider figure 7.11 for a schematic representation of a $1 \times N$ WSS. A demultiplexer situated after the input port routes the different signals to K pathways, each equipped with a variable attenuator and a $1 \times N$ optical switch. The K-th switch outputs are connected to the K-th input port of the N multiplexers serving the N output ports. This way, each channel can be routed to any output port.

The WSS may be physically assembled by resorting to the scheme illustrated in figure 7.12. Through a series of lenses, and by propagating in the free space, the multi-channel input signal strikes a diffraction grating which spatially separates the different channels. The latter impact on an array of mirrors in MEMS technology – one for each wavelength – which can be rotated on two axes so as to direct the reflected beam both vertically and horizontally. In the example from figure 7.12, the red beam is horizontally separated from the others by the grating and vertically redirected by the mirror onto the grating to be properly diffracted on the array of output fibers – which are ver-

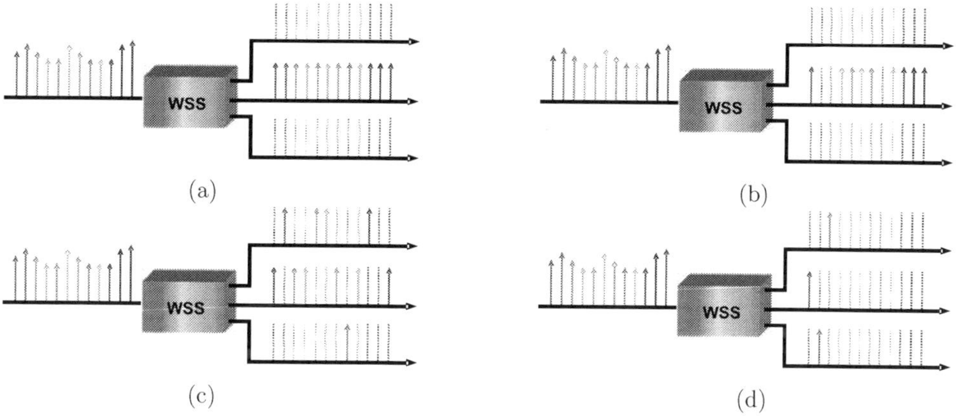

Fig. 7.10: A WSS utilised for (a) equalising different input channels, (b) selectively blocking certain channels, (c) routing signals to different output ports, (d) demultiplexing each signal on a different output port. The dotted lines indicate completely attenuated fields.

tically aligned with each other and with the input fiber. The output fiber is specifically selected by micrometrically rotating the mirror and vertically deviating the beam. In the same way as the WB, the amplitude of each channel, or its total attenuation, can be monitored by misaligning a single beam in a controlled manner with respect to the output collimating lens. All the functions from figure 7.10 can be reproduced in this case as well; besides MEMS technology, the use of digital LC is frequent when it comes to producing mirror switches. 1 × 9 WSS devices for channels spacing of 50

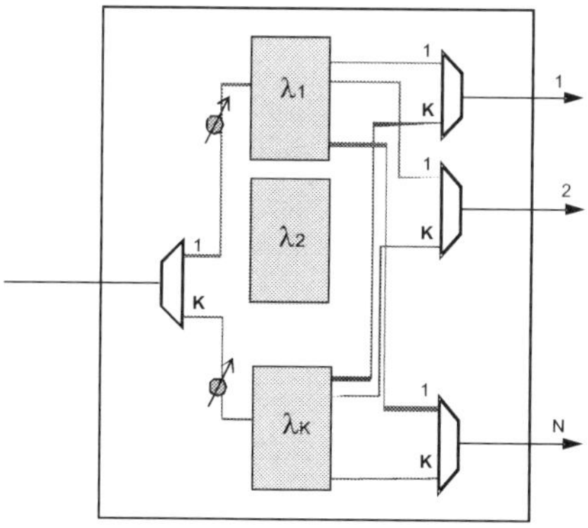

Fig. 7.11: Schematic of a 1 × N WSS.

GHz or 100 GHz are nowadays widely used in 10- and 40-$Gbit/s$ optical networks, with 100-$Gbit/s$ network being already compatible and supported.

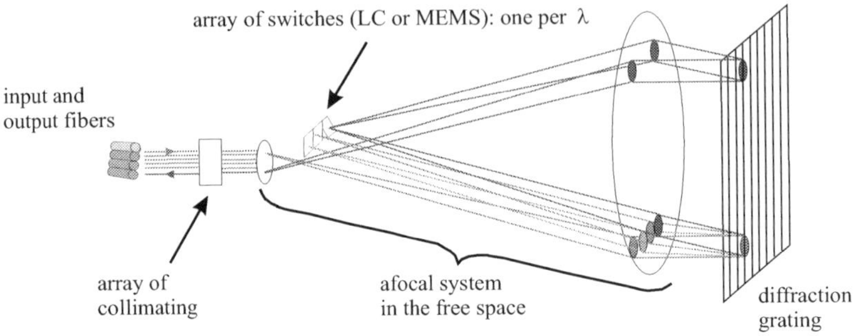

Fig. 7.12: WSS operating principle. The red signal is routed on a different fiber with respect to the input fiber.

7.4 Planar Lightwave Circuits

Planar lightwave circuits (PLCs) are integrated circuits for processing optical signals usually made in silica-on-silicon technology (SiO_2/Si) or with polymer waveguides. This is why these circuits have a high level of integration and their cost is low; however, the downside is their excessive power consumption, in the order of several dozens of watts. PLCs have a good bandwidth, with a flat transfer function at 100 GHz – though the same cannot be said at 50 GHz – and, due to the PMD

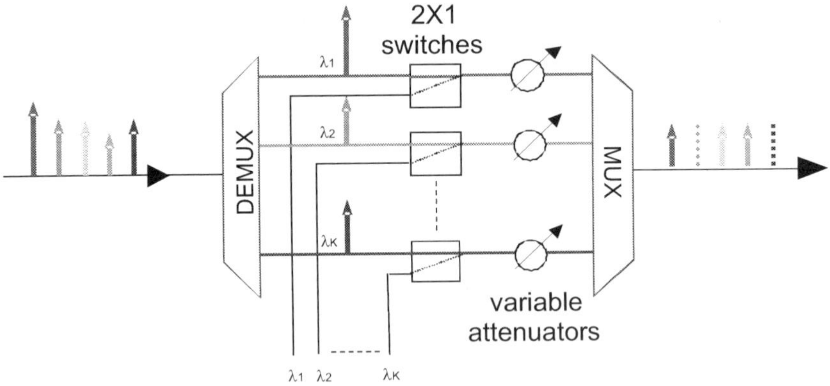

Fig. 7.13: A PLC schematic.

introduced, they cannot be easily employed at a bit rate of 40 $Gbit/s$ or over. Nevertheless, they are nowadays a mature technology, widely used in metro networks up to 10 $Gbit/s$. A PLC structure is schematised in figure 7.13: different channels reach the sole input port and are demultiplexed and routed through different paths, one per wavelength.
Once again, the variable attenuators permit to either equalise all the output channels or completely block some of them before they are multiplexed on the output port. Two AWGs perform the multiplexing and demultiplexing functions and, finally, the 2×1 switches made of simple waveguide couplers and situated on each path permit to carry out the add function at different wavelengths.

Chapter 8

The Optical Fiber

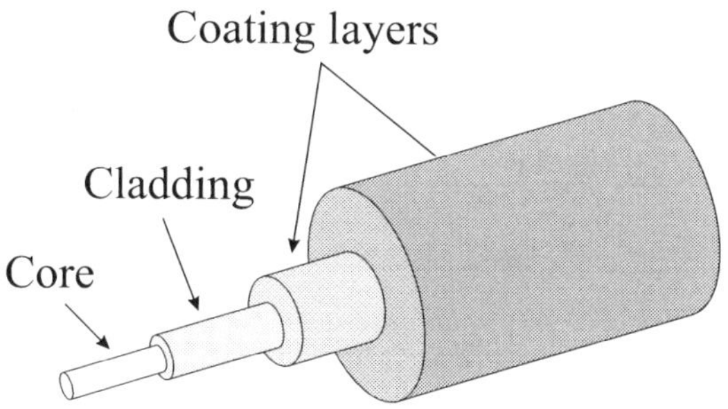

Fig. 8.1: Optical fiber.

An optical fiber is a cylindrical cable consisting of a core and a cladding layer made of different dielectric materials, and thus with dissimilar refractive indices. In practice, the extent of the cladding is finite, with current standards setting its diameter to $125\mu m$. The cladding is then usually covered by a coating material which protects the fiber and makes it flexible and mechanically strong. Figure 8.1 shows an example of an optical fiber, whereas in figure 8.2 we can see the section of a fiber which is coupled to the light from a broad-spectrum source.

8.1 The Step Fiber: Modes Calculation

As we have seen for the slab waveguide, the optical fiber also supports a finite number of guided modes as well as a continuous set of radiation modes which properly satisfy the boundary conditions. The purpose of this section is to briefly determine how to calculate the guided modes and subsequently focus on the properties of the fundamental mode – the only mode that, in practice,

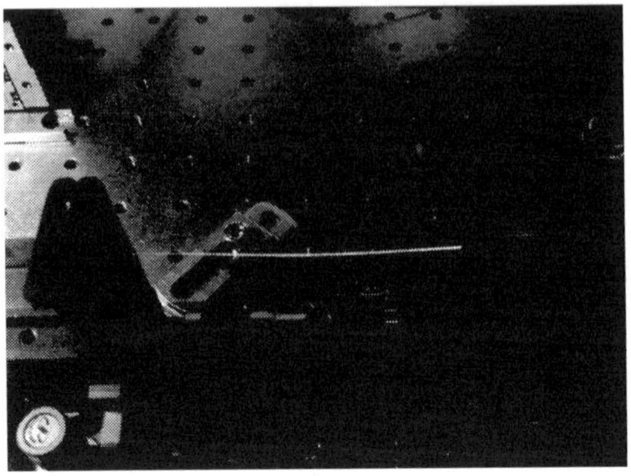

Fig. 8.2: Short section of an optical fiber coupled to the radiation from a broad-spectrum source.

can deliver information at a distance. We shall carry out our analysis by referring to the step-index (or simply, *step*) optical fiber, whose name comes from the fact that the refractive index profile as a function of the distance r from the fiber axis is assumed to be a piecewise constant function. The process is formally the same as that presented in chapter 1 and already applied to the study of the slab in chapter 2. For this reason, we will only concentrate on those aspects which are proper to the optical fiber.

Let us consider once again the Helmholtz equation for a piecewise homogeneous medium:

$$\nabla_t^2\phi + \left[\left(\frac{\omega}{c}n\right)^2 - \beta^2\right]\phi = 0$$

with ϕ indicating the longitudinal component of the mode function of the electric or magnetic field. The symmetry of the structure suggests choosing a cylindrical coordinate system, with the z-axis coinciding with the fiber axis, as outlined in figure 8.3. It follows that the interface between the core and the cladding – which must comply with the continuity conditions of the tangential components – is simply defined by the equation below:

$$r = a$$

with a being the radius of the core. In the cylindrical coordinate system, the operator ∇_t^2 becomes:

$$\nabla_t^2 = \frac{\partial^2}{\partial r^2} + \frac{1}{r}\frac{\partial}{\partial r} + \frac{1}{r^2}\frac{\partial^2}{\partial\varphi^2}.$$

Since we want to determine the guided modes for which

$$n_{cl}k_0 < \beta < n_{co}k_0,$$

it is useful to define the following quantities:

$$k_t^2 = n_{co}^2\left(\frac{\omega}{c}\right)^2 - \beta^2$$

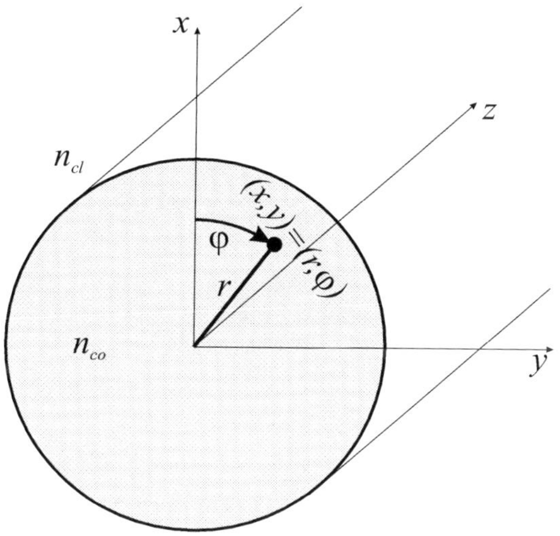

Fig. 8.3: Cylindrical coordinate system.

$$\gamma^2 = \beta^2 - n_{cl}^2 \left(\frac{\omega}{c}\right)^2,$$

where k_t and γ are two real parameters; hence:

$$k_t^2 + \gamma^2 = \left(\frac{\omega}{c}\right)^2 (n_{co}^2 - n_{cl}^2).$$

By virtue of what has just been presented, it is possible to rewrite the Helmholtz equation as follows:

$$\begin{cases} \dfrac{\partial^2\phi}{\partial r^2} + \dfrac{1}{r}\dfrac{\partial\phi}{\partial r} + \dfrac{1}{r^2}\dfrac{\partial^2\phi}{\partial\varphi^2} + k_t^2\phi = 0 & \text{if} \quad r \leq a; \\ \\ \dfrac{\partial^2\phi}{\partial r^2} + \dfrac{1}{r}\dfrac{\partial\phi}{\partial r} + \dfrac{1}{r^2}\dfrac{\partial^2\phi}{\partial\varphi^2} - \gamma^2\phi = 0 & \text{if} \quad r > a. \end{cases} \tag{8.1}$$

The solution of the equation exploits the method of separation of variables, according to which the solution can be expressed as the product of two functions, each depending on only one coordinate in the transverse plane:

$$\phi(\rho, \varphi) = R(\rho)F(\varphi).$$

By substituting this expression in (8.1):

$$\begin{cases} F\dfrac{d^2R}{dr^2} + F\dfrac{1}{r}\dfrac{dR}{dr} + R\dfrac{1}{r^2}\dfrac{d^2F}{d\varphi^2} + k_t^2 FR = 0 & \text{if} \quad r \leq a; \\ \\ F\dfrac{d^2R}{dr^2} + F\dfrac{1}{r}\dfrac{dR}{dr} + R\dfrac{1}{r^2}\dfrac{d^2F}{d\varphi^2} - \gamma^2 FR = 0 & \text{if} \quad r > a. \end{cases} \tag{8.2}$$

If we divide by FR and multiply by r^2:

$$\begin{cases} r^2 \frac{1}{R}\frac{d^2R}{dr^2} + r\frac{1}{R}\frac{dR}{dr} + \frac{1}{F}\frac{d^2F}{d\varphi^2} + r^2k_t^2 = 0 & \text{if} \quad r \leq a; \\ \\ r^2 \frac{1}{R}\frac{d^2R}{dr^2} + r\frac{1}{R}\frac{dR}{dr} + \frac{1}{F}\frac{d^2F}{d\varphi^2} - r^2\gamma^2 = 0 & \text{if} \quad r > a. \end{cases} \tag{8.3}$$

In (8.3) we have some terms which depend either on φ or r only; however, their sum needs be constant as φ and r vary. This is only possible if the two terms are both equal to a constant, indicated as ν^2. Consequently, we can obtain the following equations:

$$\begin{cases} \frac{d^2F}{d\varphi^2} + \nu^2 F = 0; \\ \\ \frac{d^2R}{dr^2} + \frac{1}{r}\frac{dR}{dr} + \left[\begin{Bmatrix} k_t^2 \\ -\gamma^2 \end{Bmatrix} - \frac{\nu^2}{r^2}\right] R = 0; \end{cases} \tag{8.4}$$

the constant ν is the *separation constant*. The terms in curly brackets refer to the core (upper term) and the cladding (lower term), respectively.

8.1.1 Azimuthal Dependence Calculation

Let us consider the equation

$$\frac{d^2F}{d\varphi^2} + \nu^2 F = 0.$$

The general expression of the solution is reported below:

$$F(\varphi) = a_1 e^{j\nu\varphi} + a_2 e^{-j\nu\varphi}.$$

The function $F(\varphi)$ is actually a periodic function with period 2π. In fact, by introducing a 2π rotation around the fiber axis we revert to the starting point; therefore:

$$F(\varphi) = F(\varphi + 2\pi),$$

from which it follows that the separation constant ν must be an *integer*.

8.1.2 Radial Dependence Calculation

We shall now consider the following equation:

$$\frac{d^2R}{dr^2} + \frac{1}{r}\frac{dR}{dr} + \left[\begin{Bmatrix} k_t^2 \\ -\gamma^2 \end{Bmatrix} - \frac{\nu^2}{r^2}\right] R = 0.$$

In order to simplify the calculations, we need to define the following normalised quantities:

$$u = k_t a;$$

$$w = \gamma a;$$

$$\rho = r/a;$$

$$v = 2\pi \frac{a}{\lambda} \sqrt{n_{co}^2 - n_{cl}^2}.$$

By doing so, the equation becomes:

$$\begin{cases} \frac{d^2 R}{d\rho^2} + \frac{1}{\rho}\frac{dR}{d\rho} + (u^2 - \frac{\nu^2}{\rho^2})R = 0 & \text{if } \rho \leq 1; \\ \\ \frac{d^2 R}{d\rho^2} + \frac{1}{\rho}\frac{dR}{d\rho} - (w^2 + \frac{\nu^2}{\rho^2})R = 0 & \text{if } \rho > 1. \end{cases} \tag{8.5}$$

The general solution of the first expression from (8.5) is a linear combination of ordinary Bessel functions of the first and second kind:

$$R(\rho) = A_1 J_\nu(u\rho) + A_2 Y_\nu(u\rho).$$

J_ν is known as Bessel function of the first kind and order ν, while Y_ν is a Bessel function of the second kind and order ν.

The general solution of the second expression from (8.5) is a linear combination of modified Bessel functions of the first and second kind:

$$R(\rho) = B_1 K_\nu(u\rho) + B_2 I_\nu(u\rho).$$

K_ν is defined as modified Bessel function of the first kind and order ν, while I_ν is known as modified Bessel function of the second kind and order ν.

The behaviour of the Bessel functions is well known. In particular, in the core the functions Y_ν diverge, hence the corresponding solution has no physical meaning. The same observation can be made for I_ν, too. For this reason it must be:

$$A_2 = 0$$

$$B_2 = 0.$$

Below is the function expressing the radial dependence of the longitudinal components:

$$R(\rho) = \begin{cases} A_1 J_\nu(u\rho) & \text{if } \rho \leq 1; \\ \\ B_1 K_\nu(u\rho) & \text{if } \rho > 1. \end{cases}$$

8.1.3 The Characteristic Equation

According to what has just been explained, the longitudinal components of the mode functions of the electric and magnetic field are thus expressed:

$$\phi(\rho, \varphi) = \left(a_1 e^{j\nu\varphi} + a_2 e^{-j\nu\varphi}\right) \left\{ \begin{array}{c} A_1 J_\nu(u\rho) \\ B_2 K_\nu(w\rho) \end{array} \right\}.$$

In this case, the continuity of E_φ, E_z, H_φ, H_z in $\rho = 1$ are the boundary conditions that the components tangential to the core-cladding interface must satisfy. This generates a homogeneous system of four equations in four unknowns: a_1, a_2, A_1 and B_1. In order to obtain a solution other than zero, by imposing that the determinant is null, we will obtain the characteristic equation:

$$[G_\nu(u) + C_\nu(w)]\,[G_\nu(u) + (1 - 2\Delta)C_\nu(w)] = q^2 \tag{8.6}$$

where:

$$\Delta = \frac{n_{co}^2 - n_{cl}^2}{2n_{co}};$$

$$G_\nu(u) = \frac{J'_\nu(u)}{uJ_\nu(u)};$$

$$C_\nu(w) = \frac{K'_\nu(w)}{wK_\nu(w)};$$

$$q^2 = \nu^2(1 - 2\Delta\frac{u^2}{w^2})\left(\frac{v^2}{u^2 w^2}\right)^2$$

and the superscript indicates the derivative with respect to the argument of the function.

The equation (8.6) can be seen as a quadratic equation in the variable $G_\nu(u)$ with two solutions. This means that the characteristic equation can actually be split into two characteristic equations:

$$G_\nu(u) + (1 - \Delta)C_\nu(w) = -\sqrt{q^2 + \Delta^2 C_\nu^2(w)} \tag{8.7}$$

$$G_\nu(u) + (1 - \Delta)C_\nu(w) = \sqrt{q^2 + \Delta^2 C_\nu^2(w)}. \tag{8.8}$$

Each characteristic equation identifies a set of modes, called $HE_{\nu,\mu}$ and $EH_{\nu,\mu}$ modes, respectively. The integer μ identifies the μ-th solution of the system determined by the corresponding characteristic equation and by

$$u^2 + w^2 = v^2. \tag{8.9}$$

If $\nu \neq 0$, all six components of the mode functions are non-zero and the modes are hybrids. On the other hand, if $\nu = 0$, the longitudinal component of the mode function – or that of the electric or magnetic field – is zero. Hence, the modes become either transverse-electric (TE) or transverse-magnetic (TM), respectively. Since $\nu = 0$, the fact that the mode functions only depend on one transverse coordinate – in this case, the radial coordinate r – seems to strengthen the analogy with the modes of the slab.

8.1.4 Cutoff and Single-mode Condition

The number of guided modes of a specific fiber at a given frequency depends on the physical and geometrical properties of the fiber itself, *i.e.* the core radius a and the core-cladding step index $n_{co}^2 - n_{cl}^2$. In order to determine the number of modes, the cutoff conditions defined by the condition $\gamma = 0$ must be known. By imposing this condition to the characteristic equations and determining the solution $u_{c\,\nu,\mu}$, it is possible to calculate the cutoff normalised frequency for the mode ν, μ, in that, if $\gamma = 0$, then $v_{c\,\nu,\mu} = u_{c\,\nu,\mu}$. The single-mode condition is particularly interesting from a practical perspective. The fundamental mode of the fiber is the HE_{11} mode, whose cutoff frequency is zero; the first higher-order mode is the TE_{01} mode, whose cut-off condition is:

$$J_0(v_c) = 0.$$

The first zero of the Bessel function of order zero corresponds to an argument equal to 2.405. The single-mode condition then becomes:

$$v < 2.405.$$

In terms of wavelength:

$$\lambda > 2.61a\sqrt{n_{co}^2 - n_{cl}^2}.$$

Let us set the cutoff wavelength at $\lambda_c = 1.2\mu m$, the index difference at $n_{co} - n_{cl} = 0.1$ and the refractive index of the cladding at $n_{cl} = 1.45$: under these circumstances, only if the radius of the core is inferior to $1.6\mu m$ will we have a single-mode fiber. This causes serious problems of alignment among different sections of a fiber or between the fiber and the source. An alignment error of just $500nm$ corresponds to approximately 30% of the core radius. Therefore, the core-cladding index difference must be reduced for the core size to be acceptable. By setting $\lambda_c = 1.2\mu m$ and a core of radius $a = 4.5\mu m$, the index difference must be less than $n_{co} - n_{cl} = 0.0035$. Fibers of this kind, having so low an index difference, are known as *weakly guiding fibers*.

Figure 8.4 illustrates the values of the cutoff normalised frequencies of the first modes of the fiber. Although the TM_{01} mode is not degenerate with the TE_{01} mode, both have the same cutoff frequency, which is just slightly lower compared to the cutoff frequency of the HE_{21} mode – deter-

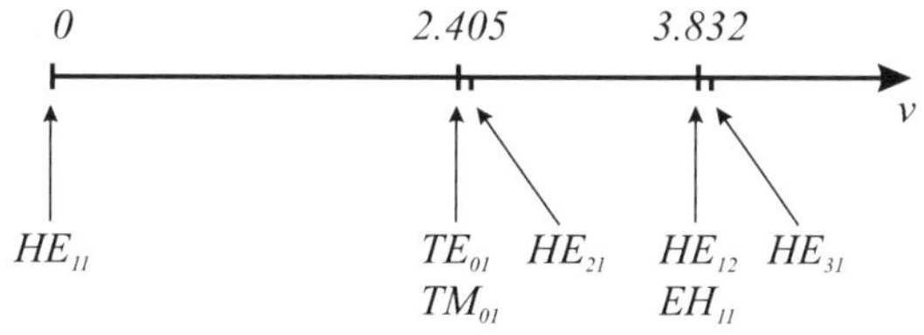

Fig. 8.4: The values of the normalised cutoff frequencies of the first modes of the fiber.

mined by the equation $J_0(v_c) = -\Delta/(1-\Delta)J_2(v_c)$. Next there are the HE_{12} and EH_{11} modes, whose cutoff frequencies correspond to the second zero of the Bessel function of order zero equal to 3.832.

8.1.5 Characteristics of the Fundamental Mode

The fundamental mode HE_{11} corresponds to the first root of the system formed by (8.7) and (8.9). Indeed, this root corresponds to two mode functions linearly independent of each other. The HE_{11} mode is actually constituted by two modes with the same phase constant at all frequencies; for this reason, they are known as *degenerate modes*. Besides, we often have one fundamental mode with two possible polarisations; figure 8.5 shows the trend of the effective index of one of these two polarisations as a function of the wavelength. Because in the calculation we have also considered the dispersion due to the silica by means of the Sellmeier equation, the refraction indices of the core and cladding also depend on the wavelength. The three components of the magnetic field of the two polarisations are shown in figure 8.6. Both polarisations are characterised by a dominant transverse component which is about two or three orders of magnitude higher than the other. The profile of the dominant component is monotonic and tends to decrease as the distance from the centre of the core increases: to a first approximation, it can be compared to a Gaussian profile. The ratio between the two transverse components depends on both the index difference and the value of the normalised frequency, as figure 8.7 illustrates. As the index difference decreases, so does the field confinement; furthermore, both the secondary transverse and the longitudinal component gradually become negligible with respect to the dominant transverse component. In the case considered, for example, the secondary transverse component is reduced from $4\cdot10^{-3}$ to $4\cdot10^{-4}$ times the dominant component. A similar approach can be adopted for the electric field.

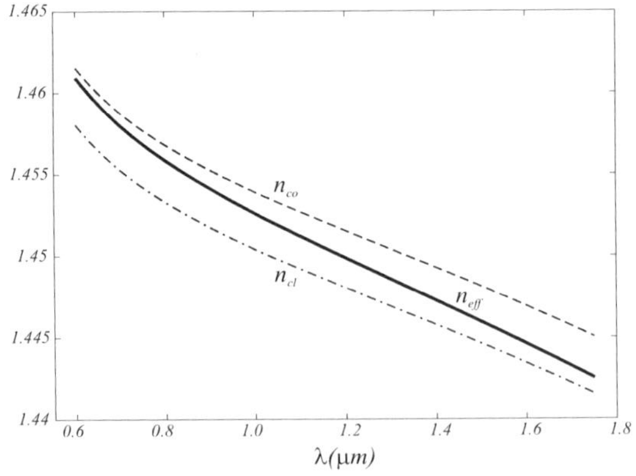

Fig. 8.5: The dispersion curve of the fundamental mode of the fiber. The core and cladding indices depend on the wavelength due to silica dispersion.

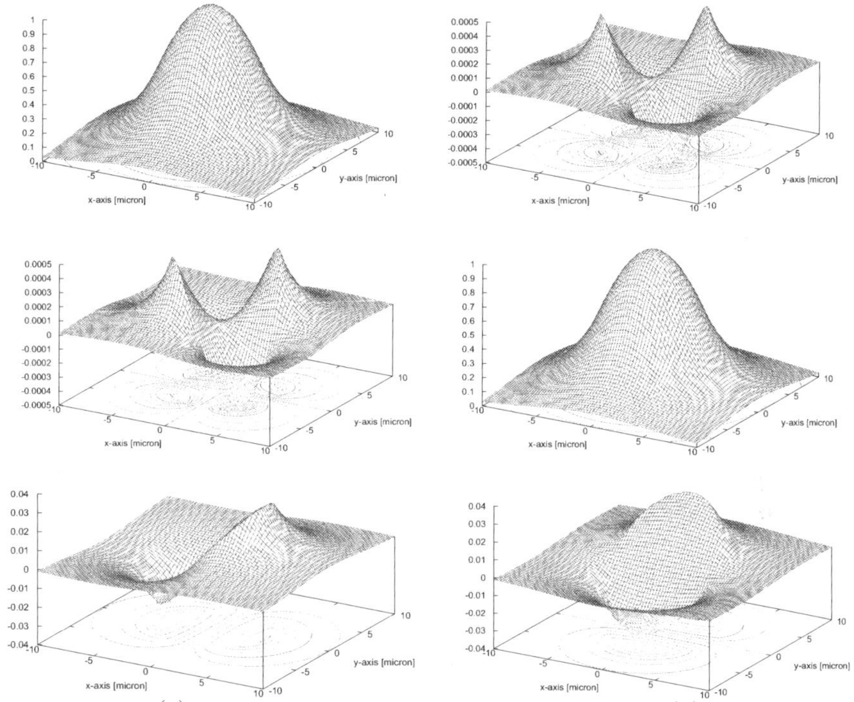

(a) (b)

Fig. 8.6: The two columns show the components of the magnetic field of the two polarisations of the $HE11$ mode. The fiber considered has a core radius equal to $r_{co} = 4.5\mu m$; the core and cladding refractive indices are $n_{co} = 1.4485$ and $n_{cl} = 1.445$, respectively; the wavelength is $\lambda = 1.55\mu m$. According to these physical and geometrical properties, and at this wavelength, the normalised frequency is $v = 1.835$.

Figure 8.8 shows the qualitative profile of the lines of force of the electric and magnetic field in the two cases. The two polarisations are discriminated according to the orientation of the electric field. We call HE_{11}^x the mode with the electric field mainly oriented along the x-axis, while the HE_{11}^y mode is mainly oriented along the y-axis. It should be noted that, for both modes, the magnetic field vector is orthogonal to the electric field vector. Thus, the E_x component of the HE_{11}^x mode is much greater than E_y, and the H_y component is much greater than H_x; the opposite is true as far as the HE_{11}^y mode is concerned. For low levels of index difference, the direction of the electric field remains the same throughout the whole transverse section. This property permits to significantly simplify the expression of the mode functions of the two polarisations of the fundamental mode:

$$\overline{E}^x(r,\varphi) = \psi(r,\varphi)\hat{\imath} \qquad \overline{E}^y(r,\varphi) = \psi(r,\varphi)\hat{\jmath};$$

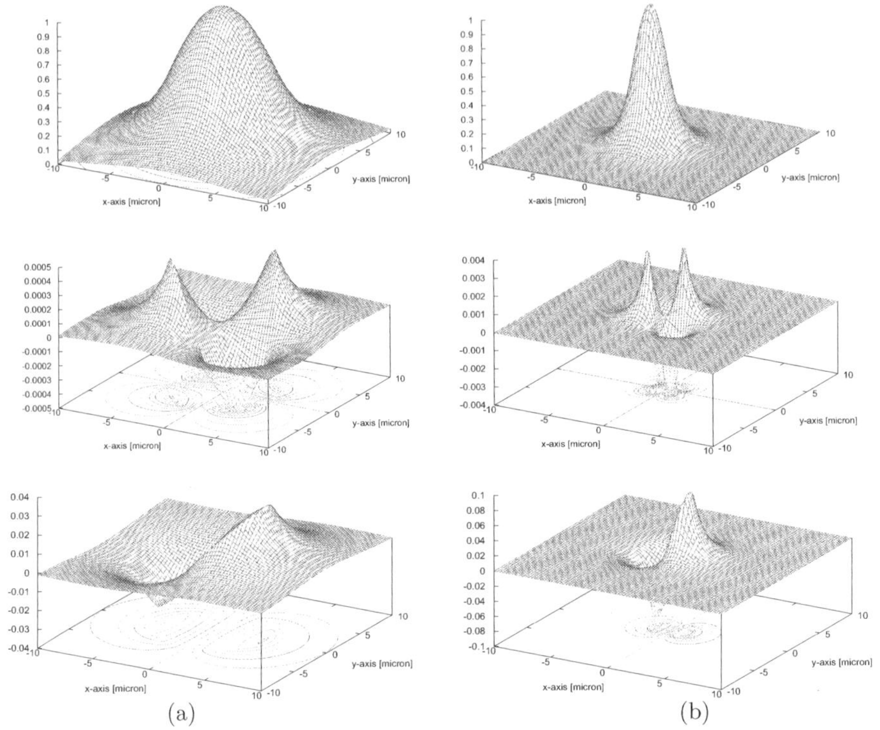

Fig. 8.7: The components of the magnetic field of the HE_{11}^{y} mode of a fiber at the wavelength $\lambda = 1.55\mu m$ and with normalised frequency $v = 1.835$. Column (a): index difference $n_{co} - n_{cl} = 3.5 \cdot 10^{-3}$, and core radius $a = 4.5\mu m$; column (b): index difference $n_{co} - n_{cl} = 30 \cdot 10^{-3}$, and core radius $a = 1.53\mu m$.

where

$$\psi(r,\varphi) = \begin{cases} J_0(r) & \text{if } \quad r \leq a \\ \dfrac{J_0(a)}{K_0(a)} K_0(r) & \text{if } \quad r > a \end{cases}$$

the superscripts x and y indicates the direction along which the mode function is polarised. In this case, the HE_{11} mode is often referred to as LP_{01} (Linearly Polarized).

In order to assess the effects linked to the presence of two polarisations of the fundamental mode, we shall consider a linear combination of two degenerate modes:

$$\overline{e}(x,y,z) = A_1 \overline{E}_1(x,y) e^{-j\beta z} + A_2 \overline{E}_2(x,y) e^{-j\beta z}$$

by extracting the phase term common to the two addends, we obtain:

$$\overline{e}(x,y,z) = \left(A_1 \overline{E}_1(x,y) + A_2 \overline{E}_2(x,y)\right) e^{-j\beta z}.$$

The $\overline{e}(x,y,z)$ mode is still degenerate to the first two. In general, if there are two degenerate modes,

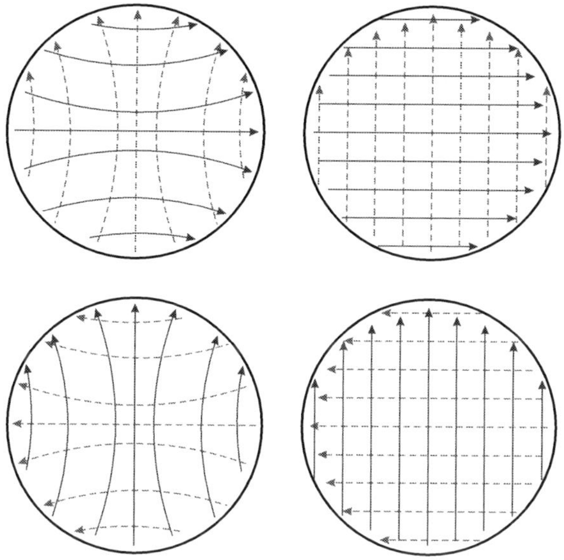

Fig. 8.8: The qualitative profile of the lines of force of the transverse component of the electric (solid line) and magnetic field (dashed line); the top images illustrate the case of the HE_{11}^x mode, while the bottom images refer to the HE_{11}^y mode of a fiber with high index difference (images on the right) and low index difference (left).

there will be infinite modes which only differ for their different mode functions:

$$\overline{E}(x,y) = A_1\overline{E}_1(x,y) + A_2\overline{E}_2(x,y).$$

This property allows us to state that, in ideal step fibers, the input field polarisation is maintained throughout the fiber by virtue of the two degenerate polarisations of the fundamental mode. This complies with the properties of rotational symmetry around the propagation axis.

Nevertheless, real fibers behave slightly differently. In fact, the interface between the core and the cladding may not be perfectly round due to some imperfections or deformations associated to the manufacturing and laying processes. On top of that, these phenomena produce a certain level of mechanical stress which makes the silica constituting the core and the cladding slightly anisotropic. As a result, we will have an overall loss in terms of degenerate polarisations of the fundamental mode. The optical fiber is thus always multimodal, even if the difference between the propagation characteristics of the two polarisations is so small that this phenomenon can almost be overlooked when compared to other causes of distortion – mainly, the intramodal and, *a fortiori*, the intermodal dispersion. Modern optical communication systems eliminate intermodal dispersion by resorting to single-mode fibers, but compensation systems for intramodal dispersion exist as well. In these cases, polarisation dispersion becomes the dominant effect – currently constituting the main cause that limits bit-rate increase in remote transmission.

8.1.6 LP Modes

We have just seen how, for low step indices, the direction of the electric field of the fundamental mode is the same throughout the entire transverse section, *i.e.* the fundamental mode is linearly polarised. The simplifications resulting from the approximation $n_{co} - n_{cl}/n_{co} \ll 1$ do not only concern the fundamental mode. In fact, it is possible to demonstrate that the $HE_{2,\mu}$ modes become degenerate with the $TE_{0,\mu}$ and $TM_{0,\mu}$ modes in weakly guiding fibers, while the $HE_{\nu+1,\mu}$ modes become degenerate with the $EH_{\nu-1,\mu}$ modes. Furthermore, we can obtain linearly polarised modes by properly combining the mode functions of the degenerate modes. These specific modes are known as *LP pseudo-modes*. The prefix "pseudo" suggests that we are not dealing with modes *sensu stricto*, since they are obtained by approximation. Nevertheless, these approximations are widely satisfied by the actual fibers where the step index is exceptionally small in order to obtain the single-mode condition with non-excessively small cores. Therefore, these pseudo-modes are of great interest from the point of view of application. In this respect, cf. the example in figure 8.9 showing the two polarisations of the LP_{11} mode obtained as linear combinations of the degenerate HE_{21} and TM_{01} modes, in the first case, and HE_{21} and TE_{01} in the second.

8.1.7 Gaussian Approximation for the Fundamental Mode

The fundamental mode LP_{01} of a step-index fiber has the shape of a Bessel function of order zero $J_0(r)$ in the core, and a modified Bessel function $K_0(r)$ in the cladding. Its profile reminds that of a Gaussian function, although it should be noted that $K_0(r)$ decays like $exp(-r)$, which is much slower than a Gaussian function. Considering the polarization along y, in the case of weak guiding conditions the so-called Gaussian approximation allows us to express the E_y component as follows:

$$E_y(r) = E_0 exp\left(\frac{-r^2}{r_0^2}\right),$$

where r_0 is known as *mode field radius* or *spot size*, while the longitudinal component and E_x are much smaller and can thus be ignored. This approximation is particularly valid in correspondence to the cutoff condition, its main advantage consisting in the definition of a specific parameter, *i.e.* the spot size: widely used in practice, it corresponds to the radial distance, from the centre of the core, for which the field amplitude is reduced by a factor of $\frac{1}{e}$ – with the power being consequently reduced by $1/e^2$. In fiber data sheets the MFD value is often reported to indicate the Mode Field Diameter, which is simply equal to twice the spot size, *i.e.* $MFD = 2r_0$. Thanks to an empirical expression, it is possible to calculate the spot size as a function of the V-number:

$$\frac{r_0}{a} \simeq 0.65 + \frac{1.619}{V^{3/2}} + \frac{2.879}{V^6}$$

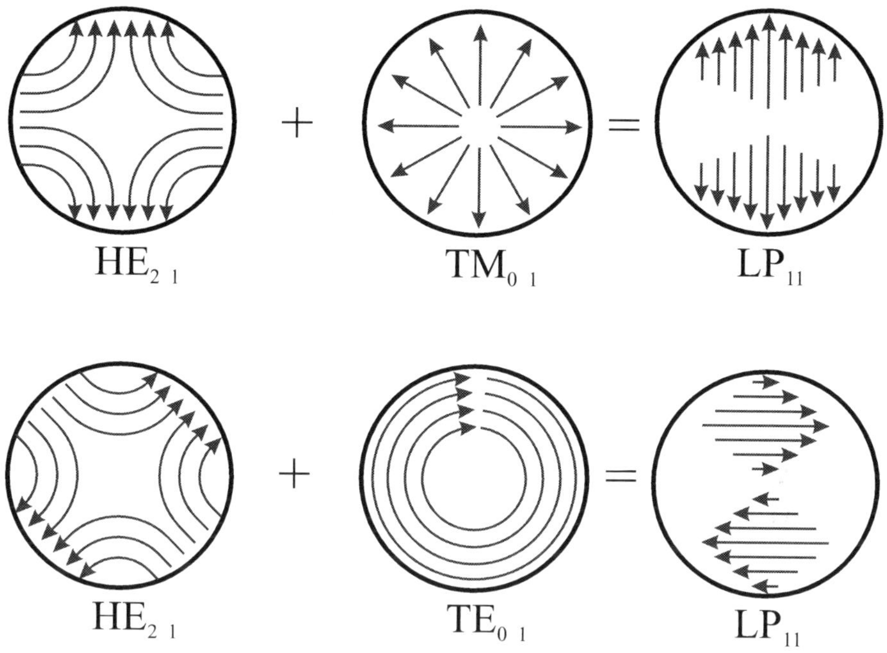

Fig. 8.9: The qualitative trend of the lines of force of the electric field transverse component of the two polarisations of the LP_{11} mode and the modes forming the said polarisations.

This formula provides accurate results with an accuracy greater than 1 %.

8.2 Birefringent or Polarization-maintaining Fibers

Polarization-maintaining (PM) fibers are optical fibers in which the two polarizations of the fundamental mode – which are the only ones above the cutoff wavelength – are no longer degenerate. This property is known as *birefringence* and it can be obtained either via anisotropic materials or elliptical cores. The most common types of birefringent fibers are schematised in figure 8.10: in the first three, silica becomes anisotropic by means of mechanical stresses induced by materials with different coefficients of thermal expansion; in the fourth and last example, the birefringence is due to ellipticity of the core. The type of polarization of the two modes depends on the characteristics of the birefringent fiber. For the sake of simplicity, from now on the polarization of the two modes will be assumed to be always linear and transverse to the direction of propagation. However, our considerations can also be valid in the general case of two modes with generic polarization.

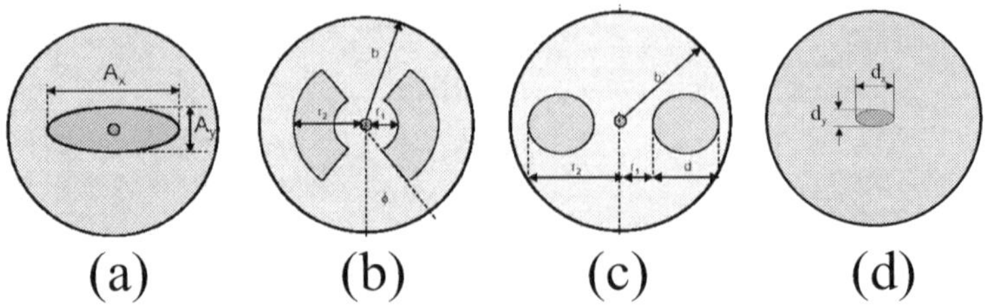

Fig. 8.10: Different examples of birefringent fibers: (a) elliptical cladding fiber; (b) bow-tie fiber; (c) PANDA fiber; (d) elliptical core fiber.

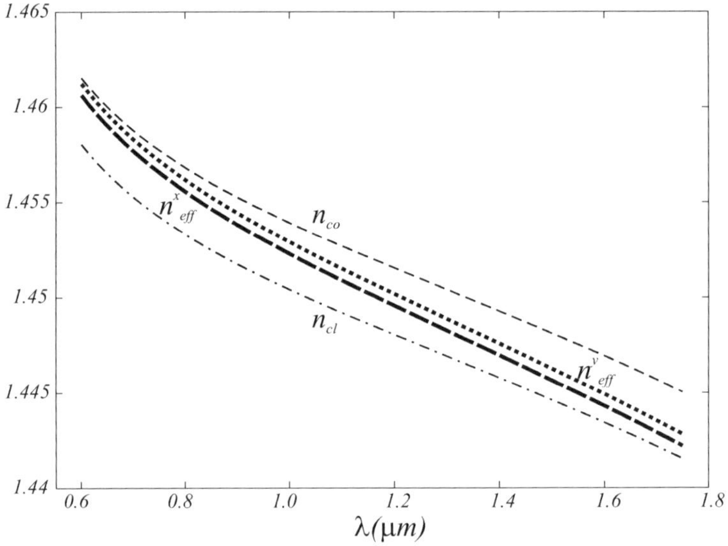

Fig. 8.11: The dispersion curves of a birefringent fiber.

According to our assumptions, we can thus write:

$$\overline{e}^{\,x} = \psi(u_1, u_2)e^{-j\beta_x z}\hat{i} \qquad \overline{e}^{\,y} = \psi(u_1, u_2)e^{-j\beta_y z}\hat{j};$$

where β_x, β_y are the phase constants of the two modes. The orientations of the electric field of the two modes identify two directions – the *main axes of the fiber* – that, in our case, coincide with the x- and y-axes of the coordinate system. Figure 8.11 illustrates the qualitative trends of the dispersion curves of the two modes. Let us suppose to work in a harmonic regime while a field excites the fiber without affecting the radiation modes:

$$\begin{aligned}\overline{e}_{in}(u_1,u_2) &= a_1\overline{E}^x(u_1,u_2)+a_2\overline{E}^y(u_1,u_2) = \\ &= \psi(u_1,u_2)\left(a_1\hat{\imath}+a_2\hat{\jmath}\right).\end{aligned}$$

The two modes are excited by a_1 and a_2 which depend on the state of polarization of the input field. Along the waveguide:

$$\begin{aligned}\overline{e}(u_1,u_2,z) &= a_1\overline{E}^x(u_1,u_2)e^{-j\beta_x z}+a_2\overline{E}^y(u_1,u_2)e^{-j\beta_y z} \\ &= \psi(u_1,u_2)\left(a_1e^{-j\beta_x z}\hat{\imath}+a_2e^{-j\beta_y z}\hat{\jmath}\right) \\ &= \psi(u_1,u_2)a_2e^{-j\beta_y z}\left(\frac{a_1}{a_2}e^{-j\Delta n_{eff}k_0 z}\hat{\imath}+\hat{\jmath}\right)\end{aligned} \quad (8.10)$$

with $\Delta n_{eff}=|n^x_{eff}-n^y_{eff}|=|\beta_x-\beta_y|/k_0$. Non-degeneration causes a progressive phase shift of the contribution of the two modes with a consequent alteration of the total field polarization, as figure 8.13 clearly demonstrates. Since the phase shift is periodic, after a specific propagation distance, known as *beat length*,

$$L_B=\frac{2\pi}{|\beta_x-\beta_y|}=\frac{\lambda}{|n^x_{eff}-n^y_{eff}|}=\frac{\lambda}{\Delta n_{eff}},$$

the polarization turns back to the input polarization state. The greater the beat length (which provides a measure of the fiber's degree of birefringence), the closer the modes are to being degenerate. In commercial birefringent fibers, the beat length is of the order of millimeters, while in standard fibers for remote transmission is of some tens-hundreds meters. In figure 8.12 we can see the example of the beat length of a fiber with elliptical core as the wavelength varies for different values of ellipticity. By virtue of the orthogonality of the modes, if the birefringent fiber is excited with a field distribution coinciding with one of the two mode functions, the state of polarization is maintained. In the ideal circular fiber, degeneration ensures that the input field polarization is maintained for any polarization, and the beat length is infinite. In reality, the imperfections of the waveguide may lead to coupling. In correspondence to the imperfection, the new boundary conditions require the presence of all the radiation and guided modes. After a reasonably small distance, the contribution of the radiation modes can be considered to be null and we are left with the contribution of the guided modes, whose amplitudes and phases will be different compared to what they were before the perturbation. This means that the perturbation has split the carried power between the different modes according to the characteristics of the perturbation itself; this partition can be calculated thanks to the coupled mode theory, which holds that the more similar the propagation constants (condition of synchronism), the higher the coupling between two modes. The perturbations in PM and circular fibers can be essentially compared, but what changes is the difference in the propagation constants; in PM fibers, this difference is high enough that the coupling phenomena between the modes can be ignored. In these fibers, if only one mode is excited – *i.e.* if the state of polarization of the input field coincides with one of the main axes of the fiber – the polarization is maintained

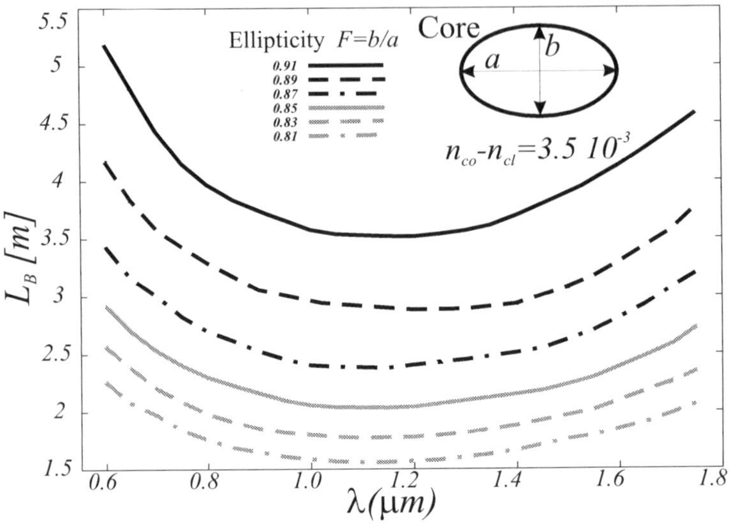

Fig. 8.12: The trend of the beat length as a function of the wavelength for a fiber with elliptical core. Cf. the different curves indicating the different values of ellipticity.

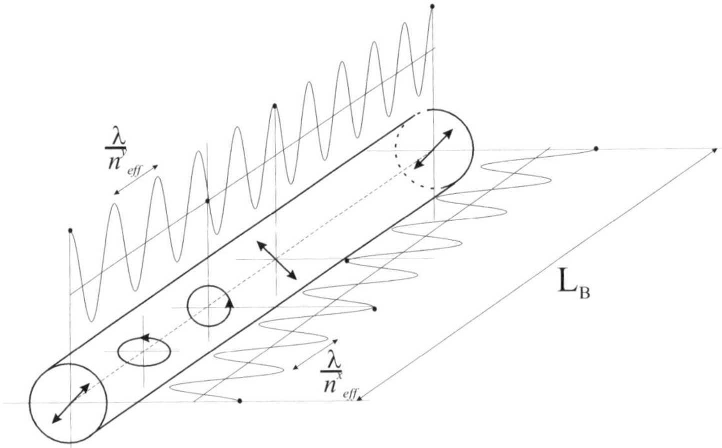

Fig. 8.13: The state of field polarization along a PM fiber.

throughout the whole waveguide even in case of perturbations. On the contrary, as far as standard fibers are concerned, since the birefringence is three or four orders of magnitude lower, a small perturbation is enough to cause a significant power transfer from one polarization to another. After a few meters of propagation, in fact, the polarization state of the signal is completely uncorrelated with the input polarization.

8.3 Transmission Characteristics of Optical Fibers

The characteristics of attenuation and dispersion affecting the performance of an optical fiber as a transmission medium are of paramount importance when it comes to determining whether or not an optical fiber can be employed in long-distance communication.

8.3.1 Optical Fiber Attenuation

The low levels of optical fiber attenuation are the main factor that led to their spread in telecommunication systems. Attenuation is usually expressed in decibels per unit of length. Let us consider a fiber of length L, with P^{in} indicating the input optical power and P^{out} the output optical power. The attenuation A is defined as follows:

$$A[dB/km] = \frac{1}{L[km]} 10Log\frac{P^{in}[mW]}{P^{out}[mW]} = \frac{1}{L}(P^{in}[dBm] - P^{out}[dBm]) \tag{8.11}$$

The technological development has made it possible to surpass the attenuation rates typical of the first optical fibers produced in the 1970s (about $20\,dB/km$) in favour of current values set at $0.2\,dB/km$. In order to measure the attenuation of an optical fiber and eliminate additional losses and errors due to connectors or variation in the launch conditions, we usually resort to the so-called *cutback technique*: first and foremost, we measure the output optical power P_1^{out} on a long length L_1 of the fiber under test; subsequently, the fiber is cut and the measurement of the output optical power P_2^{out} is repeated on the short fiber length L_2; finally, the attenuation value can be determined by calculating the difference between the two output powers and dividing the result by the length of the fiber that has been cut: $A[dB/km] = \frac{1}{L_1 - L_2}(P_2^{out}[dBm] - P_1^{out}[dBm])$.

The origins of attenuation in fiber optics can be better discussed if we differentiate between intrinsic and extrinsic causes: the first category comprises what is proper to the structure of the material itself, hence what cannot be eliminated; on the other hand, as far as extrinsic causes are concerned, the losses can be linked to the presence of impurities, imperfections or curvatures, and for this reason they can be reduced or eliminated *e.g.* by refining the production process or modifying the radius of curvature of the fiber.

Intrinsic Causes

There are mainly three causes of attenuation of this type. The first is intrinsic absorption, which ensues from electronic transitions in the ultraviolet (UV) spectral range and has a decreasing exponential trend that drops off as the wavelength increases; hence, its contribution to telecommunication wavelengths is relatively modest. Secondly, intrinsic absorption is also caused by molecular vibrations in the infrared (IR) region, becoming significant especially for wavelengths above 1660 nm. Finally, the third and main intrinsic cause is scattering, or Rayleigh scattering, which is caused by density and compositional variations in the glass: microscopic irregularities are always present and will scatter electromagnetic waves. While it is possible to reduce variation in glass composition by improving the fiber manufacturing techniques, the variations in terms of density are due to the random structure of the glass and cannot be avoided. The dimensions of these micro-irregularities are of the order of a tenth of the wavelength and constitute the spots from which radiation is spread. Electromagnetic radiation scattering occurs in all directions and entails an attenuation proportional to $1/\lambda^4$. Rayleigh scattering is thus the theoretical limit below which the attenuation of an optical fiber cannot be further reduced. As shown in figure 8.14, the minimum value of attenuation – less than $0.2dB/km$ – is reached at a wavelength of $1550nm$.

Other nonlinear scattering mechanisms exist as well, such as the so-called Brillouin and Raman scattering. However, nonlinear scattering does not contribute to the spectral attenuation of a fiber because it strongly depends on the optical power density within the fiber itself – hence, it is only significant above a certain power threshold.

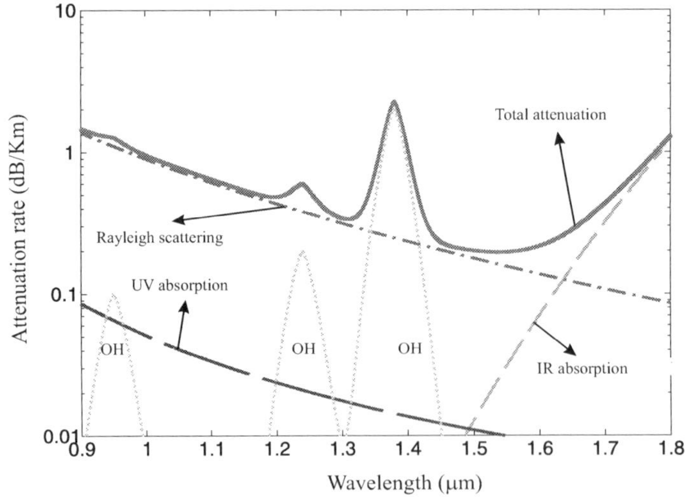

Fig. 8.14: Attenuation curves in a silica fiber.

Extrinsic Causes

Amongst the extrinsic causes of attenuation we can list the possible presence of foreign materials, extrinsic absorption caused by hydroxyl OH, and micro- as well as macro-bending losses.

Foreign materials, especially transition metals, are present due to contamination in the manufacturing process. However, current technological progresses have made it possible to completely eliminate this cause of absorption. On the other hand, absorption caused by the OH^- ion – or by water, as it is commonly said – is due to the presence of these ions in glass. Molecular vibrations in these ions give rise to *water peaks* in the absorption spectrum of a silica fiber at wavelengths of $0.72\,\mu m$, $0.88\,\mu m$, $0.95\,\mu m$, $1.13\,\mu m$, $1.24\,\mu m$ and $1.38\,\mu m$. The so-called "OH peaks" have been one of the major causes of fiber absorption for many years, and this is why, for fiber links, we chose to resort to signals situated at a wavelength distanced enough from these peaks. For this reason, we commonly distinguish between: *first window*, for wavelengths around $0.85\mu m$; *second window* at $1.30\mu m$; *third window* at $1.55\mu m$. Since the early 2000s, however, the manufacturing process has become more and more refined, which has led to the production of the first *OH-free fibers*, particularly employed in metropolitan and access networks.

Micro-bending losses are due to imperfections in the geometry of the core-cladding interface, *i.e.* fiber axis meandering or micro-deformations – note that by fiber axis we mean the boundary between the core and the cladding. Fortunately, the current manufacturing processes can produce very small micro-bending losses. Besides, other forms of microbends are possible as well: for example, they can be caused by mechanical stress, especially during the cable-laying process, resulting in micro-convexities or micro-dents. Heat stress can cause micro-bending losses as well, hence both the installation and maintenance phase must be performed with great care. The coating is the critical component of an optical fiber as far as microbend sensitivity is concerned. In this case too, high-quality coatings can be produced with today's technological standards, considerably improving this specific fiber characteristic. Generally, in single-mode fibers, micro-bending losses grow as the wavelength increases, while in multimode fibers the losses depend on the modal distribution of power and decrease as the wavelength increases. Since data on micro-bending losses are usually not specified in the data-sheets of optical fibers, we can consider these losses to be included in the total attenuation values.

The situation is different as far as macro-bending losses are concerned. One of the major advantages of optical fibers is flexibility: they can be thus employed in various places and bent if needed, though we may wonder to what extent. In terms of geometrical optics in fact, when a fiber is curved, the optical beams propagating in the core strike the core-cladding interface with an angle which is inferior to the critical angle; as a consequence, light does not undergo the process of total reflection, but is partly transmitted inside the cladding. Similarly, if we consider a single-mode fiber, when the fiber is bent, the part of the mode situated on the outside of the bend would be forced to travel at a greater speed with respect to the part of the mode on the inside of the bend; this would ensure

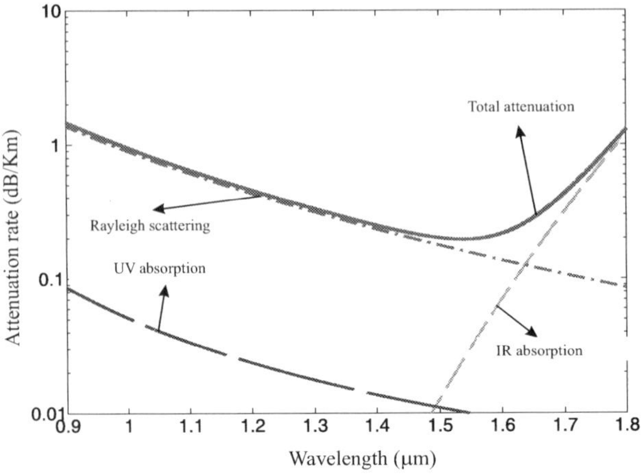

Fig. 8.15: Attenuation curves in a silica fiber without OH peaks.

that the wavefront is maintained perpendicular to the direction of propagation. As a result, part of the mode within the cladding should propagate faster than the speed of light in the medium, which is impossible, and the energy associated with this portion of the mode is lost by irradiation. Therefore, the fiber curvature leads to optical power loss. In general, macro-bending losses rapidly increase when the fiber is curved beyond a certain radius, known as critical radius of curvature. This can either be very small – a few millimetres – for strongly guiding fibers (with large numerical aperture) or of the order of tens of centimetres in the case of LMA (Large Mode Area) single-mode fibers with large effective area. Macro-bending losses increase with the wavelength, although we have a typically oscillatory dependence due to the interference between the light reflected on the cladding-coating and coating-air interface. In single-mode fibers, the losses can be reduced by decreasing the MFD, hence the wavelength. Nevertheless, the operating wavelength must be greater than the cutoff wavelength of the first higher-order mode λ_{cutoff} for the fiber to be single-mode. For an empirical performance evaluation in terms of fiber bending, we shall resort to the so-called MAC-number, which is defined as: $MAC = MFD/\lambda_{cut-off}$. The MAC-number of those fibers with reduced bending losses is less than 7, while typical values of standard fibers are usually included between 7 and 9.

Higher-order modes typically have higher bending losses compared to the fundamental mode. For this reason, the higher-order modes can be suppressed by choosing an appropriate radius of curvature, thus fabricating a mode filter. Data-sheets usually specify the values related to macro-bending losses for a certain radius of curvature at a fixed wavelength. For example, a standard single-mode fiber such as the Corning SMF28 is wound on a bobbin with a diameter of $32\,mm$, which results in a loss of approximately $0.05\,dB$ at $1550\,nm$, equivalent to 1.2% of the power lost.

It is possible to reduce the sensitivity to macro-bending losses by resorting to particular refractive index profiles, to the detriment, however, of other parameters; we will thus have to decide which feature is more relevant according to the type of application. Finally, note that bends do not only alter the optical properties of a fiber, but also its mechanical characteristics.

8.3.2 Bend-insensitive Fibers for FTTH Applications

Broadband access network technologies have seen a significant expansion recently. In particular, the increasing demand for broadband services has led to the development of media with a large transmission capacity as a result of the technology based on single-mode optical fibers. In this case we can distinguish between FTTH (fiber-to-the-home) or FTTC (fiber-to-the-cabinet) configurations. However, the implementation of fiber optics in the access network requires cables and optical fibers whose features are different from long-distance systems. The main difference is due to the large number of cables in the access network: due to the limited space and the different operations that must be performed, it is necessary to choose a fiber which is easy to handle and with reduced bending sensitivity.

In order to fabricate fibers with reduced bending losses we can resort to three main techniques. The first possibility is to employ a fiber with a slightly higher refractive index difference and a smaller core with respect to standard fibers. However, in this case we have a significant increase of the junction losses with conventional fibers. The second technique relies on the so-called trench fiber, whose name derives from the deep refractive-index trench in the area between the core and the cladding, so that the field confinement in the core is increased. As a result, bending losses are strongly reduced, though to the detriment of junction capability due to the high concentration of fluorine in the trench. The best performance can be obtained by balancing the bending-loss reduction, on the one hand, with the compatibility with the fiber installed, low splice losses and the resistance to mechanical stress due to bending, on the other. In this case, we talk about bend-optimised fibers, and a portion of their cladding, near the core, has a slightly lower refractive index; as they do not show any absorption of OH^- ions, they can be used over the entire C-WDM band.

As far as the access network is concerned, the characteristics of bend-insensitive single-mode fibers are listed in the Recommendation ITU-T G.657. This document introduces two categories of fibers: the fibers from category A are fully compliant with the G.652 standard and can also be used in other parts of the network; the other, category B, includes fibers which do not necessarily comply with ITU-T G.652, but have low values of bending losses at very low bend radii and are mainly intended for in-building use.

Main features of low bending-loss optical fibers						
Feature	Detail	Category A		Category B		
Mode Field Diameter	$\lambda = 1310\,nm$	$8.6 - 9.5\,\mu m$		$6.3 - 9.5\,\mu m$		
Cable λ_{cutoff}	max value	$1260\,nm$		$1260\,nm$		
Macro-bending loss	radius of curvature (mm)	15	10	15	10	7.5
	no. of rotations	10	1	10	1	1
	max at $1550\,nm$ (dB)	0.25	0.75	0.03	0.1	0.5
	max at $1625\,nm$ (dB)	1.0	1.5	0.1	0.2	1.0
Attenuation coefficient	max (dB/km)	0.4		0.5		

8.3.3 Dispersion

The dispersion of an optical signal causes a distortion of the signal itself both in analogue and digital transmissions. We can list several causes of fiber dispersion: in multimode fibers the main cause broadening the pulses is represented by intermodal dispersion, while in single-mode fibers this type of dispersion is eliminated and chromatic dispersion prevails; finally, polarisation-dependent dispersion becomes significant when chromatic dispersion is very low or zero.

8.3.4 Intermodal Dispersion

We shall now concentrate on a multimode fiber with a step-index profile. The optical power carried is distributed among the different modes – each travelling in the fiber with a different propagation constant and at a different speed. Hence, the receiver will be reached first by the power carried by the fundamental mode and, eventually, by the power carried by the last higher-order mode.

Let us consider a pulse of duration T entering the fiber as we try to make a rough estimate of the time delay Δt due to this phenomenon at the output end of a fiber section of length L. In ray optics approximation, the fundamental mode can be viewed as an axial beam running through the fiber in a time period equal to $t_0 = \frac{L}{c/n_1}$, where c is the speed of light in vacuum and n_1 the refractive index of the core. The last higher-order mode follows a zigzag path with an angle equal to the critical angle. The time required to cover the fiber section of length L will thus be $t_c = \frac{L/cos\alpha_c}{c/n_1}$, where $cos\alpha_c = n_2/n_1$, with n_2 indicating the cladding index. From this we can conclude that the pulse broadening is $\Delta t = t_c - t_0 = \frac{Ln_1}{cn_2}(n_1 - n_2) \simeq \frac{L}{2cn_2}(NA)^2$, where $NA = \sqrt{n_1^2 - n_2^2}$ represents the numerical aperture of the fiber. Consequently, the intermodal dispersion in a step fiber grows linearly as L increases. Furthermore, as the numerical aperture and thus the number of guided modes grow, so does the time delay.

The phenomenon of intermodal dispersion can be reduced by resorting to graded-index fibers. In these fibers, the core refractive index varies gradually, ranging from n_1 (found at the centre) to n_2 (at the core-cladding interface). This causes the ray trajectories describing the higher-order modes to bend; moreover, the rays which cover greater distances travel at higher speeds, as is clear from figure 8.17. Intermodal dispersion can be eliminated by employing single-mode fibers, which are specifically chosen in long-distance optical communication systems.

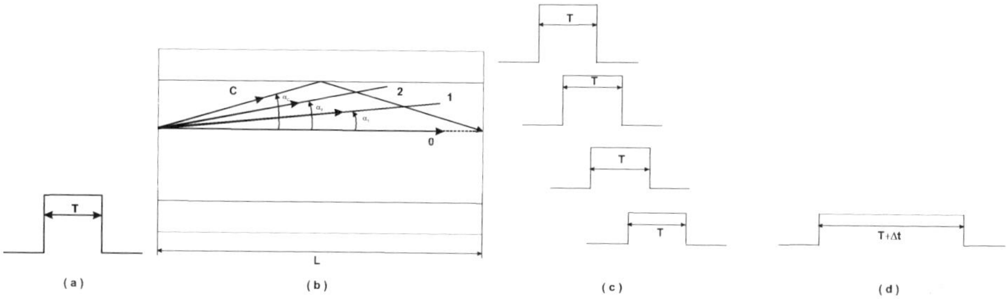

Fig. 8.16: Pulse duration broadening due to intermodal dispersion: (a) a pulse of duration T entering the fiber; (b) the optical paths of the different guided modes; (c) the output pulses for each mode; (d) the resulting pulse.

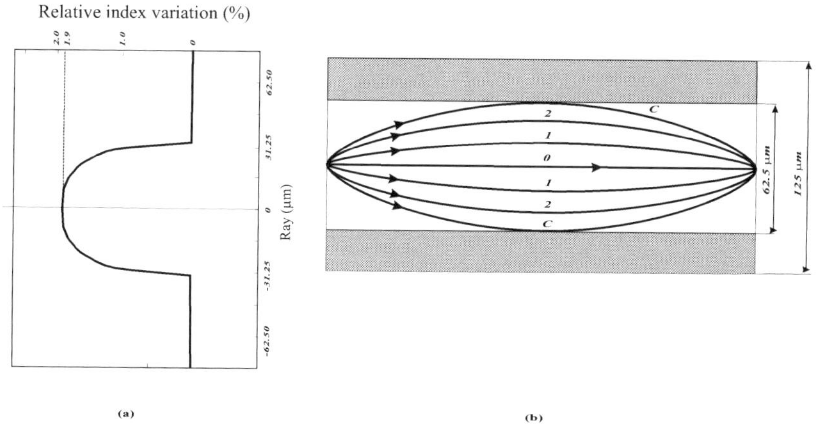

Fig. 8.17: Graded-index fiber: (a) the refractive index profile, (b) optical path of the rays.

8.3.5 Chromatic Dispersion

Considering a single-mode fiber, we have a case of chromatic, or intra-modal, dispersion when the propagation constant β does not vary linearly with frequency. In an optical fiber this effect is mainly determined by two mechanisms: the variation of the refractive index of the material with the wavelength λ, and the variation of the confinement of the mode with λ: in this case, we distinguish between material dispersion on one hand, and waveguide dispersion on the other, respectively. Although this distinction may oversimplify the discussion, we should not forget that it is not possible to isolate the two effects in a real fiber.

Material Dispersion

Material dispersion arises from the propagation of an optical signal in a bulk material – hence, not in a waveguide – for which the dispersion is only due to the refractive index variation with frequency. The propagation constant β can be written as $\beta(\omega) = n(\omega)k_0 = n(\omega)\frac{\omega}{c}$ and, if it does not vary abruptly with frequency, we can approximate, with the first three terms of the Taylor series expansion around the centre frequency ω_0, that:

$$\beta(\omega) = \beta|_{\omega_0} + (\omega - \omega_0)\frac{d\beta}{d\omega}|_{\omega_0} + \frac{1}{2}(\omega - \omega_0)^2\frac{d^2\beta}{d\omega^2}|_{\omega_0} = \beta_0 + (\omega - \omega_0)\beta_1 + \frac{1}{2}(\omega - \omega_0)^2\beta_2 \tag{8.12}$$

The group velocity v_g is equal to $v_g = \frac{1}{\beta_1} = \frac{c}{n+\omega\frac{dn}{d\omega}} = \frac{c}{N}$, where c indicates the speed of light in vacuum and N the group index of the material, defined as $N = n - \lambda\frac{dn}{d\lambda}$. At a certain wavelength, N represents the speed of light divided by the group velocity at the corresponding frequency.

The group delay for a pulse at frequency ω that propagates over a distance z is given by the propagation distance divided by the group velocity at that same frequency:

$$\tau_g = z\frac{d\beta}{d\omega} = z\frac{d(nk_0)}{d\omega} = z\frac{d\lambda}{d\omega}\frac{d}{d\lambda}(nk_0) \tag{8.13}$$

Since the wavelength in vacuum can be written as $\lambda = 2\pi c/\omega$, and then $d\lambda/d\omega = -2\pi c/\omega^2$, the group delay can be expressed in terms of wavelength:

$$\tau_g = \frac{z}{c}(n - \lambda\frac{dn}{d\lambda}). \tag{8.14}$$

By considering the group delay per unit of length $t_g = \tau_g/z$, the dispersion parameter D is defined as:

$$D = \frac{dt_g}{d\lambda} = \frac{1}{c}\frac{dN}{d\lambda} = -\frac{2\pi c}{\lambda^2}\frac{d^2\beta}{d\omega^2} = -\frac{2\pi c}{\lambda^2}\beta_2. \tag{8.15}$$

The refraction index of fused silica and silica with a percentage of germania (germanium dioxide GeO_2) is reported in figure 8.18 as a function of the wavelength. We can resort to the Sellmeier equation to analytically describe experimental data:

$$n^2 - 1 = \sum_j \frac{A_j\lambda^2}{\lambda^2 - \lambda_j^2} \tag{8.16}$$

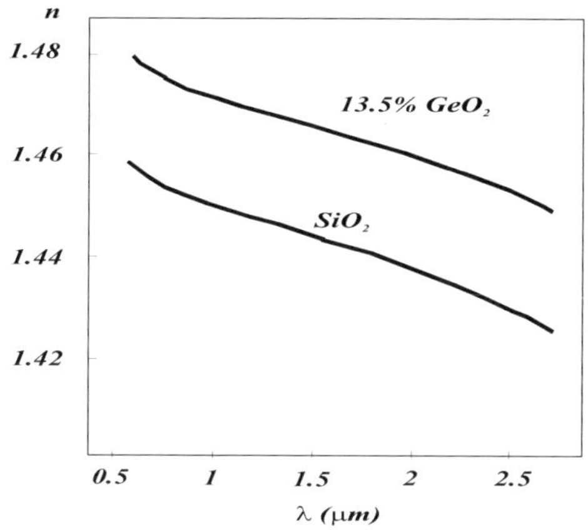

Fig. 8.18: The trend of the refractive index of pure silica and GeO_2-doped silica.

The summation must be considered for all the resonance wavelengths and is valid for wavelengths that are not close to any λ_j. As far as silica is concerned, it is enough to consider three terms adequately describing the resonances in the ultraviolet and infrared regions.

As figure 8.19 demonstrates, the group index has a minimum for wavelengths around $\lambda = 1.3\mu m$, which means that the dispersion parameter D is zero. This important feature of silica, besides the fact that losses are quite low at these wavelengths, has been crucial in the development of second-window communication systems. The value of λ for which D becomes zero depends on the

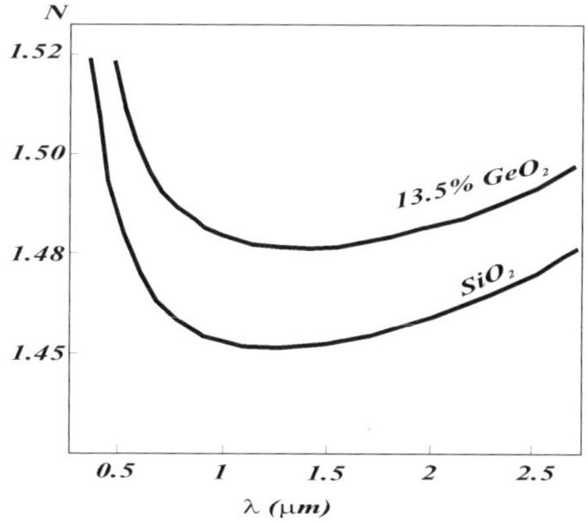

Fig. 8.19: The trend of the group index of pure silica and GeO_2-doped silica.

composition of the material and, as we shall see, it slightly changes when considering the presence of a waveguide.

Dispersion in an Optical Fiber

In an optical fiber it is necessary to consider the effects of both material and waveguide dispersion. All these properties are described by the propagation constant β or by the parameter b, which is defined as:

$$b = \frac{\beta^2 - n_2^2 k_0^2}{n_1^2 k_0^2 - n_2^2 k_0^2} \tag{8.17}$$

From this, β can be expressed as:

$$\beta = n_2 k_0 (1 + 2\Delta b)^{1/2} \approx n_2 k_0 + n_2 k_0 \Delta b \tag{8.18}$$

where $\Delta = (n_1 - n_2)/n_1$. In the equation (8.17), the term $n_2 k_0$ only describes those effects that are related to the material through the index of refraction n_2, while the term $n_2 k_0 \Delta b$, through Δ and b, depends on the properties of the waveguide, which in turn also depend on the material. Hence, the two types of dependence of the propagation constant (one on the material, the other on the waveguide) cannot be completely isolated, unless one of the two is zero. Instead of directly deriving the equation (8.17) for ω, it is perhaps more convenient to derive it for the $v - number$ or λ. Subsequently, we can obtain the following expression describing the group delay per unit of length:

$$t_g(\lambda, v) = \frac{d\beta}{d\omega} = \frac{1}{c}\{N_1 A(v) + N_2[1 - A(v)] + N_2 \Delta [A(v) - b]\} \tag{8.19}$$

where N_1 and N_2 are the core and cladding group indices respectively, and

$$A(v) = \frac{1}{2}\left[\frac{d(bv)}{dv} + b\right].$$

It is possible to demonstrate that, for a single-mode fiber, $A(v)$ represents the fraction of power of the fundamental mode LP_{01} carried in the core. This allows for an interesting physical interpretation of the equation (8.19). In fact, since the third term proportional to Δ is small, we can say that the group delay is mainly given by the weighted sum of the core and cladding group delays.

The group dispersion parameter D can be obtained by deriving (8.19). After performing some calculations, we come to the following expression:

$$D(\lambda) = D_M(\lambda) + D_W(\lambda) + D_P(\lambda).$$

The terms D_M, D_W, and D_P indicate the material, waveguide and profile dispersion, respectively; their relative formulae are thus:

$$D_M(\lambda) = \frac{1}{c}\left[\frac{dN_1}{d\lambda} A(v) + \frac{dN_2}{d\lambda}[1 - A(v)]\right]$$

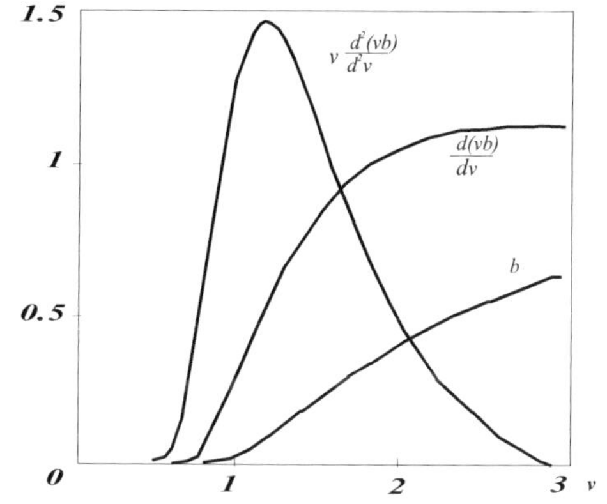

Fig. 8.20: The trend of b, $\frac{d(bv)}{dv}$ and $v\frac{d^2(bv)}{dv^2}$ for the mode LP_{01}.

$$D_W(\lambda) = -\frac{N_2^2\Delta}{n_2 c\lambda} v \frac{d^2(bv)}{dv^2}$$

$$D_P(\lambda) = -\frac{N_2^2\Delta}{n_2 c\lambda}\frac{y}{2}\left[1+\frac{y}{8}\right]\left[v\frac{d^2(bv)}{dv^2}+\frac{d(bv)}{dv}-b\right]$$

where $y = -\frac{2n_2}{N_2}\frac{\lambda}{\Delta}\frac{d\Delta}{d\lambda}$.

Material dispersion owes its name to the weighted sum of the material dispersion in the core and cladding regions; the weight factor is given by the fraction of optical power $A(v)$ in the two sections. Hence, the effects of dispersion due to guiding are implicit in D_M, even though the real mechanism of dispersion is due to the variation of the group index of the material with the wavelength. The dispersion term of the D_W waveguide is thus named because of its dependence on the second derivative of bv with respect to v. This term would exist even in case of no material dispersion, although the properties of the material are present in the refraction indices, in b and v. Finally, the term D_P is known as profile dispersion since it depends on y, or rather on the derivative $d\Delta/d\lambda$. As figure 8.20 suggests, since the terms $\frac{d(bv)}{dv}$ and $\frac{d^2(bv)}{dv^2}$ depend on v, we can conclude that waveguide dispersion is generally inferior with respect to material dispersion, though it becomes significant in silica for wavelengths around $1.3\mu m$. In the single-mode regime ($v < 2.405$), waveguide dispersion is always negative, as shown in figure 8.21. Since for $\lambda > 1.3\mu m$ material dispersion is positive, it is possible that material and waveguide dispersions cancel each other at a certain wavelength. Dispersion-shifted fibers are fabricated according to this principle. Finally, considering the trends of y, it is possible to demonstrate that, still considering silica, profile dispersion is about ten times lower than waveguide dispersion, with the consequence that profile dispersion is thus irrelevant with respect to the other two terms.

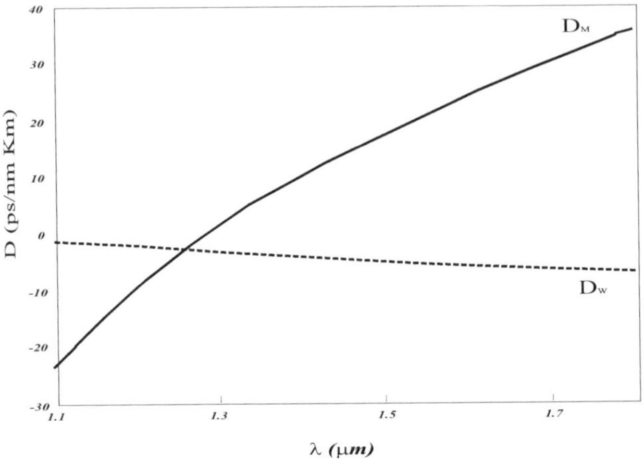

Fig. 8.21: The trend of material (D_M) and waveguide dispersion (D_W).

8.4 Fiber Classification

Optical communications standards are internationally set by the ITU (International Telecommunication Union). As far as optical fibers are concerned, they may be classified according to their dispersion.

Recommendation ITU-T G.652 describes the dispersive features of a single-mode optical fiber, often simply called standard fiber. The trend of the dispersion parameter as a function of frequency must be of the following type:

$$D(\lambda) = \frac{S_0}{4}\left[1 - \left(\frac{\lambda_0}{\lambda}\right)^4\right]$$

with λ_0 identifying the zero-dispersion wavelength ranging between $1300nm$ and $1324nm$, and $S_0 = 0.093\ (ps/nm^2km)$ designating the dispersion slope at zero dispersion. Figure 8.22 represents the trend of D in the wavelength range considered. In G.652 standard optical fibers, anomalous dispersion occurs at the working wavelength of $\lambda = 1550\ nm$, hence $D = 18\ ps/nmkm$ e $S = 0.06\ ps/(nm^2km)$.

Recommendation ITU-T G.653 defines the dispersive characteristics of a dispersion-shifted single-mode optical fiber. For this fiber, in the range $1500 - 1600\ nm$, we will have:

$$D(\lambda) = S_0(\lambda - \lambda_0)$$

with S_0 and λ_0 both equal to $1550nm$.

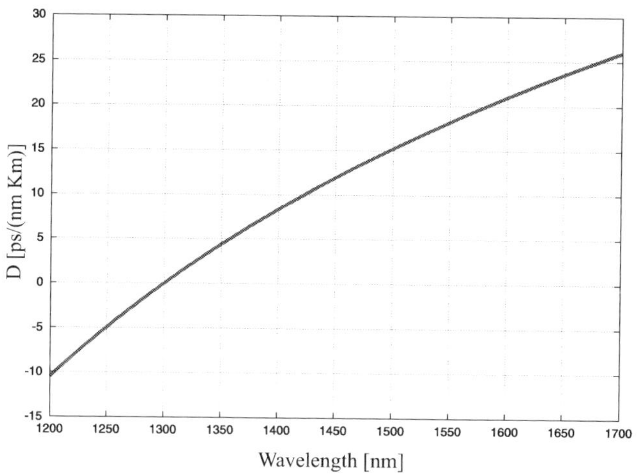

Fig. 8.22: The trend of D as a function of the wavelength as set by the Recommendation $G.652$.

As for non-zero dispersion-shifted single-mode optical fibers, the characteristics of their dispersion are reported in Recommendation ITU-T $G.655$. In this case, in the range $1500 - 1600nm$, it will be:

$$D(\lambda) = D_0 + S_0(\lambda - \lambda_0)$$

with $D_0 = 8.2ps/(nmkm)$, $S_0 = 0.043ps/(nm^2km)$ and $\lambda_0 = 1550nm$. The trend of D in the three different types of fibers is shown in figure 8.23.

8.5 Plastic Optical Fibers

Plastic optical fibers (POFs) are fibers whose core is made of acrylic resin, typically polymethyl methacrylate (PMMA), while fluorinated polymers constitute the cladding material. The cable is obtained by coating the plastic fiber with a sheath in high-density polyethylene (HDPE) measuring $2.2\,mm$ in diameter. Compared to conventional silica fibers, both materials and dimensions are different. In fact, in most of the POFs the diameter of the cladding is $1000\,\mu m$, while the core diameter is $980\,\mu m$, which means that approximately 96% of the transverse section of the fiber is constituted by the core. The refractive index of the core is typically 1.49, with a numerical aperture of 0.5. Thanks to the great dimensions of the core, light can be transmitted even if the ends of the fiber are slightly dirty or damaged, if there are strong vibrations or if the beam axis does not perfectly coincide with the centre of the fiber. Besides simplifying the installation processes, this also allows the use of low-cost optical connectors; moreover, POFs are strongly resistant and difficult to bend. As a consequence, we will have low bending losses even with radii of curvature of $20\,mm$.

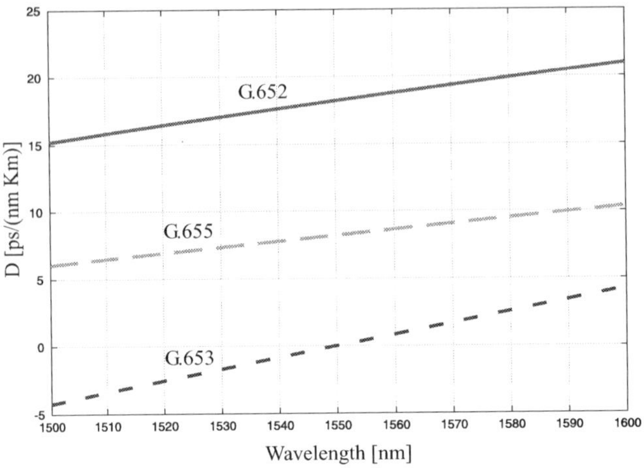

Fig. 8.23: The trend of D as a function of the wavelength according to the Recommendations ITU-T $G.652$, $G.653$ and $G.655$.

Plastic and silica optical fibers are *transparent* at different wavelengths. As figure 8.24 demonstrates, there are three transmission windows, which match the attenuation minima: 525 nm (green), 575 nm (amber), 650 nm (red). In most cases, we use a 650-nm LED as light source for an optical transceiver based on a POF. Thanks to visible light, it thus becomes immediately possible to check if the cable is working correctly. Due to high attenuation (about 150 dB/km at 650 nm), POFs cannot

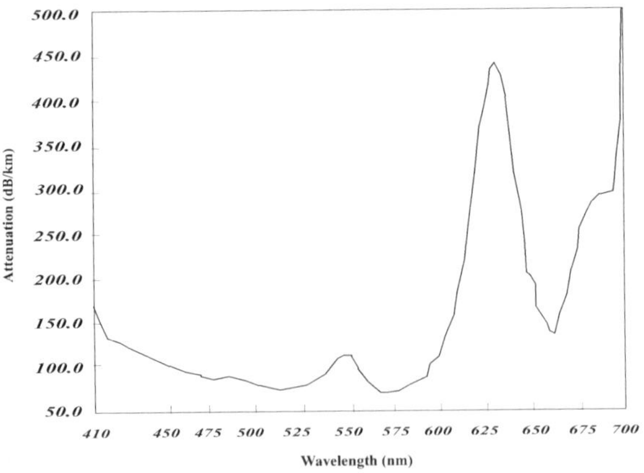

Fig. 8.24: Attenuation rates of a plastic optical fiber.

be used for long-distance communications in that the maximum distance allowed is about $400\,m$. Nevertheless, they are particularly suitable as far as short-distance applications are concerned, *e.g.* at home or in offices, where there are increasing demands for stable, simple and low-cost types of connection.

Thanks to their resilience, usability, safety, resistance against impacts, vibrations, twists and bends, POFs are better than conventional optical fibers even in automotive applications, robotics and industrial automation, video surveillance and CCTV, lighting and digital audio.

Chapter 9

Gratings

Diffraction gratings, which can be made of optical fiber or of dielectric waveguide, are exploited in a considerable amount of optical components. In fact, they are essential elements of optical communication systems because by virtue of them it is possible to produce all-optical signal processing devices.

Diffraction and interference, two typical phenomena in perturbed systems, were studied by the English physicist Sir W. H. Bragg and by his son, Sir W. L. Bragg, who formulated the well-known *Bragg's law* or *Bragg's condition* in 1913. This law was used to explain the interference pattern of an X-ray beam when it crosses the atomic structure of a crystal.

Fiber or waveguide Bragg gratings are obtained by creating a periodic variation of the refractive index of the type shown in figure 9.1, and they are based on the reflection of light at each interface, which is described by the Fresnel equation. Approximately, the process can be explained in the

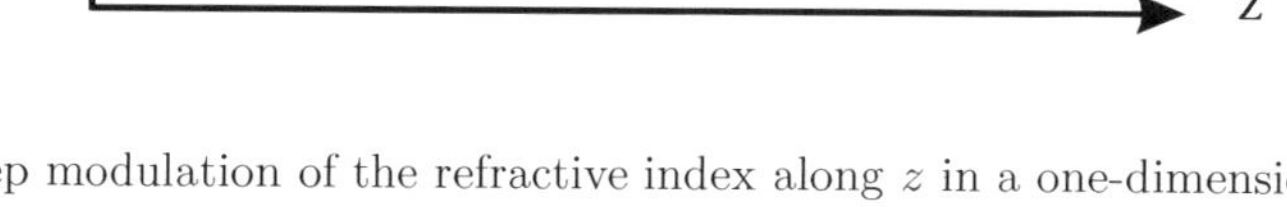

Fig. 9.1: Step modulation of the refractive index along z in a one-dimensional grating.

following terms: when light travels from a medium with low refractive index to one with higher index, a 180° phase shift occurs, whereas there is no phase shift if light travels in the opposite direction. If then we consider two optical rays reflected on two consecutive interfaces of a grating, constructive interference occurs if the extra distance covered by the second wave corresponds to half of the wavelength, so that:

$$\Lambda = \frac{\lambda}{2n}, \tag{9.1}$$

where Λ corresponds to the grating period and n to the refractive index of the optical medium. Another relevant aspect is that even when only a very small quantity of energy is reflected at each interface, with sufficiently long gratings it is possible to achieve a reflectivity close to 100%.
Guiding structures such as fibers or dielectric waveguides, however, complicate the process because they involve the modes of the waveguide instead of plane waves. For these kinds of structures, therefore the Bragg's condition, which can also be referred to as *phase-matching condition*, must be written as follows:

$$\lambda_{Bragg} = 2n_{eff}\Lambda \tag{9.2}$$

where λ_{Bragg} is the wavelength of the Bragg grating, and n_{eff} is the effective index of the guided mode considered. Since light travels from a propagating mode to a reflected mode, their coupling is called *counterpropagating.* Conversely, if the two modes travel in the same direction, the coupling is called *copropagating.* As it will be shown below, copropagating coupling requires *long-period gratings* because their period of index modulation is much larger than the wavelength.
In a coupling between two different modes, the *phase-matching* relation is usually written with propagation constants related to the wavelength and the different effective refractive indexes of the modes. It is important to remember that a perfectly uniform waveguide supports a discrete and countable set of modes which form an orthonormal basis. Since the modes are orthogonal to each other, there is no energy exchange between them. However, if the guide uniformity is perturbed, e.g. by an index modulation, the two modes are no longer orthogonal and energy exchange can therefore occur.

The variation of the refractive index can be obtained by modifying the waveguide geometry or by means of photoinduced refractive index modulation. For a long time, the only process used has been the first one, which resulted in the production of planar waveguides inscribed with techniques taken from microelectronics. Contrary to microelectronics, however, the critical feature of a waveguide is not its very small size, which can be of a few microns. In fact, the key element of waveguides is their optical quality, which is linked to their uniformity and precise geometry. Waveguides can be made of two layers of silicon or they can have a silica-on-silicon structure. They can also be made of innovative materials, such as silicon-based materials derived by a Sol-Gel process (a process through which a colloidal suspension of silicon particles is dried to obtain a porous material) doped with germanium, boron, phosphorus, or aluminum. Once the planar waveguide is ready, a corrugated grating can be grown or etched on it in order to obtain a periodic variation of the effective refractive index of the waveguide modes. Figure 9.2shows a grating under microscope.

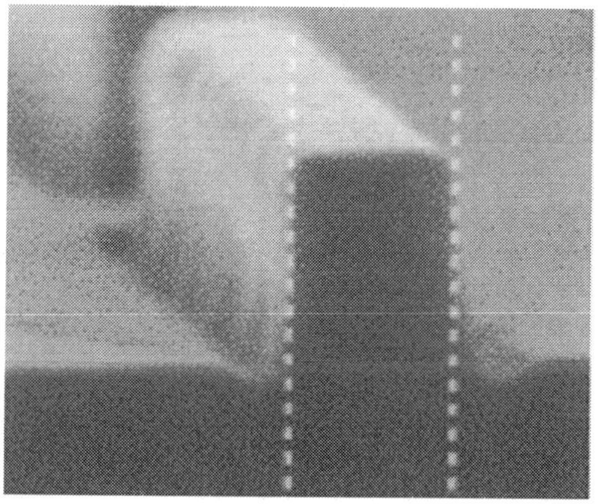

Fig. 9.2: Corrugated grating in silica-on-silicon-based planar waveguide.

This technique cannot be applied to fiber Bragg gratings, which exploit the fiber core photosensitivity instead. Photosensitivity is the ability of some materials to modify their absorption bands and refractive index when exposed to high-intensity light, such as UV light. Another aspect to consider is that germanium, typically used to dope the fiber core, is also photosensitive. Therefore, it is easy to inscribe gratings inside standard fibers for telecommunications with very few coupling problems. Nowadays, *Fiber Bragg Gratings* (FBGs) represent essential components of a wide variety of applications thanks to some of their properties. First of all, FBGs frequency response is easily verifiable. Secondly, when used as filters, FBGs can reach a reflectivity near to 100% with narrow bandwidths. Thirdly, through apodization it is possible to reduce their sidelobes, thus controlling their roll-off rate. Furthermore, it is possible to create broadband chirped FBGs, which can be used for dispersion control, that is to compensate chromatic dispersion in fiber optical transmission systems. Since FBGs are all-fiber components, their low insertion loss is another key element. Finally, considering that FBGs last approximately 25 years, their good stability must be taken into account. In fact, the changes in their spectral response linked to thermal or mechanical stress have been studied for a long while, so that today it is possible to reduce their effects or, when necessary, to exploit them for the production of optical sensors.

9.1 Photosensitivity

Photosensitivity is the ability of some materials to change permanently their physical structure when exposed to a light source. Ultraviolet radiation, for example, can be exploited to change permanently the refractive index of silicon waveguides and optical fibers. It was Ken Hill who first discovered the effect of photosensitivity in 1978, as he was studying propagation and its nonlinear effects in optical fibers. His discovery occurred accidentally while he was working on Raman lasers. Indeed, while injecting light in a special fiber, Hill realised that the light coming out from the end of the fiber tended to disappear after a few minutes of light exposure because light was back-reflected by the fiber itself. Hill and his team later discovered that this unexpected effect was caused by a grating in the fiber which had been generated by the exposure to light. After the discovery of photosensitivity in optical fibers many FBG-based components have been produced, such as filters, dispersion compensators and lasers.

Furthermore, recent studies have demonstrated that also silica-on-silicon and all-silica planar guiding structures are photosensitive. In both fibers and planar waveguides, photosensitivity is closely related to the chemical composition of the core. Dopants like cerium, europium and phosphorus, for instance, increase the degree of photosensitivity, but the best results are achieved with germanium. Germanium-doped core fibers change their refractive index when exposed to UV rays with a wavelength close to $240\,nm$ ($\sim 5.5\,eV$). This wavelength represents the absorption peak due to the defects of the crystal structure of germanium dioxide, where an atom of germanium replaces silicon. The variation of the refractive index is approximately $3 \cdot 10^{-5}$ for standard fibers doped with 3% germanium. Germanium is the standard dopant for telecommunication fibers and it is used to increase the refractive index of the core with respect to the refractive index of the cladding, which is usually doped with fluorine or phosphorus.

Complex gratings often require high photosensitivity in order to allow deep amplitude modulation and the presence of several gratings with different wavelength in the same layer of the guide. Nevertheless, the number of defects of high-quality telecommunication fibers is constantly decreasing thanks to more efficient production methods. Therefore, fibers are relatively transparent to UV light and not very photosensitive, so that new techniques have been introduced in order to increase the defects and the reactivity to UV radiation. In order to reduce the damage caused to optical fibers, the defects should be created only in the core, not in the cladding, and only where the grating must be placed. There are several techniques to do so. The first one is to increase the concentration of the dopant, germanium, since it is approximately proportional to the defect density. If single-mode fibers are considered, increasing the index difference between the core and the cladding implies reducing their core diameter, which is technically impossible below certain thresholds and always causes problems in coupling and splicing, the latter being a process for permanent joining of two optical fibers. A different approach entails the MCVD deposition of the glass matrix in a controlled environment, so that it is possible to increase the defect concentration by reducing the amount of oxygen atoms. This process increases photosensitivity while keeping the core dimension and the waveguide mode diameter compatible with standard fibers.

Another technique uses boron as a co-dopant because it reduces the refractive index of silicon. By mixing controlled quantities of boron to high germanium concentrations it is thus possible to obtain an index difference compatible with standard fibers, and at the same time a satisfactory defect concentration. Furthermore, boron co-doped fibers show a higher and faster response to UV radiation.

The most widespread technique to increase photosensitivity, however, is hydrogen loading (*H_2-loading*). If germanium-doped fibers are pre-soaked in hydrogen at high pressure and room temperature, their reactivity to UV radiation can be enhanced by two orders of magnitude. Through *H_2-loading*, hydrogen enters the glass matrix of the fiber, but once the fiber is exposed to air, hydrogen slowly comes out so that the UV radiation process must be completed before hydrogen concentration becomes too low.

9.2 Bragg gratings in dielectric waveguides

In integrated optics, gratings are widely used in several applications. Basically, they are constituted by an asymmetric planar guide where the film-cladding interface is not flat, but shows a periodic variation along the longitudinal direction. The *coupled-mode theory* (see Appendix B) can be used to describe how light behaves in grating structures.

First of all, the coupling coefficients must be calculated.

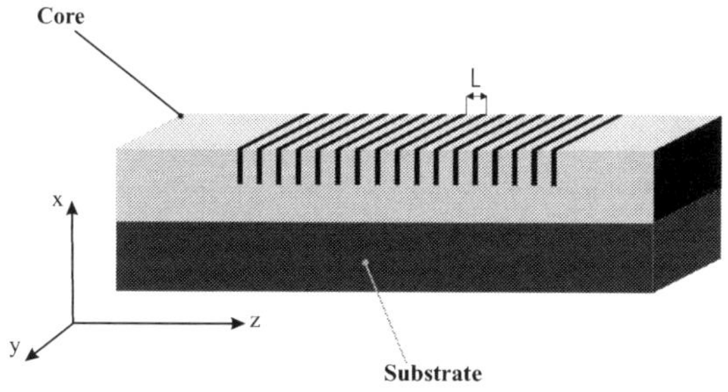

Fig. 9.3: Planar guide with grating period Λ.

Consider a planar waveguide with period Λ in which the upper layer is perturbed by means of a variation in permittivity. The variation will be written as follows:

$$\Delta\epsilon = u(x) f_\epsilon(z).$$

where $u(x)$ is a function identifying the section of the transversal plane where the perturbation occurs. Figure 9.4 shows the scheme of a grating, being d the film thickness and h the height of the

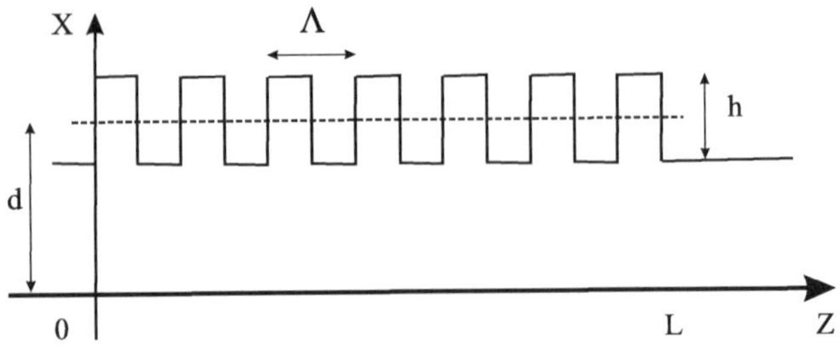

Fig. 9.4: Grating scheme.

perturbed region. In this case, $u(x)$ will be :

$$u(x) = \begin{cases} 1 \;\; se \;\; |x-d| < h/2 \\ 0 \;\; se \;\; |x-d| > h/2. \end{cases}$$

If then we consider a coordinate system where $f_\epsilon(z)$ is an even function of period Λ, the function will be written as follows:

$$f_\epsilon(z) = \Delta\epsilon_1 \cos\frac{2\pi}{\Lambda}z + \Delta\epsilon_2 \cos\frac{4\pi}{\Lambda}z + ... + \Delta\epsilon_n \cos\frac{2n\pi}{\Lambda}z + ...$$

where the coefficient $\Delta\epsilon_n$ depends on the shape of $\Delta\epsilon(z)$. Nevertheless, according to the superposition principle, each harmonic causing perturbation can be analysed individually, and then the single terms can be summed together in order to obtain the final result. Therefore, the analysis can be restricted to the effects of the fundamental wave, so that the function becomes:

$$\Delta\epsilon = u(x)\Delta\epsilon_1 \cos\frac{2\pi}{\Lambda}z = \frac{1}{2}u(x)\Delta\epsilon_1\left(e^{j\frac{2\pi}{\Lambda}z} + e^{-j\frac{2\pi}{\Lambda}z}\right).$$

Consider now TE wave propagation, in which the longitudinal electric field component is always 0. The longitudinal coupling coefficient will thus be 0, whereas the transverse one, which corresponds to both the copropagating and counterpropagating coupling coefficient, will be calculated as follows:

$$K^t_{\nu\mu} = \frac{1}{2}\frac{\omega}{4}\Delta\epsilon_1\left(e^{j\frac{2\pi}{\Lambda}z} + e^{-j\frac{2\pi}{\Lambda}z}\right)\int\int_{-\infty}^{\infty} u(x)\overline{E}_{t\nu}\cdot\overline{E}^*_{t\mu}dx\,dy \;\; =$$

$$= \;\; k_{\nu\mu}\left(e^{j\frac{2\pi}{\Lambda}z} + e^{-j\frac{2\pi}{\Lambda}z}\right) \;\; ;$$

where $k_{\nu\mu}$ is a constant independent of z. If we consider the equation in (B.27), given the condition $\delta = 0$ in accordance with the principle of stationary phase, and given β_μ and Λ as predetermined variables, the two possible solutions for co-propagating modes will be:

$$\beta_\nu = \beta_\mu + \frac{2\pi}{\Lambda};$$

$$\beta_\nu = \beta_\mu - \frac{2\pi}{\Lambda};$$

whereas these will be the solutions for the counterpropagating modes:

$$\beta_\nu = \frac{2\pi}{\Lambda} - \beta_\mu;$$

$$\beta_\nu = -\frac{2\pi}{\Lambda} - \beta_\mu.$$

In order to understand which modes can be involved in the coupling process, it is useful to consider the dispersion curves of the unperturbed guide, where the modes are arranged according to the decreasing value of β, which means that the low-order modes have higher β values and vice versa. The first type of copropagating coupling occurs through mode conversion from a high-order mode (β_ν) to a low-order one (β_μ). In the second type of coupling, the opposite occurs. The third case describes a counterpropagating coupling between ν-th and μ-th modes, and it is thus possible only if $\Lambda < 2\pi/\beta_\mu$, since β_ν must have a positive value. Finally, the required positive value of β_ν excludes the fourth type of coupling too.

In this section, the unperturbed guides will be considered single-mode. This eliminates mode-conversion phenomena between guided modes but not between or with radiation modes, which are not considered in the present discussion. Therefore, we can only consider the coupling between two counterpropagating waves of the fundamental mode, which is described by the third equation above. Since the two modes have the same propagation constant, that is $\beta_\nu = \beta_\mu = \beta$, the equation can be written as follows:

$$\beta - \frac{\pi}{\Lambda} = 0. \tag{9.3}$$

It is important to remember that it is only thanks to Λ, i.e. the periodicity along the z-axis, that it is possible to satisfy Bragg's law, thus allowing mode coupling. If the structure was invariant along the propagation direction z, that is $\Lambda \to \infty$, the above conditions would allow only μ-th mode conversion with itself, thus eliminating the possibility of counterpropagating coupling. This is obviously so, because in lack of perturbations no reflection of waves occurs.

Therefore, there are only two equations in this system of equations for mode coupling:

$$\frac{dA}{dz} = -jk_{ab}Be^{j(\beta_a+\beta_b-2\pi/\Lambda)z} = -jk_{ab}Be^{j\delta z} \tag{9.4}$$

$$\frac{dB}{dz} = jk^*_{ab}Ae^{-j(\beta_a+\beta_b-2\pi/\Lambda)z} = jk^*_{ab}Ae^{-j\delta z} \tag{9.5}$$

where the symbols A, a and B, b are used to refer to the variables related to, respectively, direct and reflected waves. In order to solve the system we need to uncouple the equations in (9.4) and (9.5). As a first step, therefore, we need to find the derivative of the first equation with respect to z, and we can then substitute the second equation into the derivative obtained. The result will be a second order equation where $A(z)$ is the only unknown element:

$$\frac{d^2A}{dz^2} - j\delta\frac{dA}{dz} - |k_{ab}|^2A = 0$$

This equation can be solved through integration, as shown in B.2. Supposing that the mode-coupling runs from $z = 0$ to $z = L$, we can take as an initial condition $A(0) = A_0$ and $B(L) = 0$. Consequently, $A(z)$ can be written as follows:

$$\begin{aligned} A(z) &= \left(\frac{A_0 e^{j\delta z/2}}{(S/2)\cosh(SL/2) + j(\delta/2)\sinh(SL/2)} \right) \times \\ &\times \left\{ \left[(S/2)\cosh(SL/2) + j(\delta/2)\sinh(SL/2) \right] \cosh(Sz/2) \; + \right. \\ &\left. - \left[(S/2)\sinh(SL/2) + j(\delta/2)\cosh(SL/2) \right] \sinh(Sz/2) \right\}; \end{aligned}$$

where S is set as:

$$S = 2\sqrt{|k_{ab}|^2 - (\delta/2)^2}$$

On the other hand, $B(z)$ can be written as:

$$\begin{aligned} B(z) &= \frac{j}{k_{ab}} \left(\frac{|k_{ab}|^2 A_0 e^{-j\delta z/2}}{(S/2)\cosh(SL/2) + j(\delta/2)\sinh(SL/2)} \right) \times \\ &\times \left[\cosh(SL/2)\sinh(Sz/2) - \sinh(SL/2)\cosh(Sz/2) \right] ; \end{aligned}$$

Figure 9.5 shows the trends of $|A(z)|^2$ and $|B(z)|^2$, which form direct waves progressively decreasing their amplitude along the grating. At the same time, however, the amplitude of the reflected wave increases. If the grating is long enough, a total reflection of the incident wave can be obtained, as it

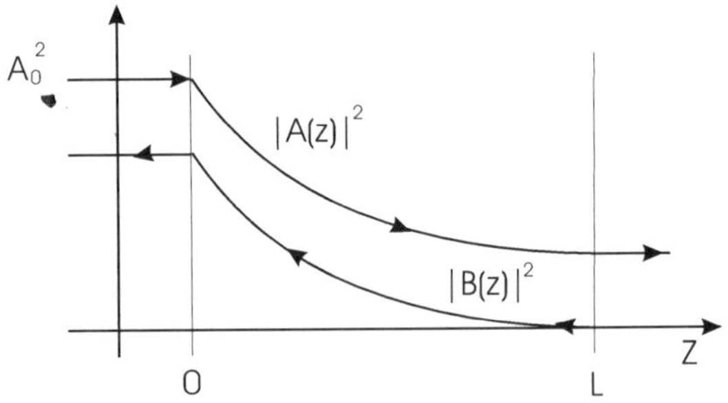

Fig. 9.5: Trends of $|A(z)|^2$ and $|B(z)|^2$ along an L-long grating.

occurs in distributed reflectors.

Therefore, it is now useful to introduce the reflection coefficient:

$$\Gamma(z) = \frac{B(z)}{A(z)}.$$

When $z = 0$ the reflection coefficient will be:

$$\Gamma(0) = -j\frac{|k_{ab}|^2}{k_{ab}} \frac{\sinh(SL/2)}{(S/2)\cosh(SL/2) + j(\delta/2)\sinh(SL/2)}.$$

As it can be seen, there are only two parameters in the equation, namely $\delta/|k_{ab}|$ and $k_{ab}L$. The first one refers to the progressive variation of synchronism with respect to the coupling coefficient, whereas the second one refers to the total coupling. The trend of $|\Gamma(0)|^2$ is shown in figure 9.6. Note that the reflection coefficient is approximately constant in proximity of the synchronism condition, but it drops abruptly when δ is high. This determines high reflectivity only for the waves within the band represented by the first lobe of the curve $\Gamma(0)$. Furthermore, reflectivity increases for higher values of the coupling coefficient, which in turn depends on the refractive index and on the spatial distribution of the modes involved in the coupling.
As it is shown in figure 9.7, the band becomes increasingly narrow as the grating length L increases, thus making the grating more selective. Moreover, since the phase constants of Bragg's law are directly related to the frequency of propagation modes, the grating selectivity influences the grating spectral bandwidth. That is the reason why gratings are often used for wavelength selection and stabilization in a lot of optical devices, such as in Distributed Feed-Back (DFB) lasers.
One last remark should be made about situations of perfect synchronism ($\delta = 0$), when the reflection

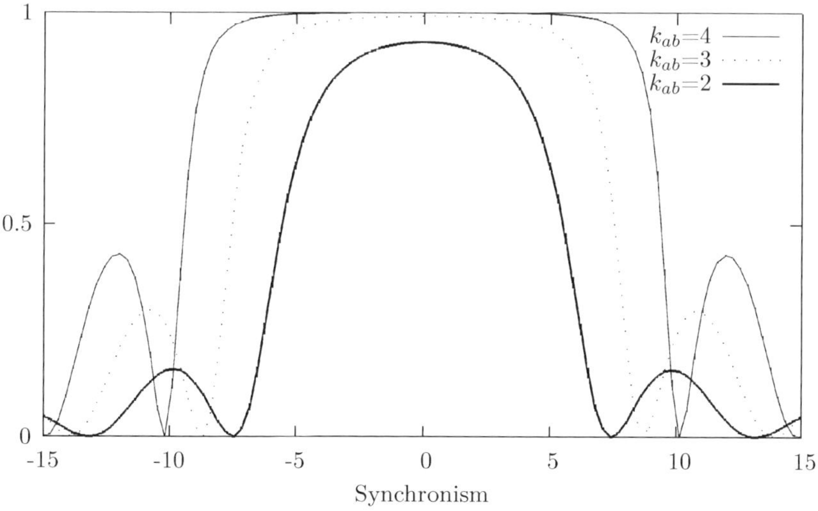

Fig. 9.6: Trend of $|\Gamma(0)|^2$ in a grating with $\Lambda = 0$. The trend is expressed as a function of $\delta[m^{-1}]$ and according to different coupling coefficients $k_{ab}[m^{-1}]$.

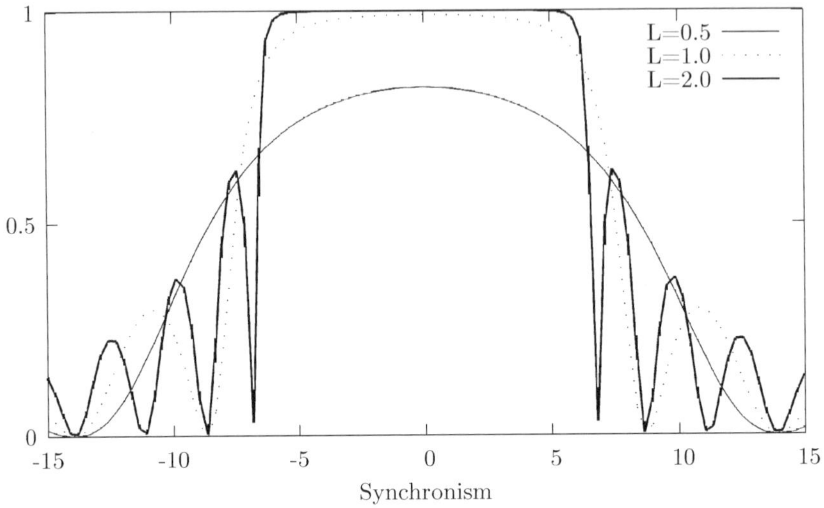

Fig. 9.7: Trend of $|\Gamma(0)|^2$ according to different lengths of $L[m]$, for $k_{ab} = 3m^{-1}$.

coefficient is reduced to:

$$\Gamma(0) = -j\tanh(|k_{ab}|L).$$

Since it is purely imaginary, the phase difference between the incident wave, $A(z)$, and the reflected one, $B(z)$, is $\pi/2$.

It is also important to notice that in the case of perfect synchronism the coupled direct and reflected modes have the same phase constants ($\beta_\mu = \beta_\nu = \beta = n\,2\pi/\lambda$), so that Bragg's law can be written as follows:

$$2\,n\,\frac{2\pi}{\lambda} = \frac{2\pi}{\Lambda}$$

From this equation, we derive the one seen at the beginning of this chapter (see (9.1)):

$$\Lambda = \frac{\lambda}{2n}.$$

Therefore, the grating period corresponds to half wavelength of the radiation which propagates in the guide. This condition, however, can only be achieved by means of high quality fabrication methods.

9.3 Fiber Bragg Gratings

In general, the fiber core refractive index modulation is expressed by a $\delta n_{core}(z)$ variation:

$$\delta n_{core}(z) = \overline{\delta n_{core}}(z)\left\{1 + \upsilon(z)\cos\left(\frac{2\pi z}{\Lambda(z)} + \varphi(z)\right)\right\} \tag{9.6}$$

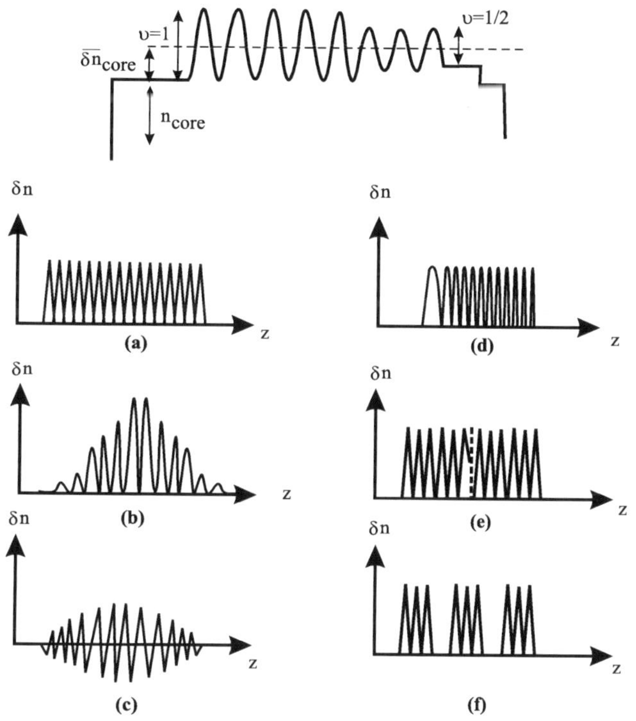

Fig. 9.8: Examples of index modulation, all described by equation (9.6).

where $\overline{\delta n_{core}}(z)$ is the average index modulation, $\upsilon(z)$ is the visibility, that is, the amplitude of the modulation, $\Lambda(z)$ is the grating period, and $\varphi(z)$ is the phase (see figure 9.8). This equation can be used to describe different types of gratings. The δn of **Uniform Bragg Gratings** is shown in figure 9.8(a) and is characterised by the constant values of the three parameters υ, $\overline{\delta n_{core}}$ and Λ. This is the simplest kind of grating, but its spectral response shows many undesired sidelobes, so that uniform Bragg gratings are not used in complex devices such as filters for WDM (Wavelength Division Multiplexing) transmission systems. For more complex systems, *apodized gratings* have been developed, in which the value of the modulation index can be controlled. Apodization acts on two parameters of the index modulation, namely the average index modulation and its visibility.

DC apodization refers to the average index modulation, whereas AC apodization refers to visibility. **Intensity-modulated fiber Bragg gratings**, shown in figure 9.8(b), have a constant visibility (v) and grating period Λ, but show average index variation $\overline{\delta n}_{core}(z)$. Therefore, they are also called gratings sensitive to the average modulation. Figure 9.8(c)shows gratings with varying visibility. In these kinds of gratings $v(z)$ is controlled, while the average index modulation $\overline{\delta n}_{core}$ and the period Λ are constant. They are also called gratings apodized in visibility. **Chirped Gratings** (figure 9.8(c)) are characterised by variation in the grating period Λ. They are used especially for the chromatic dispersion compensation which occurs in optical fiber communication. In these gratings, visibility (v) and average $\overline{\delta n}_{core}$ are constants. Finally, **complex gratings**, shown in figure 9.8(e)/(f), vary in each parameter, $\varphi(z)$, $\overline{\delta n}_{core}(z)$ and $\Lambda(z)$, so that they can be considered as a combination of an apodized and a chirped grating. This kind of structures is used for applications with complex functions and specific requirements.

The general index modulation model might seem difficult, but in fact it allows calculating the formulas for the coupling coefficients which are required to solve the coupled-mode equations. Before going into detail, it is useful here to introduce some simplifications which seem justifiable for the purpose of the present book, and which will simplify the analysis. To start with, wave propagation in optical fibers is usually single-mode. Secondly, the fundamental mode is the LP_{01}, where $E_z \approx 0$, so that $K^z_{\nu\mu}$ can be ignored while calculating the coupling coefficients. Furthermore, since the variation $\Delta\varepsilon$ is not a function of x or y, it can be extracted from the integral which defines $K^t_{\nu\mu}$. Since we know that:

$$\begin{aligned}
\frac{\varepsilon + \Delta\varepsilon}{\varepsilon_0} &= (n + \delta n)^2 \\
\frac{\varepsilon + \Delta\varepsilon}{\varepsilon_0} &= n^2 + 2\,n\,\delta n + \delta n^2 \\
\varepsilon + \Delta\varepsilon &= \varepsilon_0\,n^2 + 2\,\varepsilon_0\,n\,\delta n + \varepsilon_0 \delta n^2
\end{aligned}$$

by ignoring the second order perturbation δn^2 we thus obtain:

$$\Delta\varepsilon = 2\,\varepsilon_0\,n\,\delta n. \tag{9.7}$$

Therefore, the $K^t_{\nu\mu}$ coefficient can finally be written as follows:

$$
\begin{aligned}
K^t_{\nu\mu} &= \frac{\omega}{4}\Delta\varepsilon \iint_{-\infty}^{+\infty} \overline{E}_{t\nu}\overline{E}^*_{t\mu}\,dx\,dy \\
K^t_{\nu\mu} &= \frac{\omega}{2}\varepsilon_0\, n_{core}\, \delta n \iint_{-\infty}^{+\infty} \overline{E}_{t\nu}\overline{E}^*_{t\mu}\,dx\,dy \\
K^t_{\nu\mu} &= \frac{\omega}{2}\,\varepsilon_0\, n_{core}\, \overline{\delta n}_{core}\,(z)\left\{1+\upsilon(z)\cos\left(\frac{2\pi}{\Lambda(z)}z+\varphi\,(z)\right)\right\}\iint_{-\infty}^{+\infty} \overline{E}_{t\nu}\overline{E}^*_{t\mu}\,dx\,dy \\
K^t_{\nu\mu} &= \frac{\omega}{2}\,\varepsilon_0\, n_{core}\, \overline{\delta n}_{core}(z)\iint_{-\infty}^{+\infty} \overline{E}_{t\nu}\overline{E}^*_{t\mu}\,dx\,dy\; + \\
&+ \frac{\omega}{2}\,\varepsilon_0\, n_{core}\, \overline{\delta n}_{core}(z)\upsilon(z)\cos\left(\frac{2\pi}{\Lambda(z)}z+\varphi(z)\right)\iint_{-\infty}^{+\infty} \overline{E}_{t\nu}\overline{E}^*_{t\mu}\,dx\,dy \\
K^t_{\nu\mu} &= \sigma_{\nu\mu}\,(z) + 2\kappa_{\nu\mu}\,(z)\cos\left(\frac{2\pi}{\Lambda(z)}z+\varphi\,(z)\right)
\end{aligned}
$$

that is:

$$K^t_{\nu\mu}\,(z) = \sigma_{\nu\mu}\,(z) + \kappa_{\nu\mu}\,(z)\left(e^{j\left(\frac{2\pi}{\Lambda}z+\varphi\right)} + e^{-j\left(\frac{2\pi}{\Lambda}z+\varphi\right)}\right) \tag{9.8}$$

where:

$$\sigma_{\nu\mu}\,(z) = \frac{\omega}{2}\varepsilon_0 n_{core}\overline{\delta n}_{core}\,(z)\iint_{core} \overline{E}_{t\nu}\overline{E}^*_{t\mu}dx\,dy; \tag{9.9a}$$

$$\kappa_{\nu\mu}\,(z) = \frac{\upsilon\,(z)}{2}\sigma_{\nu\mu}\,(z)\,. \tag{9.9b}$$

In the previous equations, $\sigma_{\nu\mu}\,(z)$ describes an average coupling, which is constant if $\overline{\delta n}$ does not depend on z, and which is usually called *DC coupling coefficient.* Conversely, $\kappa_{\nu\mu}(z)$, which multiplies the cosine function, is called *AC coupling coefficient.* Note that in (9.9a) the integration is limited to the core because it is supposed that the cladding permittivity ε is not perturbed.

Therefore, the mode coupling between a forward fundamental mode of $A(z)$ amplitude and a backward fundamental mode of $B(z)$ amplitude, which share the same phase constants ($\beta_\nu = \beta_\mu = \beta$), can be reduced to a system of two equations in two variables. Following the coupled-mode equations in (B.27), we will thus obtain:

$$
\begin{aligned}
\frac{dA}{dz} &= -j\left(K^t_{aa}A + K^t_{ab}Be^{j2\beta z}\right) \\
\frac{dB}{dz} &= j\left(K^t_{bb}B + K^t_{ba}Ae^{-j2\beta z}\right)
\end{aligned}
\tag{9.10}
$$

and according to the definition given in (9.8) of the coupling coefficients, we obtain:

$$K^t_{ab} = \left(K^t_{ba}\right)^*. \tag{9.11}$$

If we substitute the calculated $K^t_{\nu\mu}(z)$ to the first equation we then find:

$$\begin{aligned}\frac{dA}{dz} = & -j\left\{\sigma_{aa} + \kappa_{aa}\left[e^{j\left(\frac{2\pi}{\Lambda}z+\varphi\right)} + e^{-j\left(\frac{2\pi}{\Lambda}z+\varphi\right)}\right]\right\}A+ \\ & -j\left\{\sigma_{ab} + \kappa_{ab}\left[e^{j\left(\frac{2\pi}{\Lambda}z+\varphi\right)} + e^{-j\left(\frac{2\pi}{\Lambda}z+\varphi\right)}\right]\right\}e^{j2\beta z}B\end{aligned} \tag{9.12}$$

and if we do the same for the second equation we find:

$$\begin{aligned}\frac{dB}{dz} = & +j\left\{\sigma_{bb} + \kappa_{bb}\left[e^{j\left(\frac{2\pi}{\Lambda}z+\varphi\right)} + e^{-j\left(\frac{2\pi}{\Lambda}z+\varphi\right)}\right]\right\}B+ \\ & +j\left\{\sigma_{ba} + \kappa_{ba}\left[e^{j\left(\frac{2\pi}{\Lambda}z+\varphi\right)} + e^{-j\left(\frac{2\pi}{\Lambda}z+\varphi\right)}\right]\right\}e^{-j2\beta z}A.\end{aligned} \tag{9.13}$$

At this stage, by means of stationary-phase approximation and the result in (9.3), which has already been found for gratings in dielectric waveguides, we obtain two equations where the terms have no oscillation, or they have a pulse oscillation of $\beta - \frac{\pi}{\Lambda} = 0$. By substituting these terms in (9.12) and (9.13) we thus obtain:

$$\begin{aligned}\frac{dA}{dz} &= -j\sigma_{aa}A - j\kappa_{ab}Be^{2j\left(\beta-\frac{\pi}{\Lambda}\right)z-j\varphi} \\ \frac{dB}{dz} &= j\sigma_{bb}B + j\kappa_{ba}Ae^{-2j\left(\beta-\frac{\pi}{\Lambda}\right)z+j\varphi}\end{aligned} \tag{9.14}$$

and finally, using the equation in (9.11):

$$\begin{aligned}\frac{dA}{dz} &= -j\sigma_{aa}A - j\kappa^{*}Be^{2j\left(\beta-\frac{\pi}{\Lambda}\right)z-j\varphi}; \\ \frac{dB}{dz} &= j\sigma_{bb}B + j\kappa Ae^{-2j\left(\beta-\frac{\pi}{\Lambda}\right)z+j\varphi};\end{aligned} \tag{9.15}$$

The two equations can be solved by analytic or numerical integration. In both cases, the solutions of this coupled equations are written as functions of the detuning (δ):

$$\delta = \beta - \frac{\pi}{\Lambda} = 2\pi n_{eff}\left(\frac{1}{\lambda} - \frac{1}{2n_{eff}\Lambda}\right) = 2\pi n_{eff}\left(\frac{1}{\lambda} - \frac{1}{\lambda_{Bragg}}\right) \tag{9.16}$$

where the Bragg wavelength, λ_{Bragg}, refers to the wavelength of a grating of Λ period and can be written as:

$$\lambda_{Bragg} = 2n_{eff}\Lambda. \tag{9.17}$$

Detuning, instead, refers to the shift from the phase-matching condition $\delta = 0$ and varies according to the frequency for counterpropagating modes in a grating with predetermined parameters.

9.4 Properties of the Uniform Bragg Gratings

The most common parameters to define grating properties are: reflectivity, transmittivity, bandwidth, phase response, and grating dispersion, which are all functions of the detuning δ. These parameters are usually written as functions of the *normalised detuning*, that is:

$$\frac{\delta}{2\pi n_{eff}/\lambda} = \frac{\lambda_{Bragg} - \lambda}{\lambda_{Bragg}}.$$

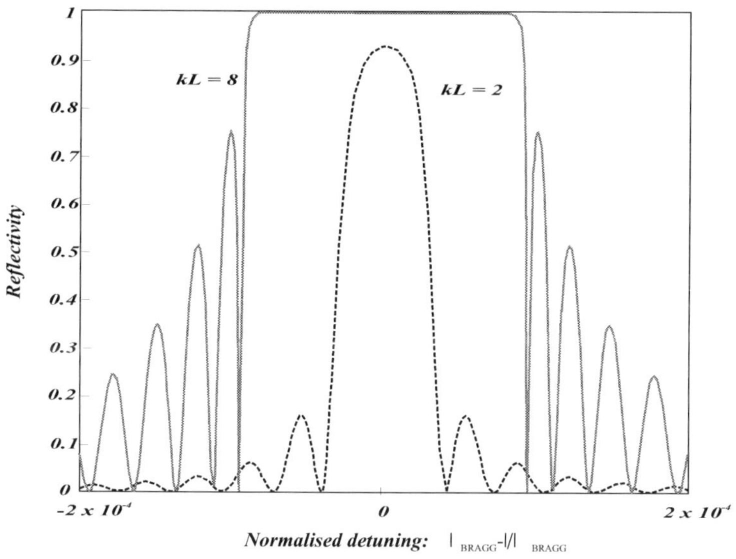

Fig. 9.9: Reflectivity coefficient of uniform gratings written as a function of the normalised detuning $\frac{\lambda_{Bragg}-\lambda}{\lambda_{Bragg}}$ for $\kappa L = 2$ e $\kappa L = 8$.

Reflection Coefficient

The amplitude reflection coefficient is always expressed as:

$$\Gamma(z) = \frac{B(z)}{A(z)}. \tag{9.18}$$

It is a complex mathematical function which can be expressed according to its magnitude and phase. Its magnitude squared for $z = 0$ represents the reflectivity, which is shown in figure 9.9 as a function of the normalised detuning $\frac{\lambda_{Bragg}-\lambda}{\lambda_{Bragg}}$ for two gratings with a different κL product. As the figure shows, in uniform gratings the central peak is always surrounded by smaller lateral peaks. This behaviour in the so-called *roll-off band* is caused by the limited length of the grating and by the constant index modulation. Both features determine an abrupt interface with the uniform unmodulated regions at the beginning and at the end of the grating. Furthermore, they provoke a Fabry-Perot effect. Apodization, however, can reduce the oscillations.

Bandwidth

A convenient definition of the bandwith can be obtained by considering the first minimums of reflection on both sides of the grating central peak. Therefore, bandwidth can be mathematically

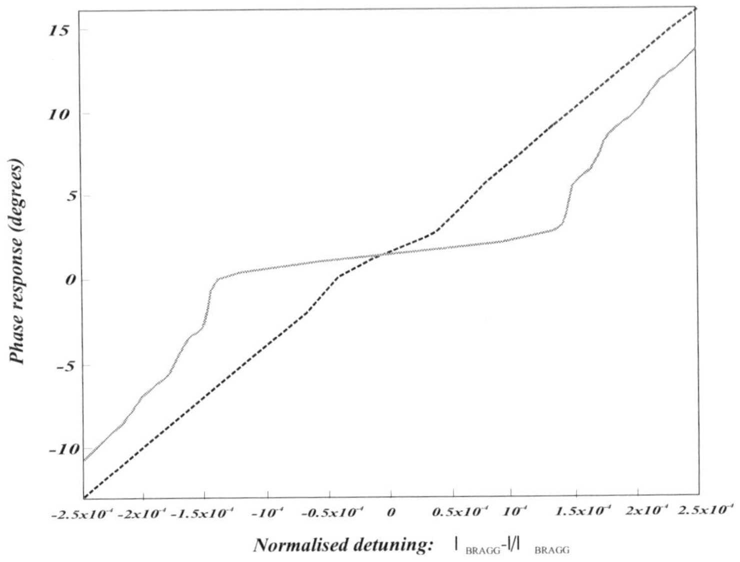

Fig. 9.10: Uniform grating phase response as a function of normalised detuning for $\kappa L = 2$ and $\kappa L = 8$.

calculated by searching for the first reflectivity zeros, thus obtaining:

$$B = \frac{\lambda^2}{2\pi n_{eff} L}\sqrt{(\kappa L)^2 + \pi^2}. \tag{9.19}$$

Note that if the AC coupling coefficient must be low so that $(\kappa L)^2 \ll \pi^2$, it follows also that the bandwidth becomes an inverse function of the grating length L:

$$B = \frac{\lambda^2}{2 n_{eff} L}$$

Conversely, if $(\kappa L)^2 \gg \pi^2$ the band does not depend on the grating length because the grating is able to prevent the light from travelling towards the end of the grating itself. In this case the bandwidth will be:

$$B = \frac{\lambda^2 \kappa}{2\pi n_{eff}}$$

Phase Response and Group Delay

The amplitude reflection coefficient has its own phase, which is illustrated in figure 9.10. Note that the out-of-band phase response tends to be linear, similarly to a phase delay caused by light propagation in fibers. Indeed, the frequencies which are distant from the phase-matching condition

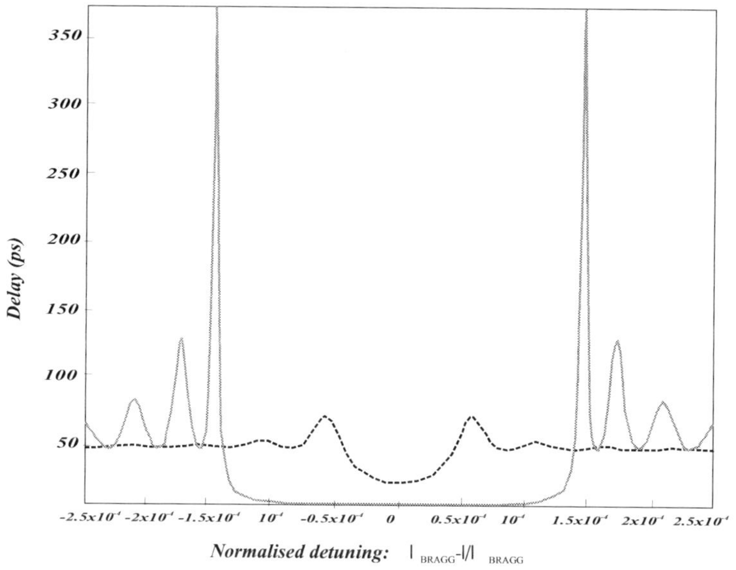

Fig. 9.11: Group delay as a function of normalised detuning for $\kappa L = 2$ and $\kappa L = 8$.

act as if the grating did not exist.
The phase response helps to define the group delay for the light reflected by the grating, which is usually measured in picoseconds:

$$\tau = \frac{d\phi}{d\omega} = -\frac{\lambda^2}{2\pi c}\frac{d\phi}{d\lambda} \tag{9.20}$$

Figure 9.11 shows the group delay of two gratings for $\kappa L = 2$ and $\kappa L = 8$.

Chromatic Dispersion

Chromatic dispersion d (expressed in ps/nm) varies according to the wavelength and describes the group velocity variation which determines the group delay. It can be written as follows:

$$d = \frac{d\tau}{d\lambda} = -\frac{2\pi c}{\lambda^2}\frac{d^2\phi}{d\omega^2} \tag{9.21}$$

Figure 9.12 shows its behaviour in a uniform grating for $\kappa L = 8$.
Chromatic dispersion is important in gratings and has been studied to develop special gratings with controlled dispersion. In uniform gratings, dispersion is equal to zero around synchronism condition and becomes appreciable only near the sidebands of the reflection spectrum. This phenomenon can be explained thanks to the already mentioned Fabry-Perot effect, which occurs at the ends of the

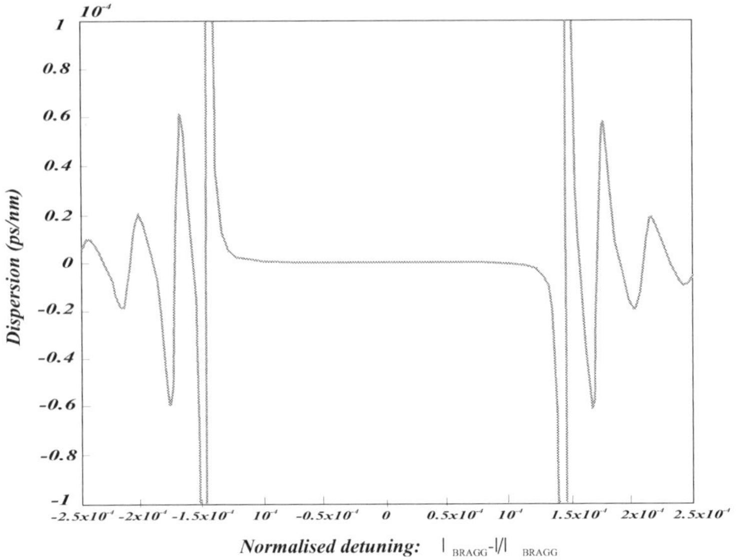

Fig. 9.12: Grating dispersion as a function of normalised detuning for $\kappa L = 8$.

device. The zeros in the reflection spectrum are resonance frequencies of the Fabry-Perot cavity which are somehow trapped into the grating.

9.5 Nonuniform Gratings

The majority of gratings used for practical applications are nonuniform, because nonuniformity improves their frequency response and phase response. As already mentioned above, a key feature of passive components used in WDM technologies is their squared spectrum with a sharp roll-off and no sidelobes. It is possible to obtain such features and to adapt the grating to specific functions by regulating the grating uniformity (and nonuniformity). Furthermore, it is possible to introduce localised phase shifts inside the gratings in order to create bandpass filters with an extremely narrow transmission peak. Recent studies have also focused on superstructure gratings, in which some parameters, such as visibility or the mean value of the modulation index, are periodically varied along the grating with much larger periods than that of the grating itself.

Speaking of apodization, AC apodization is used to reduce the Fabry-Perot effect and the ripples which may appear on the dispersion curve. As a matter of fact, in these cases DC apodization is usually not effective, since the frequency for the maximum reflectivity of the grating depends on the average value of the apodization. In a Gaussian apodized grating, for example, the peak reflectivity is shifted to shorter wavelengths at the beginning and at the end of the grating. This creates a sort of Fabry-Perot cavity at shorter wavelengths, which results in minor peaks in the spectral response. Therefore, apodization is more effective when it affects only visibility. Mathematically, this implies varying $v(z)$ in equation (9.6).

9.6 FBG Applications

Add-drop Multiplexer for WDM Systems

Nowadays, all single-carrier systems have been replaced by WDM systems. *Optical add-drop multiplexers* (OADM) represent the most effective method to control multi-carrier systems because they allow selective adding and dropping of channels. In order to build an OADM, a variety of devices can be exploited, such as arrayed waveguide gratings, Fabry-Perot filters, or thin film filters. Nevertheless, it is with fiber Bragg gratings and fiber optic circulators that the lowest insertion losses (which can be less than $2\,dB$ for added or dropped channels and $3\,dB$ for pass-through channels) and the highest channel isolation ($> 50\,dB$) are obtained. As shown in figure 9.13, a multi-channel optical signal is routed through the circulator and while travelling along the fiber meets an FBG which reflects a single channel at a given wavelength. The channel is then dropped from the circulator. In the same way, a wavelength channel is launched into the second circulator and is then added to the signals which are travelling along the fiber after being reflected by the same FBG.

It is clear that in these devices the features of the filters play a key role because, for example, for a $40\,GBit/sec$ modulation the usable bandwidth is limited to a $0.32\,nm$ and the nearest channel is just $0.4\,nm$ away, so that only $0.08\,nm$ of the bandwidth can be used to allow the filter to achieve an adequate roll-off, which is necessary to isolate the adjacent channels without affecting the one considered. It is essential not to attenuate the channel, because it may travel through a considerable number of filters but its spectrum should not be altered. This is the reason why the requirements for the bandwidth of WDM systems are usually given at $-0.5\,dB$ instead of $-3\,dB$.

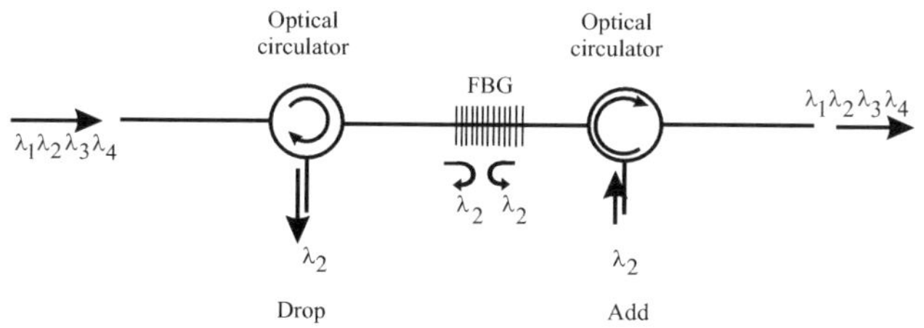

Fig. 9.13: Add-drop multiplexer using an FBG and two circulators

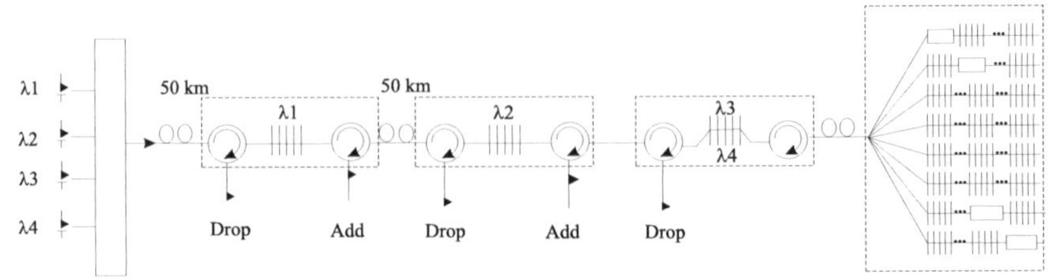

Fig. 9.14: Demultiplexer for WDM systems.

Similarly, the requirements for the out-of-band dispersion are stringent in order not to modify the waveform of the signal travelling through the filter, while at the same time the in-band group delay ripple of the filter must be lower than $10\,ps$ for $10\,GBit/sec$ modulations and $2.5\,ps$ for $40\,GBit/sec$ modulations. Positive results are achieved with both apodized fiber Bragg gratings and with demultiplexers similar to those in figure 9.14, which are used at the end of the fiber to split the channels and are in fact formed by cascaded gratings. In the case of demultiplexer the in-band features are particularly relevant because the signal must travel through a series of filters before coming out. In particular, as it has already been highlighted, the filter should not reduce the spectral bandwidth of the signal, but at the same time it should isolate the adjacent channel in order to avoid crosstalk.

FBG-assisted Couplers

The major disadvantage of the add-drop filters described in the previous section is that they contain circulators, which are complex devices and are not made of fiber. In fact, researchers aim at producing all-fiber components in order to reduce insertion loss. It is possible to produce all-fiber add-drop filters by using a fused coupler whose behaviour is partially modified by a fiber Bragg grating in the fusion region. The coupler is obtained by gradually stretching the fibers until the core is no longer recognisable, and then by fusing the two fibers in the stretched regions. The device functions as a

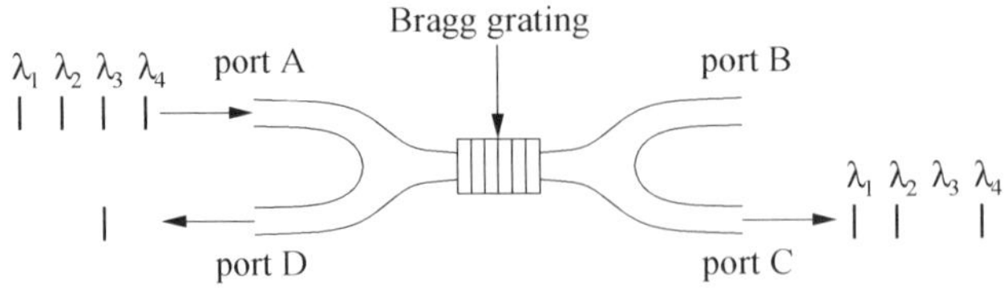

Fig. 9.15: FBG-assisted coupler for add-drop filters.

directional coupler. By means of the FBG it is possible to modify the behaviour of the channel so that the frequency for which the grating was devised is reflected on port D, while the other channels can travel through the coupler with standard behaviour, which means that they exit through port C. Since the component is symmetrical, by using port B it is also possible to add a channel to the signal travelling from port A to port C.

Dispersion Compensators

Optical fiber transmission systems rely on OOK modulation to transfer information, thus acquiring an impulse shape similar to a Gaussian function. It is mathematically shown that chromatic dispersion on Gaussian pulses leads to a pulse broadening, so that the energy is distributed in larger intervals. Consequently, there is a degradation of the eye diagrams which are used to study the bit error rate (BER) in digital transmission.

Researchers have been able to produce special fibers - the *dispersion-shifted fibers* - which reduce dispersion to $1550\,nm$ in C band. Low dispersion, in turn, minimises the non linear effects of four-wave mixing, which are more difficult to control. Bragg gratings can be used in different configurations to produce multichannel compensators.

To start with, dispersion compensation can be obtained with uniform Bragg gratings in transmission, as shown in figure 9.16. There are two main advantages with this kind of configuration: firstly, no further component is required; secondly, the compensator can be introduced in a pre-existing system by simply photoinducing the Bragg grating in the optical fiber. This solution works because the filter is used in a band where the group delay has a normal dispersion which compensates for the anomalous dispersion occurring in the fiber. The major disadvantage, however, is that this technique can compensate only for a single channel, even if it has been shown that it covers long propagation distances.

Figure 9.17 shows a much more versatile technique for dispersion compensation, namely chirped FBG, which usually have a linear variation of the grating period. This kind of configuration is in reflection, so that it requires the introduction of a circulator, which causes an insertion loss of $2-6\,dB$. This technique rests on the principle that different wavelengths are reflected from different

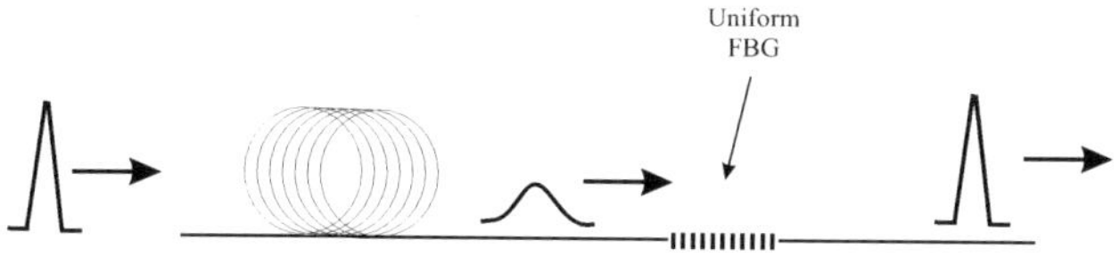

Fig. 9.16: Dispersion compensation with uniform fiber Bragg grating in transmission.

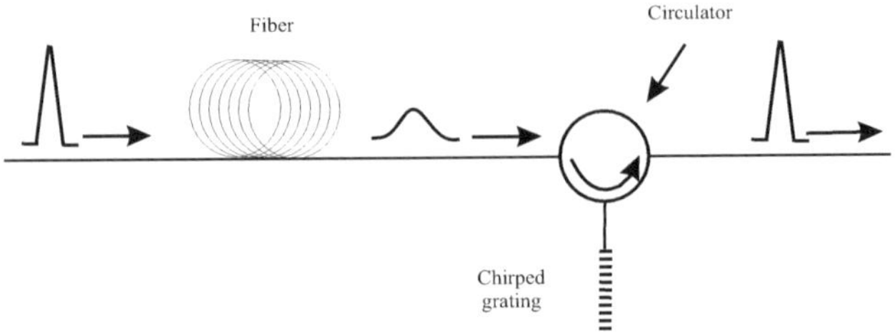

Fig. 9.17: Dispersion compensation with chirped fiber Bragg grating in reflection.

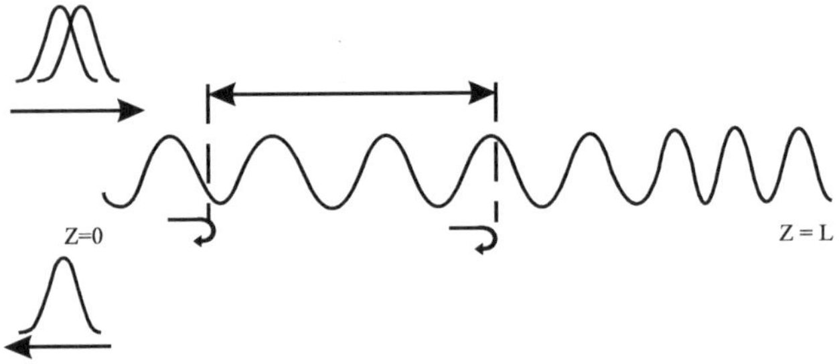

Fig. 9.18: Scheme of the principle of compensation with chirped FBG.

parts of the grating, so that the short wavelengths can be delayed in order to restore the original pulse. Figure 9.18 illustrates how compensation with chirped FBG works. The major problem of this technique is that the grating length limits the delay which can be introduced between long and short wavelengths, that is, the result of the band-dispersion product is limited. A $10\,cm$ grating, for instance, can compensate for $100\,km$ of standard fiber on a $0.5\,nm$ band. The length is limited by the chirped phase mask commonly used for these gratings. Nevertheless, new techniques have been developed for gratings which can be up to $1\,m$ long.

Gain Flattening Filters

The gain of optical amplifiers with no filters depends on the wavelength. In modern WDM systems it is important to have a broad band with homogeneous gain, because otherwise the usable bandwidth of a series of amplifiers is reduced to approximately $10\,nm$ in proximity of the peak of maximum gain. Therefore, the most suitable solution is represented by gain flattening filters, which have the further advantage of having a minimum effect on the total noise figure of the amplifier.
These devices work by attenuating selected wavelength bands in order to reduce the gain peak. To

do so, tilted FBGs can be used. These kinds of short-period gratings are inscribed at an angle with the optical axis, so that light is coupled in the cladding modes, thus attenuating the bands. Another technique to obtain gain flattening filters is based on long-period gratings, which also allow the coupling of guided modes to the cladding modes.

Tilted gratings are less efficient because part of the intensity of the reflected energy remains inside the amplifier, so that insulators are needed, although they increase the filter insertion losses and worsen the signal-to-noise ratio. Furthermore, several gratings with controlled intensity are required, so that the production process becomes longer and more difficult. That is the reason why long-period gratings are more common. Figure 9.19 shows the effects of FBG long-period gratings on filters. As it can be seen, the gain is attenuated at its peak, so that it is possible to obtain a 30 nm band with an oscillation of about 1 dB.

Equalising filters are usually inserted between two sections of the amplifier to improve the impact on the noise figure. An equalising filter acts on the noise figure and the amplified spontaneous emission (ASE). More specifically, it reduces the counterpropagating ASE coming from the second section. Moreover, a reduced counterpropagating ASE in the first layer increases the pump power available to amplify the signal. The device obtained is constituted by a first section - the low-noise amplifier (LNA) - isolated from a second section - the power amplifier - by means of the filter. It is important to notice that since long-period gratings couple light to the cladding modes, they show high sensitivity to optical fiber curvature. Consequently, an accurate *packaging* is essential for their proper operation. Another popular device are filters and gratings which work as mirrors to reflect the pump or the signal (or both). That is because if the pump power is not completely exploited by the signal, it is better to send it back into the active medium in order to maximise the total gain.

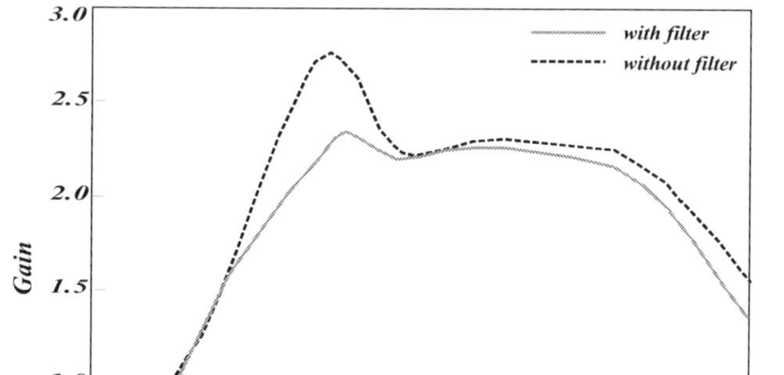

Fig. 9.19: Effect of a long-period FBG filter on the optical gain of an EDFA.

FBG Lasers

As previously mentioned, Bragg gratings can be used to optimise laser diodes performances in a wide variety of applications, such as DFB and DBR lasers with single-mode emission, mode-locked lasers, or coherence collapsed lasers. The underlying principle is that gratings written directly in the pigtail or in the laser cavity represent a simple and efficient way to control the wavelength and the other emission parameters. As a matter of fact, it is relatively easy to produce gratings with a precise peak position of the FBG reflection. It is also possible to ensure high thermal stability by controlling the grating temperature or by embedding it into a compensating package.
Although lasers will be analysed in detail in chapter 12, it is important here to remember a fundamental application of gratings to lasers, namely their use in *high power cladding pumped fiber lasers*, shown in figure 9.20 . High power cladding pumped fiber lasers increase the power required, for

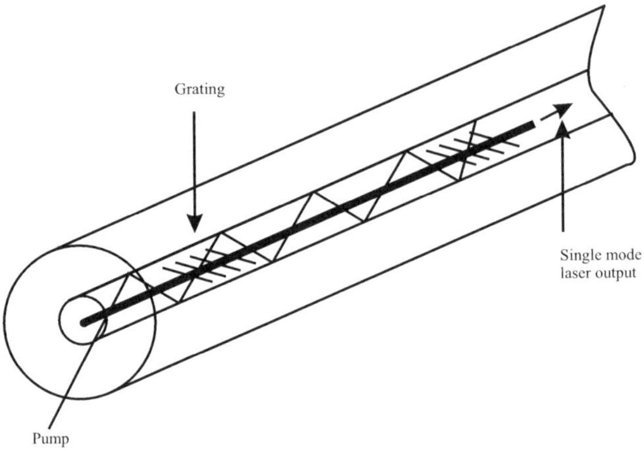

Fig. 9.20: Structure of a high power cladding pumped fiber laser.

instance, to pump an optical amplifier, because they are pumped themselves in a more efficient and gradual way by means of a power laser diode array. A cladding pumped fiber laser is constituted by an erbium-ytterbium doped core which guides the emission, together with an outer core which surrounds the first one and which is able to provide multimode guidance of the pump wavelengths. The outer core is in turn surrounded by a polymer cladding, whose large numerical aperture allows the coupling of pump diodes. Finally, Bragg gratings are inscribed in the cladding, so that the pump power is repeatedly reflected into the active inner core. For the most part, energy is converted to an output laser emission with power of $9\,W$ or more, which means that the laser is more than two orders of magnitude more powerful than a simple diode. Similarly to DBR, FBG can be inscribed also in the inner core as feedback elements in order to improve the frequency stability of the lasers.

FBG Sensors

If the fiber properties are varied in order to modify the refractive index or the period of the inscribed gratings, it follows that the Bragg wavelength of the grating itself changes. By monitoring the reflected spectrum it is possible to obtain sensors which are sensitive to thermal and mechanical stress, pressure sensors, or even sensors with sensitivity to low-frequency acoustic signals. A considerable advantage of FBG sensors is that their response is wavelength-coded, which implies that the spectral variations associated with the monitored quantities do not depend on the incident intensity and are related to each single grating, so that different sensors can be inscribed at different points of the same optical fiber without interference.

FBG sensors can be used to monitor passive optical networks (PON), which are designed to deploy broadband networks with diffused coverage. By means of optical fibers, PON users are connected to routers where different channels are converted into a WDM signal, which is transferred to a central station. Since routers are temperature-sensitive, however, there might be problems of misalignment between WDM channels and router passbands, thus generating, in turn, power loss or crosstalk. An FBG sensor on a sensing channel provides a temperature-dependent feedback, so that the wavelengths of WDM channels can be adjusted according to the drift of the router.

Chapter 10

Optical Modulators

Modulated optical signals are important in communication systems because they are used to encode and carry information. Although it is possible to produce directly modulated lasers, they are usually avoided because they cannot achieve both a high extinction ratio and a high frequency modulation at the same time, since the frequency of the relaxation oscillations sets an upper limit to the frequency modulation. Furthermore, the constant variation of the injection current, which is required to modulate the laser beam, varies the absorption coefficient and the refractive index. Therefore, chirping occurs and causes the spectral instability of the laser output. This effect is particularly relevant in WDM systems, because an interference between the adjacent channels due to a low frequency stability can seriously damage the system. There is another considerable disadvantage of direct modulation. As a matter of fact, in order to cover the distance between two amplifiers, long distance communication systems require high power signals, which can even reach some tens of mW. Therefore, the laser current must reach $100\,mA$ or more, which is difficult to achieve at high bit rates with direct modulation. Consequently, external modulators have been developed, so that the modulation of the light beam occurs outside the laser cavity. Figure 10.1 shows an example of a commercially available modulator. For external modulation, it is necessary to coordinate the interaction between the signal oscillating at optical carrier frequency and the modulating signal, either electrical or acoustic. At the same time, the available bandwidth of the fiber should be fully exploited. Consequently, the quality of modulation depends on either the electro-optic or the electro-acoustic characteristics of the modulators. The most important parameters of measurement can be summarised as follows. The *insertion loss* refers to the loss of power of the optical beam while travelling through the device. This may occur, for example, when the guide and the fiber meet, or due to in-guide absorption, or to tail absorption where the metal electrodes meet. The *polarization dependent loss* indicates the different attenuation of the optical beam according to the input polarization. The *extinction ratio* is the ratio between the maximum and the minimum optical intensity. It describes the modulation depth, i.e. the ability of a device to separate two different output states, the dark and the bright states. The *frequency* of the modulator is, however, the most important parameter for a modulator, because it indicates the maximum speed required to modulate the input signal, which is also the speed at which the communication system operates. The *bandwidth* or *modulation bandwidth*, B, is the switching speed between the dark and the bright states of a modulator and it is related to the switching time, T, through the equation $B = 2\pi/T$. In order to carry the signals at a given frequency, or bit rate, the modulator must show an adequate frequency response, that is, it must have sufficient bandwidth. Since it is assumed that the required bandwidth is $B \geq 3/4$ bit rate, it follows that in order to have a bit rate of $10\, Gb/s$ the bandwidth B must be of $8\, GHz$ at least. Finally, the *driving voltage* indicates the voltage required to set the modulator at its operating point.

This chapter describes how electro-optic modulators work, with particular reference to Mach-Zehnder modulators, acousto-optic modulators, and electro-absorption modulators. Since these products have different performances depending on how they are produced and how they work, the advantages and disadvantages of each will be described in the following sections.

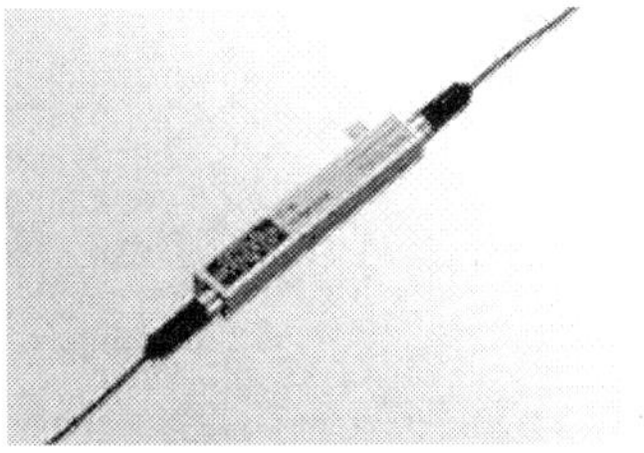

Fig. 10.1: Example of a commercially available modulator

10.1 Electro-optic Modulators

Electro-optic modulators (EOMs) depend on the *Pockels effect*, according to which the refractive index n of some materials changes when they are exposed to low-frequency or radiofrequency electric fields E_{RF}, so that:

$$n = n_0 - \frac{1}{2} n_0^3 r E_{RF}(x, y), \tag{10.1}$$

where n_0 represents the refractive index under no influence of an external electric field and r is the Pockels coefficient of the material considered. The Pockels effect is one of the most effective way to modify the refractive index by means of an external field, and it is thus widely used in both phase and amplitude modulation.

10.1.1 Phase Modulators

Figure 10.2 shows the structure of a phase modulator. According to the equation in (10.1), the electric field E_{RF} generated by the voltage V applied to the electrodes varies the refractive index of the material as follows:

$$\Delta n(x,y) = -\frac{1}{2} n_0^3 r E_{RF}(x,y). \tag{10.2}$$

The electric field $E_{RF}(x,y)$ is related to the voltage V by means of the following equation:

$$E_{RF}(x,y) = \frac{V}{d} e_{RF}(x,y), \tag{10.3}$$

being d the distance between the electrodes and e_{RF} a normalized function, $0 \leq |e_{RF}| \leq 1$, which considers the spatial distribution of the electric field and basically depends on the electrode configurations. The variation of the refractive index also influences the effective index n_{eff} of the waveguide mode propagating along the optical guide. By developing a perturbation theory, it is possible to demonstrate that a Δn perturbation of the refractive index changes the effective index as follows:

$$\Delta n_{eff} = \frac{\int\int \Delta n |E|^2 dS}{2P}, \tag{10.4}$$

where E is the electric field at the optical frequency of the mode of the dielectric guide, and P is the power carried by the field. Therefore, by substituting (10.3) to the equation in (10.2) and then again the latter in (10.4), we obtain:

$$\Delta n_{eff} = -\frac{1}{2} n_0^3 r \frac{V}{d} \Gamma, \tag{10.5}$$

where Γ is the overlap integral between the optical field and the radio-frequency (RF) field:

$$\Gamma = \frac{\int\int e_{RF} |E|^2 dS}{2P}.$$

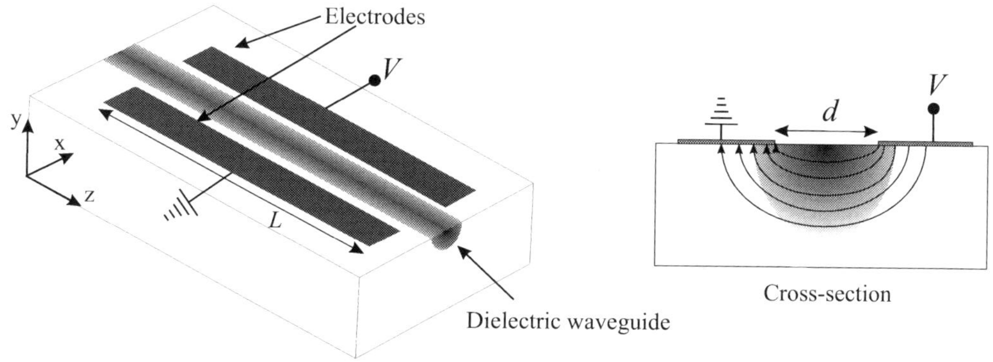

Fig. 10.2: Schematic representation of a phase modulator.

By setting L as the electrode length, the instantaneous phase shift caused by the varying voltage $V(t)$ is:

$$\Delta\phi(t) = \frac{2\pi}{\lambda}\Delta n_{eff}(t)L = -\frac{\pi}{\lambda}n_0^3 r\frac{V(t)}{d}\Gamma L. \tag{10.6}$$

The V_π voltage, also called *half-wave voltage*, is the voltage required in order to obtain a phase shift of π:

$$V_\pi = \frac{d}{L}\frac{\lambda}{n_0^3 r\Gamma}. \tag{10.7}$$

Therefore, by substituting (10.7) in (10.5) and (10.6), we obtain:

$$\Delta n_{eff} = -\frac{\lambda}{2L}\frac{V(t)}{V_\pi}, \tag{10.8}$$

$$\Delta\phi = -\pi\frac{V(t)}{V_\pi}. \tag{10.9}$$

where:

$$\eta = \frac{\Delta\phi}{V(t)} \tag{10.10}$$

is called *modulation efficiency* and represents the phase shift obtained when unit voltage is applied. As it will be shown below, η depends on the frequency of the modulating signal, $V(t)$. By introducing V_π, the modulation efficiency becomes $\eta = \pi/V_\pi$. Furthermore, given the same bandwidth for different modulators, according to the equation in (10.7), the lower the product $V_\pi L$ is, the more efficient the modulator. It follows that it is essential to use materials with high electro-optic coefficients and to improve the design accuracy of the guide structures and the electrodes in order to maximise the overlap between the RF and the optical fields. Consider, for example, a lithium niobate modulator where $n_0 = 2.14$ and $r = 30.8\,pm/V$. Suppose, then, that the distance between the electrodes is $d = 10\,\mu m$ and the overlap is $\Gamma = 0.6$. We thus obtain: $V_\pi L = d\lambda/(n_0^3 r\Gamma) = 85\,V\,mm$. It follows that the electrodes length must be $L = 2.125\,cm$ in order for the driving voltage to be approximately $4\,V$.

10.1.2 Mach-Zehnder Modulators

The electro-optic effect can only act on phase modulation. By introducing a phase modulator into a Mach-Zehnder interferometer, however, the phase modulation is converted into an amplitude modulation. Figure 10.3 shows a schematic representation of a Mach-Zehnder modulator. In order to understand how these devices work, consider a Mach-Zehnder modulator using single-mode guides. If we apply a voltage to the electrodes, thanks to the electro-optic effect the electric fields are able to modify the phase shifts in the two arms, thus changing their interference pattern when the light is recombined at the output of the modulator. If the relative phase shift is a multiple of 2π, the output power is the sum of the powers of the two electric fields and the fundamental mode of the output guide will be excited. On the other hand, if the relative phase shift is an odd multiple of π,

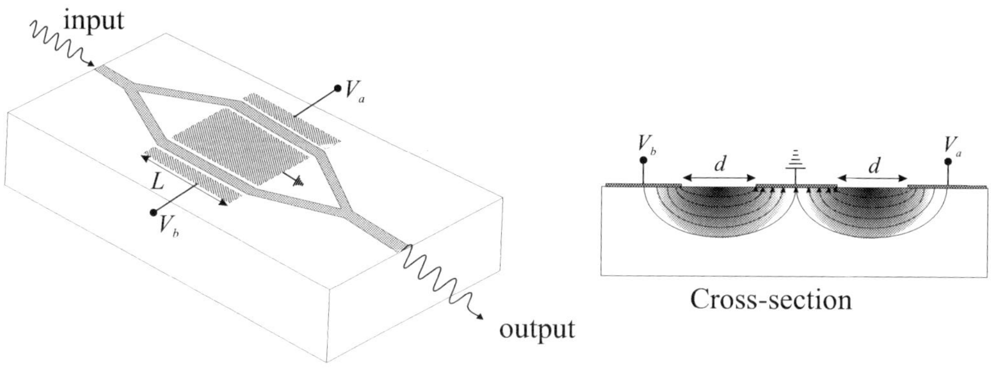

Fig. 10.3: Schematic representation of an amplitude modulator based on a Mach-Zehnder interferometer. V_a and V_b are the voltages applied to the modulator arms.

the contributions will have opposite sign, so that the output field will show odd symmetry, thus in fact exciting only the radiation modes and allowing no power transmission along the guide. Figure 10.4 shows the two cases. Nevertheless, the output power never equals the input one, even when it is given by the sum of the contributions. This is due to the insertion loss and the Y-junction loss which occur when the radiation modes are excited. Although Y-junctions introduce higher loss compared to couplers and MMI, they operate on larger bandwidths. The insertion loss will typically be of about $5\,dB$, whereas the extinction ratio will be of about $20\,dB$.
In Y-junctions, the transfer matrix of the input and output ports is, respectively:

$$[T]_{in} = \frac{1}{\sqrt{2}} \begin{bmatrix} 1 \\ 1 \end{bmatrix}; \qquad [T]_{out} = \frac{1}{\sqrt{2}} \begin{bmatrix} 1 & 1 \end{bmatrix}$$

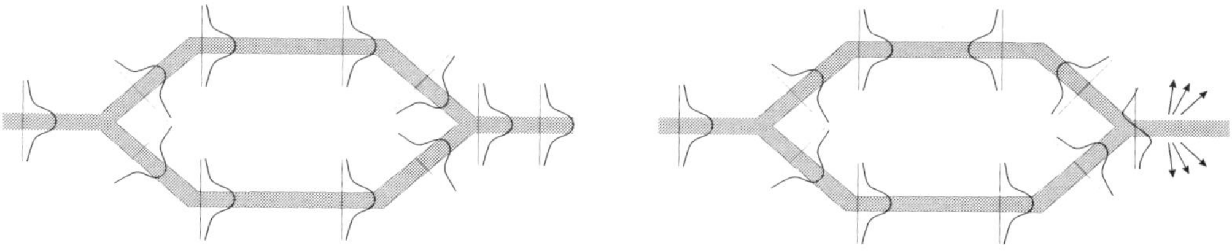

Fig. 10.4: Interferences in the Mach Zehnder modulators. On the left, sum of the contributions. On the right, contributions in opposite phase.

As it was shown in chapter 5, in phase shifters:

$$[T]_{shif} = \begin{bmatrix} e^{-j\frac{2\pi}{\lambda}n_{eff_a}L} & 0 \\ 0 & e^{-j\frac{2\pi}{\lambda}n_{eff_b}L} \end{bmatrix},$$

so that:

$$\begin{aligned} [T]_{mach} &= [T]_{out} \cdot [T]_{shif} \cdot [T]_{in} \\ &= \frac{1}{2}\begin{bmatrix} 1 & 1 \end{bmatrix} \cdot \begin{bmatrix} e^{-j\frac{2\pi}{\lambda}n_{eff_a}L} & 0 \\ 0 & e^{-j\frac{2\pi}{\lambda}n_{eff_b}L} \end{bmatrix} \cdot \begin{bmatrix} 1 \\ 1 \end{bmatrix} \\ &= \cos\left[\frac{\pi}{\lambda}(n_{eff_a} - n_{eff_b})L\right] e^{-j\frac{2\pi}{\lambda}\frac{n_{eff_a}+n_{eff_b}}{2}L}. \end{aligned}$$

According to (10.5) and (10.8):

$$\begin{aligned} n_{eff_a} &= n_{eff_0} - \frac{1}{2}n_0^3 r\frac{V_a}{d}\Gamma = n_{eff_0} - \frac{\lambda}{2L}\frac{V_a}{V_\pi}, \\ n_{eff_b} &= n_{eff_0} - \frac{1}{2}n_0^3 r\frac{V_b}{d}\Gamma = n_{eff_0} - \frac{\lambda}{2L}\frac{V_b}{V_\pi} \end{aligned}$$

so that:

$$[T]_{mach} = \cos\left(\frac{\pi}{2}\frac{V_b - V_a}{V_\pi}\right) e^{j\frac{\pi}{2}\frac{V_b+V_a}{V_\pi}} e^{-j\frac{2\pi}{\lambda}n_{eff_0}L}.$$

Supposing that the optical input power P_{in} of a Mach-Zehnder modulator has a constant value, the output electric field has the following amplitude:

$$A_{out} = \sqrt{P_{in}}\cos\left(\frac{\pi}{2}\frac{V_b - V_a}{V_\pi}\right) e^{j\frac{\pi}{2}\frac{V_b+V_a}{V_\pi}} e^{-j\frac{2\pi}{\lambda}n_{eff_0}L}.$$

Therefore, the output signal is both amplitude- and phase-modulated, with values depending on the applied voltages V_a and V_b. It is only $e^{-j\frac{2\pi}{\lambda}n_{eff_0}L}$ which does not affect the modulation because it is constant.

As for amplitude modulation, the power is:

$$\begin{aligned} P_{out} &= P_{in}\cos^2\left(\frac{\pi}{2}\frac{V_b - V_a}{V_\pi}\right) \\ &= P_{in}\frac{1}{2}\left[1 + \cos\left(\pi\frac{V_b - V_a}{V_\pi}\right)\right] = \frac{1}{2}P_{in}\left[1 + \cos\left(\pi\frac{V_D}{V_\pi}\right)\right] \end{aligned} \quad (10.11)$$

which is shown in figure 10.5 as a function of $V_D = V_b - V_a$. It is thus clear that, given a fixed input voltage, in order to maximise the output power we need to operate at the point of maximum slope on the curve. If we consider noninverting inputs, the slope of the curve must be positive, so that the first operating point is $V_{D_{pol}} = (3/2)V_\pi$. This value corresponds to the bias voltage of the modulator and indicates the operating point on the output power curve. Another value must be added to this, namely peak-to-peak voltage, $V_{D_{pp}}$, whose maximum variation is V_π. Note that if the waveform of the modulating signal $V_D(t)$ is trapezoidal, the output power is a raised-cosine function,

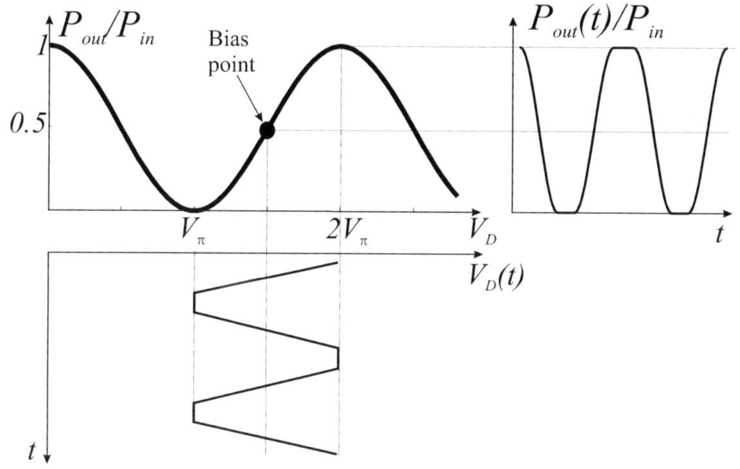

Fig. 10.5: Input-output characteristics of a Mach-Zehnder Modulator.

whose roll-off varies according to the rise time of the voltage ramp. By reducing the voltage shift $V_{D_{pp}}$ around the bias point, we operate in fact on a linear characteristic curve, although the output extinction ratio is reduced.

As for the modulation phase, the instantaneous phase of the field corresponds to:

$$\theta(t) = \frac{\pi}{2} \frac{V_a(t) + V_b(t)}{V_\pi}. \tag{10.12}$$

This kind of modulation, however, should be limited because it is a disadvantage for the amplitude modulation. This is because it increases the signal degradation when the signal travels through a dispersive medium, such as the optical fiber (see chapter 14). The parameter used to measure this phenomenon is called *chirp coefficient*:

$$C_0 = \frac{\dfrac{d\theta}{dt}}{\dfrac{1}{2P}\dfrac{dP}{dt}}.$$

By substituting (10.11) and (10.12), we obtain:

$$C_0 = -\frac{\dfrac{dV_a}{dt} + \dfrac{dV_b}{dt}}{\left[\dfrac{dV_a}{dt} - \dfrac{dV_b}{dt}\right] \tan\left(\dfrac{\pi}{2}\dfrac{V_b - V_a}{V_\pi}\right)}.$$

If we consider the expression for V_D and consider $V_S = V_a + V_b$, it follows that $V_b = (V_S + V_D)/2$ and $V_a = (V_S - V_D)/2$, so that we obtain:

$$C_0 = -\frac{\frac{dV_S}{dt}}{\frac{dV_D}{dt}\tan\left(\frac{\pi}{2}\frac{V_D}{V_\pi}\right)}$$

Together with (10.11) this shows that the amplitude and phase modulations can be controlled by means of V_S and V_D. For example, by applying to the two electrodes a dual drive scheme in which $V_a(t) = -V_b(t)$, it follows that $V_S \equiv 0$ and also that $C_0 = 0$, which means that the signal is not chirped. Furthermore, the output power and the chirp coefficient can be controlled independently through, respectively, the V_D and V_S voltages. Therefore, it follows that the chirp coefficient of the signal can be selected, thus compensating the dispersion along the optical fiber.
Apart from avoiding phase modulation, in analog modulation it is essential that the device keeps a linear regime. Thus, given $V_D = V_{D_{pol}} + \Delta V_D = 3/4\, V_\pi + \Delta V_D$, if we sum the terms the equation in (10.11) becomes:

$$\begin{aligned} P_{out} &= P_{in}\frac{1}{2}\left[1 + \cos\left(\frac{3}{2}\pi + \pi\frac{\Delta V_D}{V_\pi}\right)\right] \\ &= P_{in}\frac{1}{2}\left[1 + \sin\left(\pi\frac{\Delta V_D}{V_\pi}\right)\right]. \end{aligned}$$

If $\Delta V_D << V_\pi$, then $\sin(\pi \Delta V_D/V_\pi) \simeq \pi \Delta V_D/V_\pi$, so that:

$$P_{out} = \frac{P_{in}}{2}\left[1 + \frac{\pi}{V_\pi}\Delta V_D\right] = \frac{P_{ing}}{2}\left[1 + \eta \Delta V_D\right]$$

This equation illustrates the variation of the output power as a function of a ΔV_D with η slope, being η the modulation efficiency of a single arm defined in (10.10).

10.1.3 Travelling-wave Modulators

In modulators, bandwidth is one of the most important characteristics. As a matter of fact, if the time variation of the RF signal is too fast, the modulator is not able to transfer the bandwidth to the optical signal, thus reducing the modulation efficiency. This phenomenon basically depends on the response time of the electro-optic material, i.e. the time needed by the medium in order to adapt to the applied field, and on the parasitic capacitance of the voltage driver. In ferroelectric materials, e.g. lithium niobate, and in a wide series of semiconductors (except for silicon and germanium) the delay related to the electro-optic tensor can be ignored, since the medium has a time response of a few tens of femtoseconds. Nevertheless, in macromolecular materials, such as polymers and liquid crystals, the response varies from a few milliseconds to a few seconds. Therefore, when the modulation frequency is low (lower than $1GHz$), it is the time constant of the driver which determines the maximum operating frequency of the device. As for higher frequencies, the wavelength of the RF

signal can be compared to the length of the electrodes, L, so that the electrodes must be analysed in terms of distributed constants. This means that the electrodes form a transmission line with its own phase velocity v_{pe} and impedance Z_0. When propagating along the interferometer arms, the voltage that is seen by the optical field is no longer constant, but varies along the propagation direction, so that its variation reduces the modulation efficiency. The modulator, therefore, has a limited bandwidth due to the different velocity of propagation between the optical and the RF fields.
This phenomenon can be explained by supposing that the RF and the optical signals enter the modulator from the same end, with the electrodes being connected at the other end with a matched load $Z_L = Z_0$. The phase modulation will thus be between two copropagating waves, that is, between the optical and the RF signals.

Suppose to introduce a sinusoidal RF signal with frequency ν_e into the modulator. The voltage between the electrodes will thus be a wave of amplitude V_0 which propagates at velocity v_{pe}:

$$V(z,t) = V_0 \cos\left[2\pi\nu_e\left(t - \frac{z}{v_{pe}}\right)\right]. \qquad (10.13)$$

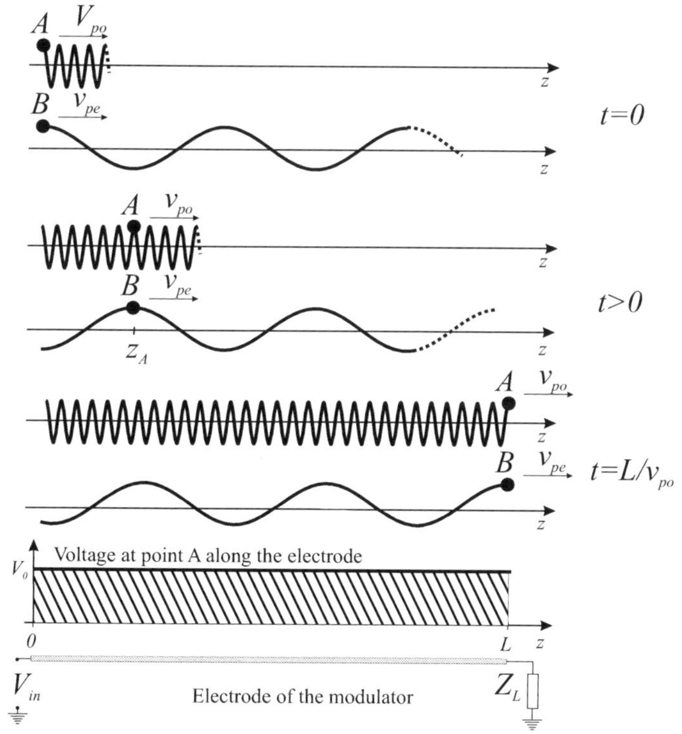

Fig. 10.6: RF and optical waves propagating along an arm of the modulator for $v_{pe} = v_{po}$.

Consider now the equiphase point of the optical signal, which is illustrated as A in figure 10.6. The equiphase point propagates at velocity v_{po} and while propagating undergoes a phase modulation which varies according to the voltage along the electrodes. For $t = 0$, the equiphase point A is at the input side and coincides with the equiphase point B of the RF wave, so that:

$$V(0,0) = V_0.$$

If $v_{pe} = v_{po}$, the propagation of the two points always coincides along the length L of the electrodes. Since the voltage at A is constant at V_0 during propagation, according to (10.9) the equation becomes:

$$\Delta\phi = -\frac{\pi}{L\,V_\pi}\int_0^L V_0\,dz = -\pi\frac{V_0}{V_\pi}.$$

However, the two velocities do not in fact coincide. Therefore, supposing that $v_{po} > v_{pe}$, A propagates at a relative velocity $v_r = v_{po} - v_{pe} > 0$ with respect to B. It follows that at each instant t, A can be found in $z_A = v_{po}t$, whereas B in $z_B = v_{pe}t < z_A$. Figure 10.7 graphically shows this situation. Since $t = z_A/v_{po}$, according to (10.13) the voltage at point A is:

$$V(z_A, \frac{z_A}{v_{po}}) = V_0 \cos\left[2\pi\nu_e\left(\frac{z_A}{v_{po}} - \frac{z_A}{v_{pe}}\right)\right] = V_0\cos\left[2\pi\nu_e\frac{v_{po}-v_{pe}}{v_{po}v_{pe}}z_A\right]. \qquad (10.14)$$

This means that during propagation, at the equiphase point A there is a voltage between the electrodes which varies according to the law in (10.14), so that:

$$\begin{aligned}\Delta\phi &= -\frac{\pi}{LV_\pi}\int_0^L V_0\cos\left[2\pi\nu_e\frac{v_{po}-v_{pe}}{v_{po}v_{pe}}z\right]dz \\ &= -\frac{\pi}{LV_\pi}V_0\frac{v_{po}v_{pe}}{v_{po}-v_{pe}}\frac{1}{2\pi\nu_e}\sin\left[2\pi\nu_e\frac{v_{po}-v_{pe}}{v_{po}v_{pe}}L\right] \\ &= -\frac{\pi}{V_\pi}V_0\,\mathrm{sinc}\left[2\pi\nu_e\frac{v_{po}-v_{pe}}{v_{po}v_{pe}}L\right].\end{aligned}$$

Therefore, according to the equation of modulation efficiency in (10.10), assuming $V(t)$ is at the peak value V_0, we obtain:

$$\eta(\nu_e) = \frac{\pi}{V_\pi}\mathrm{sinc}\left[2\pi\nu_e\frac{v_{po}-v_{pe}}{v_{po}v_{pe}}L\right].$$

The value of its squared magnitude, normalised with respect to the one at zero frequency, represents the normalised power response:

$$H(\nu_e) = \frac{\eta(\nu_e)\eta^*(\nu_e)}{\eta(0)\eta^*(0)} = \mathrm{sinc}^2\left[2\pi\nu_e\frac{v_{po}-v_{pe}}{v_{po}v_{pe}}L\right].$$

The frequency response is thus constant only for $v_{pe} = v_{po}$, i.e. when the waves propagate at the same velocity. In all other cases, the modulation efficiency is reduced for increasing frequency. In particular, the higher the difference between the two velocities, the higher the reduction of modulation efficiency, as shown in figure 10.8. Considering $sinc^2(2\pi/5) \simeq 0.5$, the product between

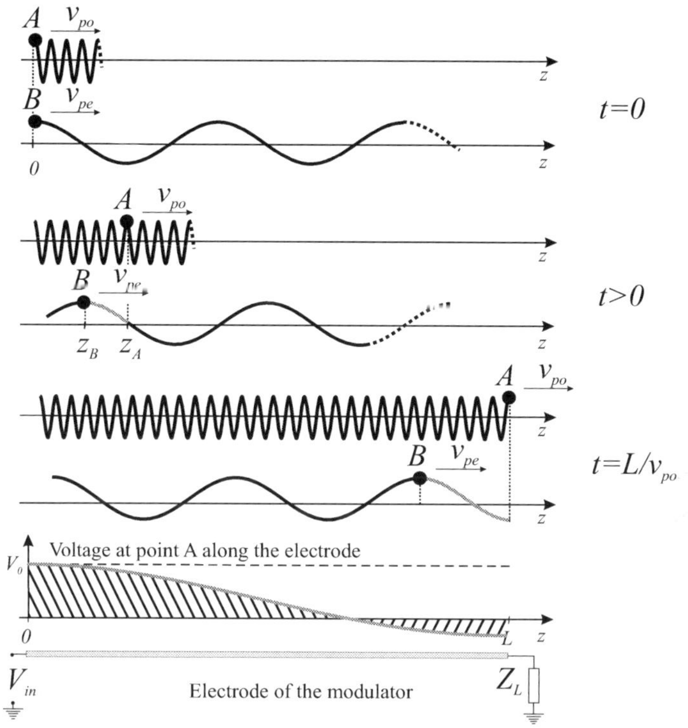

Fig. 10.7: Propagation of RF and optical waves along an arm of the modulator for $v_{pe} < v_{po}$.

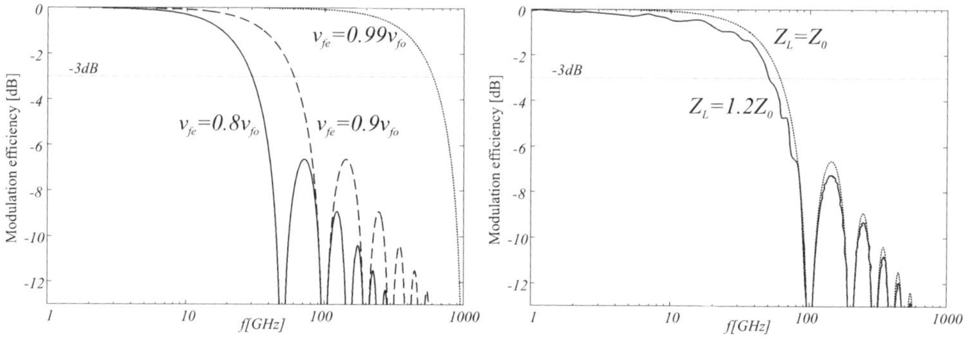

Fig. 10.8: Modulation efficiency at different frequencies. On the left, the modulation efficiency is related to the velocity difference. On the right, to the load mismatch.

the $3-dB$ bandwidth of the B_{3dB} modulator and the length of the electrodes becomes:

$$B_{3dB}L \simeq \frac{v_{po}\, v_{pe}}{5\,(v_{pe}-v_{po})}.$$

Therefore, reducing the length of the electrodes permits to increase the bandwidth, although the driving voltage between electrodes increases too. Consider a single RF wave which is copropagating with the optical signal. If the load to which the electrodes are connected, Z_L, is not perfectly matched, it produces a counterpropagating wave which interferes with the copropagating one, thus generating a standing wave along the electrodes. This, in turns, causes a perturbation of the frequency response (see figure 10.8), which damages the efficiency of the modulator. Figure 10.9 shows the schematic representation of a dual-drive travelling-wave modulator. The electrodes, which are usually built with gold to minimise the loss, form a coplanar transmission line whose geometry is optimised in order to match the velocities of the optical and RF frequencies as much as possible. The lines are terminated on matched loads. It is thus clear that modulators for high bit rates are not produced in the same way of modulators for lower bit rates. In fact, their production requires knowledge and competence about both microwave and optical transmission lines.

10.1.4 Lithium Niobate Modulators

Mach-Zehnder Lithium Niobate ($LiNbO_3$) modulators are nowadays the most widely used in communication systems because they operate over large bandwidths and minimise the effects of dispersion (see section above).

Furthermore, they can be chirped, and their chirp value can be adjusted in order to compensate the dispersion introduced by the transmission system. In fact, thanks to integrated optics and $LiNbO_3$ technology, lithium niobate modulators have become fast, reliable over time, stable for large temperature intervals and operable with low driving voltages. Furthermore, the introduction of titanium diffused lithium niobate technology has overcome the problem of modulators built with proton-exchange waveguides which only guide a single polarization. Devices in titanium diffused

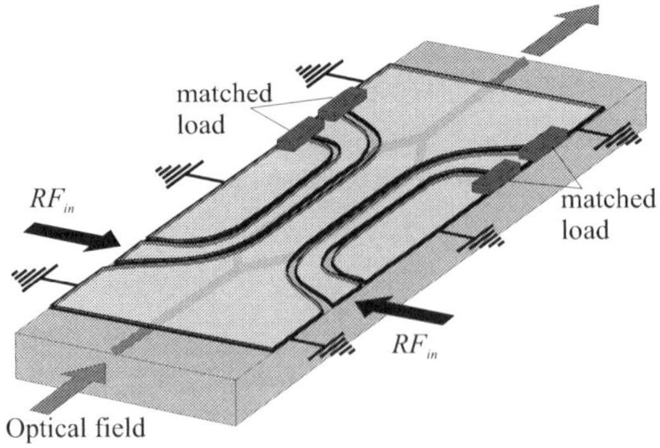

Fig. 10.9: Schematic representation of a dual-drive Mach-Zehnder travelling-wave modulator with two drivers.

lithium niobate, in fact, show a polarization-dependent loss lower than half dB. The package of a Mach-Zehnder interferometer usually includes fiber connectors, two ports for the input electrical signals (one for the bias voltage, which sets the bias point, the other for the modulating voltage). Table 10.1 illustrates the basic characteristics of a Mach-Zehnder modulator. The effect of Mach-

Parameter	Value	Unit of measurement
Package size	$11 \times 0.9 \times 1.3$	cm
Operating wavelength	1525 - 1605	nm
Insertion loss	< 5.5	dB
Extinction ratio	> 20	dB
Return loss	-40	dB
V_π voltage	< 4	V

Tabella 10.1: Technical characteristics of a commercially available $LiNbO_3$ Mach-Zehnder interferometric modulator using Ti-diffused channel waveguides. The fiber used for the optical input is generally a panda-type polarization-maintaining fiber (PMF), whereas single-mode fibers (SMF-28) can be used for the output.

Zehnder modulators on the system performance can be described by means of an eye diagram in which the eye opening slowly deteriorates at increasing modulation rates.

X-cut and *Z-cut* Configurations

In developing lithium niobate modulators it is essential to consider the orientation of the crystal axes with respect to the waveguide and the electrodes, because the crystal cut affects both the modulator efficiency and the chirp introduced by it. In fact, lithium niobate is an anisotropic material with respect to both the refractive index and the electro-optic coefficient. Therefore, if we consider a coordinate system with the z-axis oriented along the optic axis of the crystal, (10.1) becomes:

$$\begin{aligned}
n_x &= n_o - \frac{1}{2} n_o^3 r_{13} E_{RF} \\
n_y &= n_o - \frac{1}{2} n_o^3 r_{13} E_{RF} \\
n_z &= n_e - \frac{1}{2} n_e^3 r_{33} E_{RF}
\end{aligned}$$

where n_e and n_o refer to the extraordinary and ordinary refractive indexes, whereas $r_{33} = 30.0\,pm/V$ and $r_{13} = 8.6\,pm/V$ are the Pockels electro-optic coefficients along the optic axis and in its corresponding orthogonal plane. Due to anisotropy, $LiNbO_3$ response to the E_{RF} field depends on the electrodes orientation. Figure 10.10 shows the four most common configurations for Mach-Zehnder modulators. The external electric field of the waveguide must be z-oriented because $LiNbO_3$ shows

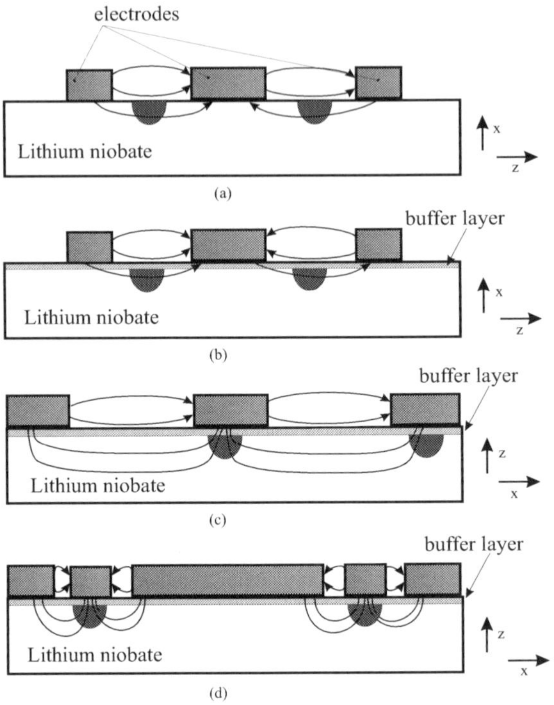

Fig. 10.10: Schematic representation of the four most common electrodes configurations for a) *nonbuffered x-cut*, b) *buffered x-cut*, c) *buffered single-drive z-cut*, and d) *buffered dual-drive z-cut*

a higher electro-optic coefficient along this axis. It follows that the waveguide must be placed between the electrodes for an $x-cut$ configuration, and below them for a $z-cut$ configuration. Since the electrodes are always placed over the waveguides, the $z-cut$ devices require a buffer layer in order to minimise the attenuation of the optical mode due to metal absorption. Nevertheless, the buffer layer reduces the value of the E_{RF} field at the waveguide. The $x-cut$ devices do not require buffer layers because the electrodes are not placed over the waveguides. They can use buffer layers, however, in order to reach rates as high as a few GHz. In fact, buffer layers are able to adapt the propagation velocity of the RF and the optical waves.
Furthermore, the $x-cut$ configuration shows a chirp-free modulation due to the symmetry of fields applied to the electrodes. In figure 10.10(c) the central waveguide shows a good overlap between the RF field and the optical field, so that the value of V_π is lower than for the $x-cut$ configuration. Nevertheless, the asymmetry of the waveguides with respect to the RF signal determines a non-zero chirp, which can be eliminated by means of a dual drive configuration. The $x-cut$ devices without a buffer layer show the best efficiency per unit length, since the $V_\pi L$ product is 30% lower than the

$z - cut$ devices shown in figure 10.10(c). Unfortunately, however, the velocity mismatching between the RF and the optical waves limits their bandwidth.

10.2 Electro-Absorption Modulators

Electro-Absorption Modulators (EAMs) are able to vary the absorption coefficient in semiconductors through the driving voltage, exploiting the so called *Franz-Keldysh* effect. As it is known, the

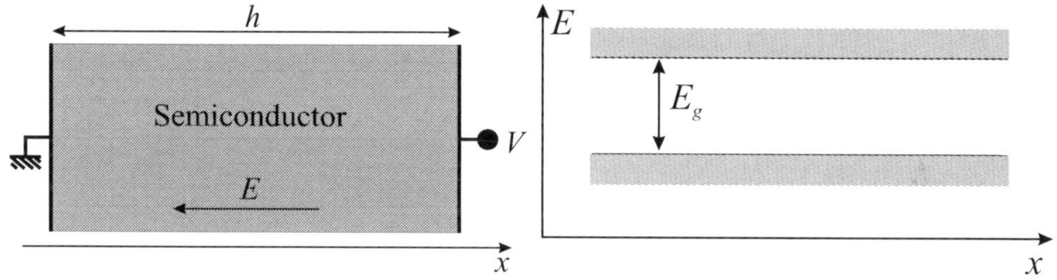

Fig. 10.11: Semiconductor and band structure for $V = 0$.

attenuation of the electromagnetic field propagating in a semiconductor highly depends on the possibility of photons to excite an electron transition from the valence band to the conduction band. Consider the case represented in figure 10.11. If the driving voltage is zero, the energy levels of the energy bands are constant along x. In this case, the electron transition is allowed only if the energy

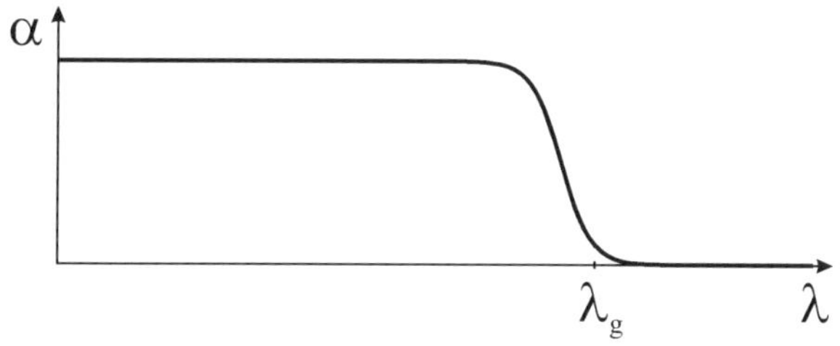

Fig. 10.12: Absorption spectrum of a semiconductor in the absence of an external electric field.

of the photon, $E_{ph} = h\nu = hc/\lambda$, is greater than the energy gap, E_g, so that $E_{ph} \geq E_g$. Therefore, by making the wavelength explicit, we obtain:

$$\lambda \leq \lambda_g,$$

where $\lambda_g = hc/E_g$. The absorption spectrum of an α material as a function of λ is thus represented as shown in figure 10.12.

If we apply a voltage $V \neq 0$ to the semiconductor, however, the electric field $\overline{\mathcal{E}} = V/h$ generates a linear band-bending of the type shown in figure 10.13. In such a situation, an electron of the valence band which absorbs a photon of E_{ph} energy lower than E_g, has to overcome a further energy gap $\Delta E = E_g - E_{ph}$ in order to reach the conduction band. However, the band-bending progressively reduces the residual gap to zero while moving towards increasing values of x. The electron, therefore, is faced with a potential barrier of the type shown in 10.13, and it has a non-zero probability of crossing the barrier (tunnelling process). The probability is inversely proportional to the barrier height ΔE and width $\Delta x = \Delta E / q\overline{\mathcal{E}}$. To sum up, when a voltage is applied to a semiconductor, the photons may be absorbed with less energy than the band-gap. The absorption coefficient behaves as shown in figure 10.14. Given a wavelength λ_0 slightly longer than λ_g, it is possible to increase or decrease α by controlling the voltage V. In order to maximise the variation of the absorption coefficient while limiting the variation of applied voltage, it is essential that the curve shows a steep slope $\alpha(\lambda)$. Electroabsorption modulators, consequently, only exploit direct band gap semiconductors. In this kind of modulators, it is important to consider that it is possible to fabricate a semiconductor waveguide directly on the laser sublayer, or that the light source and the external modulator can be integrated into one chip whose length is typically of a few hundred microns.

A simple scheme of an electroabsorption modulator consists of an inversely polarised PIN junction placed along the path of the signal which must be modulated with the junction interface parallel to the direction of the optical beam, as shown in figure 10.15. When the field crosses the depletion region of the junction while the voltage is varying, there is a modulation of the signal amplitude. In particular, if no voltage is applied, the wave does not change, whereas it is absorbed if the junction is biased. Consider α_1 and α_0 as the absorption coefficients in absence and presence of a bias voltage, respectively. For the desired wavelength λ_0, they must be:

$$\alpha_1 << \frac{1}{d}, \qquad \alpha_0 >> \frac{1}{d},$$

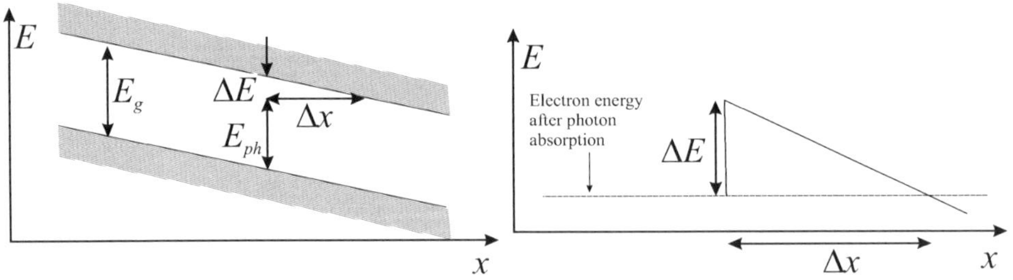

Fig. 10.13: On the left: band structure for $V \neq 0$; on the right: potential barrier.

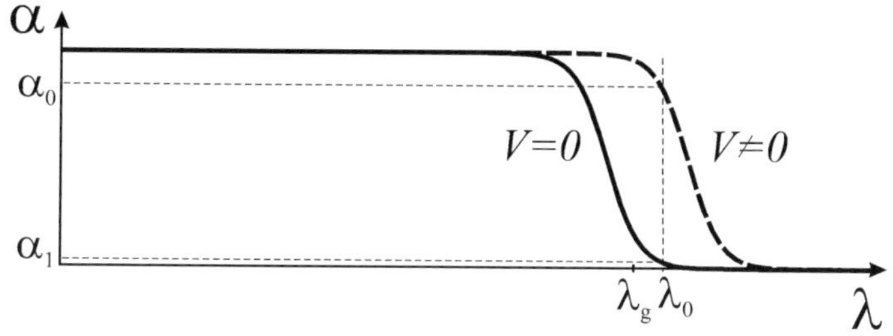

Fig. 10.14: Absorption spectrum of a semiconductor in the absence of an external electric field.

so that:

$$P_{out_0} = P_{in g} e^{-\alpha_0 d} << P_{ing}, \qquad P_{out_1} = P_{ing} e^{-\alpha_1 d} \simeq P_{ing}.$$

The Extinction Ratio (ER) is expressend in dB and corresponds to:

$$ER = 10 \log \frac{P_{out_1}}{P_{out_0}}.$$

By substituting the equations above we thus obtain:

$$ER = 10 \log \frac{e^{-\alpha_1 d}}{e^{-\alpha_0 d}} = (\alpha_{1_{dB}} - \alpha_{0_{dB}})\, d$$

with $\alpha_{dB} = 10 \log(e)\alpha = 4.343\alpha$.

The first electroabsorption modulators worked at wavelengths of about $10\,\mu m$, and were then substituted by devices working in second window with modulation rates of tens of MHz. Nowadays, electroabsorption modulators work in C-band $(1530 - 1565\,nm)$, having both excellent optical and electric performance.

Table 10.2 shows the basic parameters of a modulator for an OOK-NRZ optical transmission system. The driving voltage is an asset for this kind of modulators, because it is considerably lower

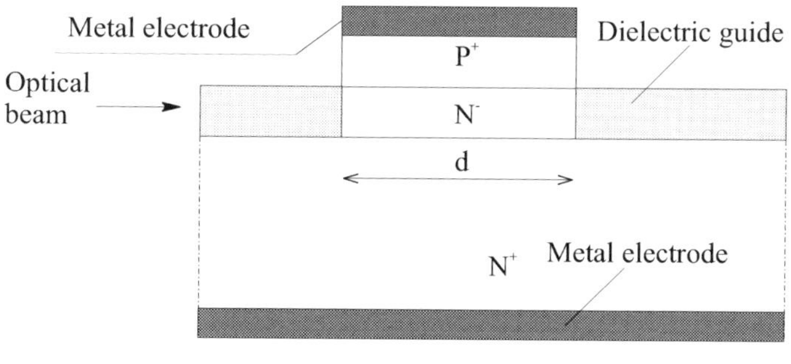

Fig. 10.15: Principle of operation of an Electroabsorption modulator (EAM) .

Parameter	Value	Unit of measurement
Insertion loss	9	dB
Polarization-dependent loss	0.5	dB
Extinction ratio	20	dB
Modulation rate	40	Gb/s
Electric bandwidth (at -3 dB)	30	GHz
Operating temperature	from 0 to 70	$^{\circ}C$
Driving voltage	from 1 to 3	V
Optical input power (peak value)	13	dBm

Tabella 10.2: Basic parameters of an electroabsorption modulator

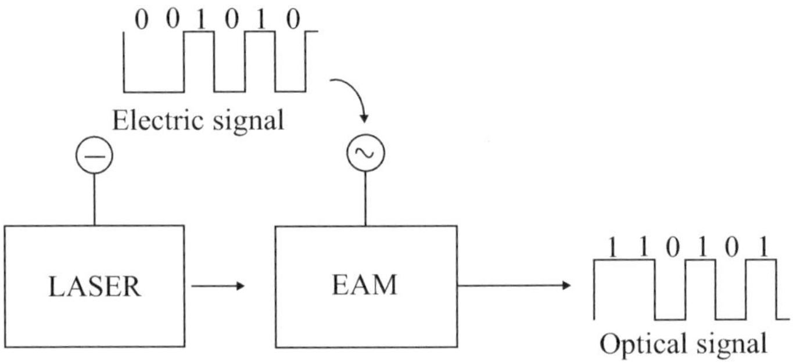

Fig. 10.16: EAM in a transmission system, at the output of the laser source.

than that required, for instance, by Mach-Zehnder modulators. The main disadvantage, however, is the signal chirp. Furthermore, in semiconductors the variation of the absorption coefficient also implies the variation of the refractive index, which in turn determines a phase modulation of the signal. Contrary to what happens in Mach-Zehnder modulators, however, the chirp coefficient cannot be controlled by means of the other parameters and can hardly be suppressed using the available fabrication techniques.

10.2.1 Saturable Absorbers (SA)

Modulators can be used to modulate a laser output or they can work as demultiplexers, as shown in figures 10.16 and 10.17. More specifically, figure 10.16 shows a modulator in a standard transmission setup, placed at the output of the laser source, where it is able to modulate the electric signal on the optical carrier. Figure 10.17, on the other hand, shows a modulator driven by a periodic electric clock signal, which is used to produce electric time-division-multiplexed (ETDM) demultiplexers.

Similarly to the modulator shown above, these devices use electric signals which open time windows in order to transfer selected bits, which can be either high level or low level bits. All the signals outside the windows are absorbed.
So far, two basic applications of electroabsorption modulators have been described. Nevertheless, electroabsorption modulators can also be used as Saturable Absorbers (SA), that is, devices in which all the electrons from the valence band are excited by a strong optical signal, so that all possible transitions to the conduction band are saturated. Consequently, the weaker input signals cannot be attenuated because their photons cannot be absorbed, so that they encounter no obstacle and they can be transferred as if the material were transparent. This effect can be exploited for demultiplexing in a SA with optical control. Figure 10.18, for example, shows the case in which the signal varies but the polarization of the junctions is kept constant. The optical clock signal is absorbed, thus opening time windows through which the signal can be transferred.
Furthermore, optical control also allows wavelength conversion, as shown in figure 10.19. The information of the input signal is impressed on a continuous-wave optical signal at a different wavelength. The continuous-wave signal must be weak. The advantage of this type of converters is that the signal polarity is preserved, so that the data are not inverted, as it occurs with semiconductor optical amplifiers.

10.3 Acousto-optic Modulators

Optical properties of materials may depend on external fields different from the electric one. The *photoelastic*, or *elasto-optic effect*, for example, describes a change of the tensor of dielectric permittivity, which implies a change of the optical properties of the medium, due to the introduction of a pressure field. The present discussion will focus on the specific case of the interaction between acoustic waves and a light beam entering the medium. This effect is generally referred to as the *acousto-optic effect*. The term refers to the possibility of varying the direction of an incident optical wave by means of an acoustic wave generated by an electro-acoustic transducer, usually a piezoelec-

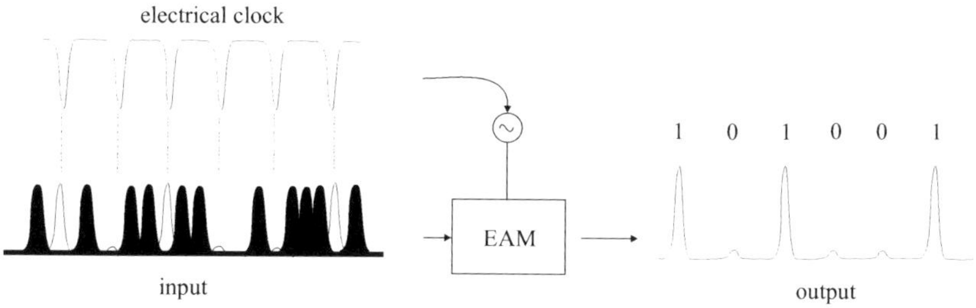

Fig. 10.17: EAM-based electrically controlled demultiplexer.

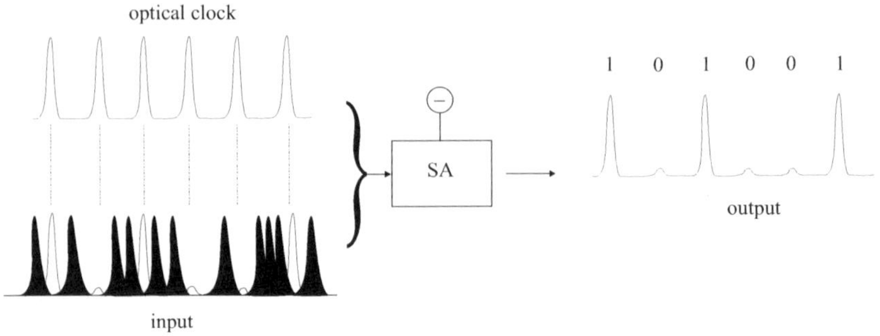

Fig. 10.18: SA-based optically controlled demultiplexer.

tric transducer. The phenomenon can be qualitatively explained by considering an acoustic wave applied to an optical medium sensitive to elastic effects. The acoustic waves generate a series of compressions and rarefactions which are caused by the vibration of molecules deviating from their equilibrium position. This, in turn, varies the density and the refractive index of the material, as shown in figure 10.20, so that the incident light beam crosses layers with different refractive indexes. The perturbation caused by the acoustic wave is a function of both time and space. Nevertheless, since the speed of sound in the medium is lower than the speed of light by around 5 orders of magnitude, the incident light beam is affected by a stationary perturbation and by optical phenomena similar to those caused by diffraction gratings. For the sake of simplicity, consider an acoustic plane wave, $s(x,t)$, propagating in an isotropic crystal in the x direction. Λ is the wavelength, $v_s = \Omega\Lambda/2\pi$ the velocity, and Ω the angular frequency. By taking S_0 as the amplitude of the wave, the strain at position x and time t is:

$$s(x,t) = S_0 \cos\left(\Omega t - \frac{2\pi}{\Lambda}x\right). \tag{10.15}$$

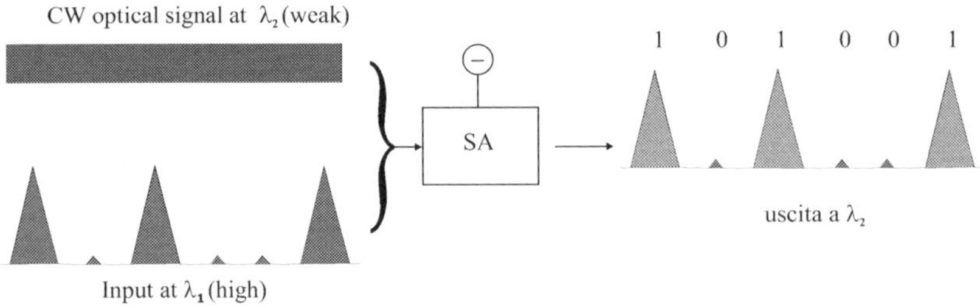

Fig. 10.19: SA for wavelength conversion.

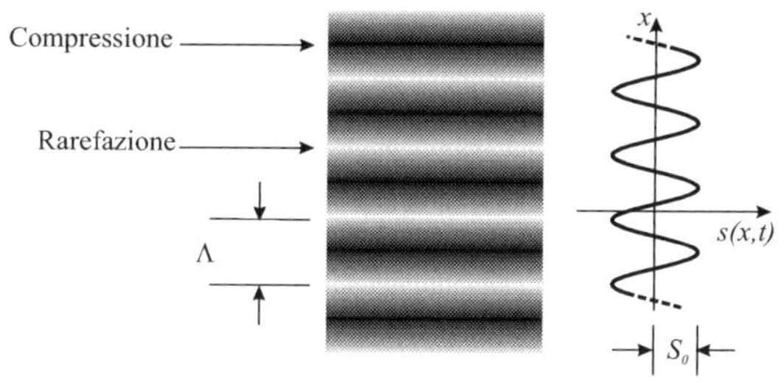

Fig. 10.20: Change in the refractive index of the medium as a function of the deformation caused by an $s(x,t)$ acoustic wave.

If we consider an optically transparent medium characterised by a refractive index n in absence of sound, the strain $s\,(x,t)$ creates a perturbation Δn of the refractive index which corresponds to:

$$\Delta n\,(x,t) = -\frac{1}{2} p n^3 s\,(x,t)\,, \tag{10.16}$$

where p is the photoelastic constant. Figure 10.20 illustrates this situation. The minus sign in (10.16) indicates that positive strain, i.e. dilatation, reduces the refractive index. Consequently, the medium has a time-varying and inhomogeneous refractive index in the form of a wave: $n\,(x,t) = n - \Delta n_0 \cos\left(\Omega t - \frac{2\pi}{\Lambda} x\right)$. The amplitude of the wave is:

$$\Delta n_0 = \frac{1}{2} p n^3 S_0. \tag{10.17}$$

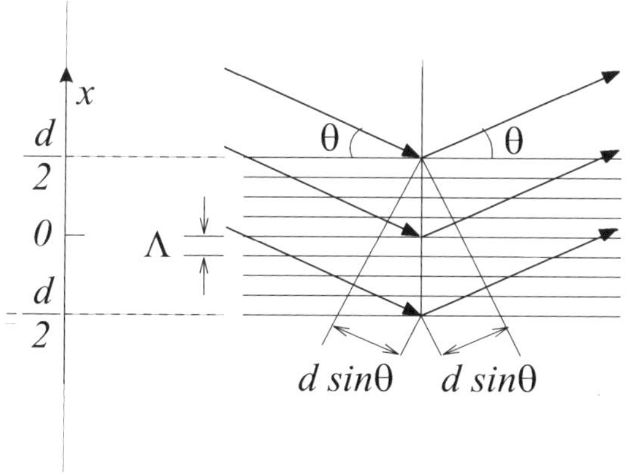

Fig. 10.21: Interaction between optical radiation and an inhomogeneous medium on which an acoustic wave has been applied.

Parameter	Value	Unit of measurement
Spectral Range λ	600 - 1600	nm
Rise Time	1.4	ns
Acoustic Velocity	6.31×10^3	m/sec
Modulation Bandwidth	400	MHz
Max. Power	630	mW
Input Impedance	50	Ω

Tabella 10.3: Technical parameters of a TeO_2 (Tellurium Dioxide) acousto-optic modulator. The modulator is horizontally polarised, and the polarization is parallel to the sound direction.

The intensity of the sound wave, I_s, corresponds to:

$$I_s = \frac{1}{2}\varrho v_s^3 S_0^2, \tag{10.18}$$

where ϱ represents the density of the medium. By substituting I_s inside the expression in (10.17) we thus find that the maximum variation of the refractive index is proportional to the square root of the acoustic intensity:

$$\Delta n_0 = \sqrt{\frac{1}{2}\mathcal{M} I_s}, \tag{10.19}$$

where

$$\mathcal{M} = \frac{p^2 n^6}{\varrho v_s^3}, \tag{10.20}$$

is a material parameter representing the effectiveness of sound in altering the refractive index. $\mathcal{M}$ represents a figure of merit for the strength of the acousto-optic effect of the material.
Consider an incident plane wave in the acousto-optic crystal, as shown in figure 10.21. We divide the medium into different layers along the x direction. The thickness of each layer equals the period of the acoustic wave Λ, so that light is partially reflected and partially diffracted by the layers.

As it is known from the theory of diffraction gratings, the diffracted beam reaches its maximum intensity at $\theta^{(m)}$ angle, which in case of normal incidence corresponds to:

$$\sin\theta^{(m)} = m\frac{\lambda}{\Lambda}, \tag{10.21}$$

where m is an integer and λ corresponds to the optical wavelength of the radiation within the acousto-optic medium. The angle $\theta^{(m)}$ is the angle at which the layers which are divided by a period of the acoustic wave contribute to diffraction with a phase shift of 2π, which means that they interfere constructively. The intensity of the diffracted light on the output beam depends on the amplitude of the perturbation on the medium caused by the acoustic wave, which, in turn, ultimately depends on the electric signal. By modulating the acoustic intensity it is possible to turn the diffracted light beam on and off, that is, to modulate its intensity, as shown in figure 10.22. In

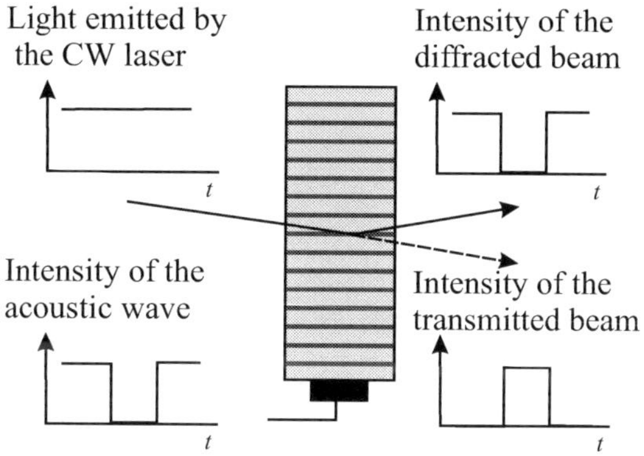

Fig. 10.22: Schematic representation of an acousto-optic modulator.

terms of bandwidth, the major disadvantages of this kind of modulators depend on the time required by the acoustic wave to travel through the part of the crystal crossed by the light wave. Generally speaking, these modulators have poorer performances than electro-optic devices also in terms of the quality of signals at the same velocity and bit rate. Table 10.3 shows some typical technical parameters of acousto-optic modulators.

Chapter 11

Optical Amplifiers

Although optical signals can propagate over long distances without amplification, both wide area and local area networks benefit from optical amplifiers.

Optical amplifiers widely differ from optoelectronic devices because they work without converting optical signals into electric signals. It follows that optical amplifiers are transparent to modulation format. Furthermore, since they are based on stimulated emission, they do not depend on the shape of the optical signal or on its lifetime. Optoelectronic devices, however, cannot always be replaced by optical amplifiers, especially in long distance transmission. Optoelectronic devices, for instance, are necessary to compensate for nonlinear distortion and noise caused, respectively, by the fibers and the amplifiers themselves. They do so because, apart from amplifying the signal, they allow its re-shaping and re-timing, thus being commonly referred to as 3R regenerators (Re-amplifying, Re-shaping, Re-timing). Since with modern technologies 3R regeneration achieves high-quality and cost-effective results only in the electrical domain, it is thus necessary to use optoelectronic devices. As for optical amplification, modern communication systems usually exploit three types of amplifiers: Semiconductor Optical Amplifiers (SOAs); Doped Fiber Amplifiers (DFAs) or, less often, Doped Waveguide Amplifiers (DWAs); and Raman Amplifiers (RAs), which are made of fiber, too.

11.1 Characteristics of Optical Amplifiers

The fundamental parameters of optical amplifiers are: gain, gain bandwidth, gain saturation, polarization sensitivity, and noise.

The gain G represents the ratio of the output power P_{out} of a signal to its input power P_{in}:

$$G = \frac{P_{out}}{P_{in}},$$

which expressed in dB becomes:

$$G_{dB} = P_{dB_{out}} - P_{dB_{in}} = 10\log(G).$$

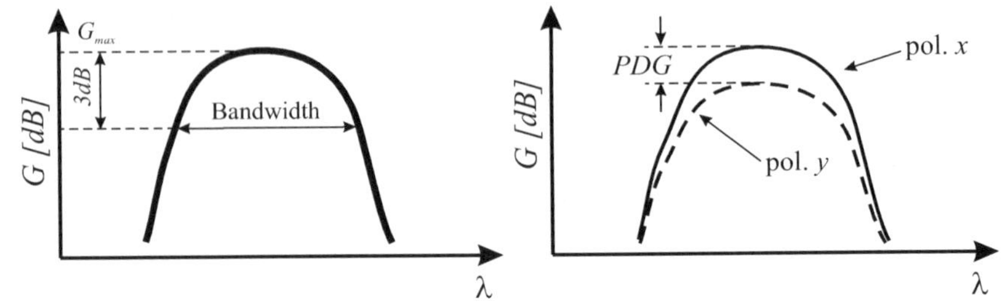

Fig. 11.1: Left: gain spectrum of an optical amplifier. Right: gain variation as a function of the polarization of the input signal.

Figure 11.1 shows the gain spectrum of an optical amplifier. The gain bandwidth indicates the range of frequencies or wavelengths in which the gain is no less than $3dB$ below the peak value. Since the range of frequencies is much wider than the bandwidth of a single signal, once the spacing between channels has been selected, the gain bandwidth limits the number of channels which can be amplified rather than the transmission speed.

Polarization sensitivity describes the dependence of the gain on the polarization of the signal, as shown in figure 11.1 (right). It is usually referred to as Polarization Depending Gain (PDG), it is measured in dB and represents the gain difference between the two polarizations.

Saturation refers to the dependence of the gain on the power of the signal, and it is shown in figure 11.2. As the input power increases, the gain decreases until it reaches the saturation point, at which point the output power does not vary with an increase of the input power. The saturation power is typically defined as the input power required in order for the gain to drop by $3\,dB$ from the small-signal gain G_0. Beyond the saturation point, the gain falls abruptly and tends towards unity, for which $P_{out} = P_{in}$.

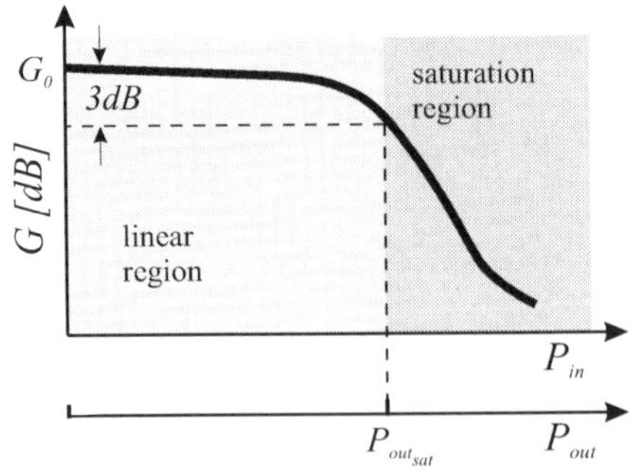

Fig. 11.2: Saturation in an optical amplifier.

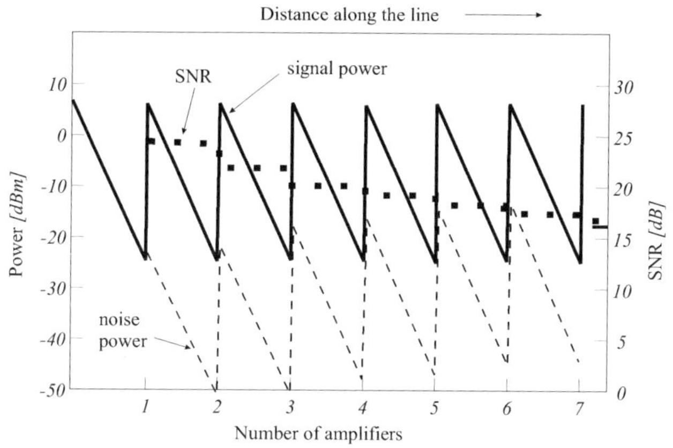

Fig. 11.3: Signal power and signal-to-noise ratio in a multistage system with seven optical amplifiers.

In optical amplifiers, the main source of noise is the *amplified spontaneous emission*(ASE), which originates from the spontaneous emission in the active region of the amplifier. The amount of noise depends on various factors, such as the probability of spontaneous emission, the amplifier gain spectrum and the population inversion parameter, which measures the degree of population inversion between two energy levels.

In optical amplifiers, the parameter relative to noise is the Noise Figure (NF), which corresponds to:

$$NF = \frac{(OSNR)_{in}}{(OSNR)_{out}}$$

where $(OSNR)_{in}$ and $(OSNR)_{out}$ are, respectively, the input and output optical signal-to-noise ratios. The NF is a number greater than 1 which indicates the degradation of the signal-to-noise ratio caused by the amplifier and it is an essential element to determine the maximum distance between two optoelectronic regenerators. Figure 11.3 illustrates the behaviour of the signal power, the noise power, and the signal-to-power ratio in a multistage system with seven optical amplifiers. At each stage, both the signal power and the noise power decrease at a constant rate, which is determined by the attenuation coefficient and which keeps the signal-to-noise ratio constant. After each amplifier, the added noise reduces the $OSNR$. No more amplifiers can be added when the $OSNR$ reaches the minimum value for which the BER is kept below a given value specified for the system. In this case it is in fact necessary to add optoelectronic regenerators.

Appendix A summarises the physical principles and the related mathematical models which are at the basis of optical amplification and are essential to understand how optical amplifiers work.

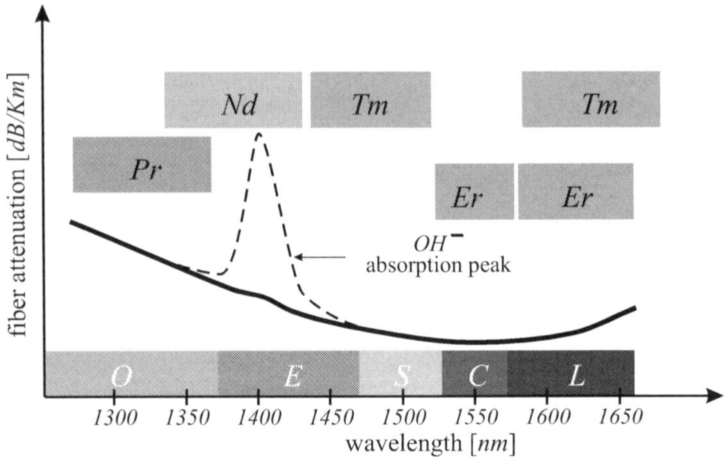

Fig. 11.4: Spectral attenuation of an optical fiber. Bandwidths of the most common doped-fiber amplifiers.

11.2 Doped-fiber Amplifiers

Doped-fiber amplifiers use as an active medium an optical fiber in which the core is typically doped with a rare-earth element. The energy levels of the dopants also depend on the composition of the host glass, because its local electric field modifies them. The most common doped-fiber amplifiers are made of erbium-doped silica glasses, because of their gain in both the C- $(1530 - 1565\,nm)$ and the L-band $(1565 - 1625\,nm)$. As for the S-band $(1460 - 1530\,nm)$, it includes amplifiers made of thulium-doped tellurite or fluoride glasses. Finally, in the $1300\,nm$ band amplifiers are usually based on praseodymium-doped fluoride fibers. Figure 11.4 shows the spectral attenuation of a standard silica fiber and of a silica fiber without OH^- ions, responsible for the absorption peak at 1390 nm. Different types of dopants and the gain bandwidths of the corresponding fiber-doped amplifiers are illustrated.

Figure 11.5 shows an example of doped fiber. The fiber core is doped only in part in order to reach a good overlap of the pump beam with the signal beam and the doped core. The pump beam travels through a WDM coupler. The way in which the pump laser injects light into the doped fiber is called *pumping scheme*. Figure 11.6 shows the most common pumping schemes. When the pump and the signal beams propagate along the same direction, the pumping scheme is *co-propagating*, whereas it is *counter-propagating* when they travel in opposite direction. The *bi-directional* pumping scheme, finally, requires both a copropagating and a counterpropagating pump.

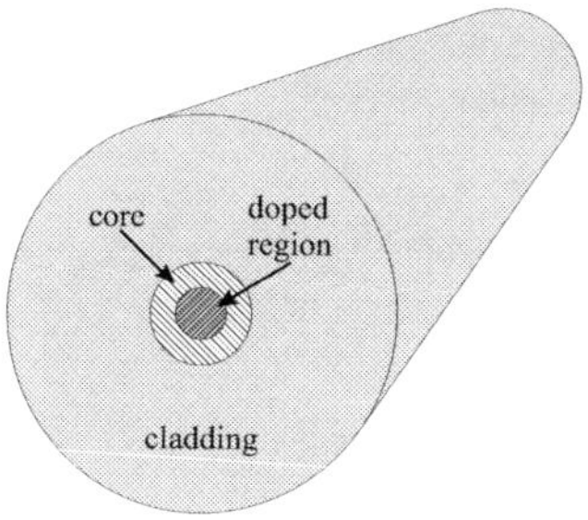

Fig. 11.5: Doped fiber cross section.

11.3 EDFA

Erbium Doped Fiber Amplifiers (EDFA) provide gains in the band around 1550 nm, i.e. where silica fiber have the minimum attenuation. Erbium shows excellent spectroscopic properties concerning both its lifetime and energy levels. Consequently, EDFAs achieve low noise figures and high pump conversion efficiency. EDFAs are essential components of modern optical communication systems because they have low costs of production and at the same time they have introduced a significant

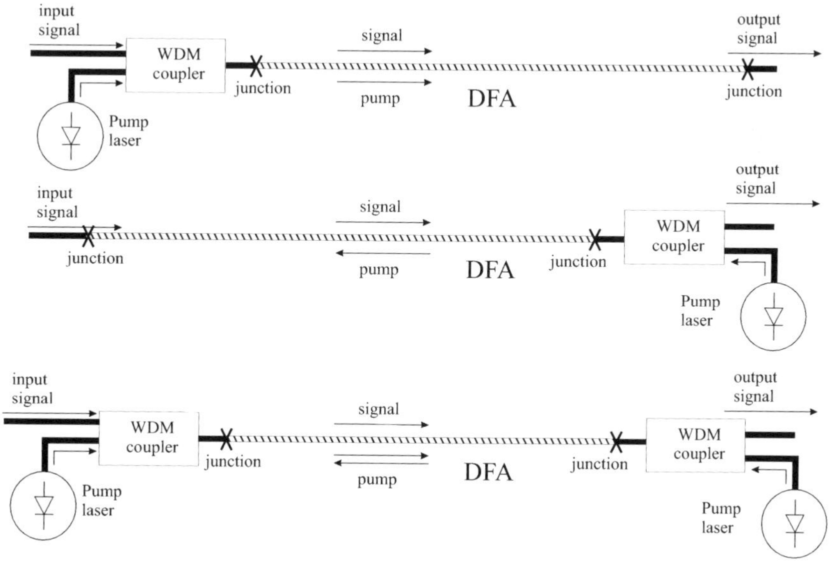

Fig. 11.6: Schematic diagram of a DFA. Copropagating pumping scheme (above); counterpropagating pumping scheme (centre), bi-directional pumping scheme (below).

increase in the transmission capacity of optical systems. Research on EDFAs started in 1987 but it was only five years later that they became commercially available thanks to the growing demand for network services.

Figure 11.7 illustrates the basic components of an EDFA, which consists of an erbium-doped fiber, a pump laser, an electronic control unit and a WDM (Wavelength Division Multiplexer) coupler which combines the signal and the pump power. An erbium-doped fiber contains a fiber core doped with a few hundred parts per million of erbium, so that the final erbium ion density is about $10^{25}\, ions/m^3$. The control unit typically consists of a current source for the pump laser and a temperature controller, which is used to avoid damaging the pump laser and in order to control the emission wavelength, which depends on temperature.

Other components of EDFAs are optical insulators and tap couplers. Optical insulators are non-reciprocal components based on the Faraday effect. Since they show low optical loss in one direction of transmission and high optical loss for signals travelling in the opposite direction, they are used to suppress the reflected waves and the counter-propagating ASE. Tap couplers are used to monitor the amplifier gain or its output power, so that the pump power can be adjusted to keep the amplifier gain constant.

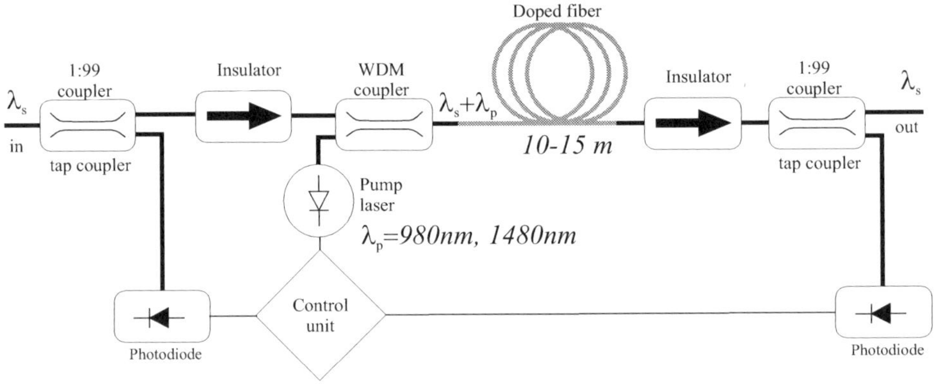

Fig. 11.7: Basic components of an EDFA.

11.3.1 Erbium cross-sections and pump wavelengths

In glass matrices (e.g. silica), erbium forms ionic bonds, thus taking the form of the trivalent ion Er^{3+}. Figure 11.8 shows the energy levels of Er^{3+}. Optical amplification occurs when the pump power is enough to generate a population inversion of the erbium ions from the ground state, ${}^4I_{15/2}$,

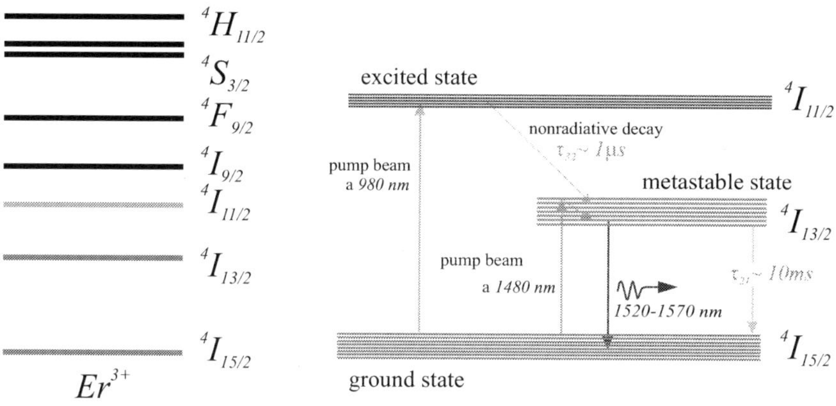

Fig. 11.8: (a) Energy levels of Er^{3+}. (b) Energy levels used in EDFA. Each level is split into sublevels due to the Stark effect.

to the metastable state [1], $^4I_{13/2}$. [2] In communication systems, EDFAs are pumped by 980 nm semiconductor laser diodes of the InGaAs type. Otherwise, they are pumped by 1480 nm InGaAsP laser diodes. 980 nm pumping causes the transition from the $^4I_{15/2}$ state to the $^4I_{11/2}$ state and then the immediate non-radiative decay from the $^4I_{11/2}$ state to the $^4I_{13/2}$ state. On the other hand, 1480 nm pumping causes direct transition from the $^4I_{15/2}$ state to the $^4I_{13/2}$. Theoretically, it would also be possible to use 800 nm or 665 nm laser diodes, but their pumping efficiency is degraded by their excited state absorption (ESA) from the $^4I_{13/2}$ state to higher states.

The degeneracy of energy levels which would occur with isolated Er^{3+} ions is removed in fibers due to the crystal field surrounding the ions which are doped into the glass matrix. This phenomenon, called the Stark effect, is exploited for wide bandwidth amplification because it allows 56 different transitions between the 8 sublevels of the ground state and the 7 sublevels of the metastable state. Figure 11.9 shows the absorption and emission cross-sections of Er^{3+} as functions of the wavelength. In each level, the population distribution follows the Boltzmann distribution and the transition to equilibrium occurs much faster than the 10 ms lifetime of a metastable state.

[1] The metastable state is a particular excited state of an atom or an ion which has a longer lifetime than ordinary excited states but which usually has a shorter lifetime than the ground state. It can thus be considered as a temporarily stable state.

[2] The rules for the nomenclature of the energy levels are the following: the angular momentum, L, is expressed by a letter, following the spectroscopic notation $L = 0, 1, 2, 3, 4, 5, 6 \rightarrow X = S, P, D, F, G, H, I$; the superscript on the left $2S + 1$ indicates the spin, S; the total angular momentum, J, is indicated by a subscript on the right of the letter. Therefore, we finally obtain $^{2S+1}L_J$.

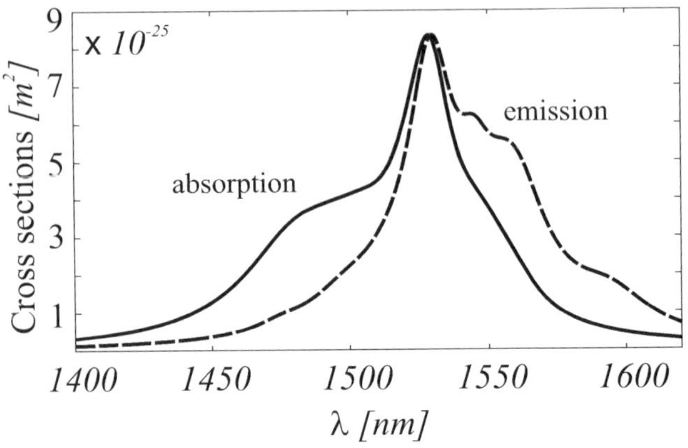

Fig. 11.9: Absorption and emission cross-sections of the transition from the $^4I_{13/2}$ to the $^4I_{15/2}$ state in an Er^{3+} doped silica glass.

11.3.2 EDFA Equations

The gain and noise performances of EDFAs can be described by means of a set of coupled differential equations containing population and propagation equations. Since at $980\,nm$ and $1480\,nm$ pump wavelength there is no ESA and the $10\,\mu s$ lifetime of the $^4I_{11/2}$ state is much lower than that of the $^4I_{13/2}$ metastable state, the population of the $^4I_{11/2}$ state can be ignored, and all the ions can be considered in the $^4I_{13/2}$ or $^4I_{15/2}$ state. For the sake of simplicity, therefore, the following discussion will consider EDFAs as a simple two-level system.

EDFAs properties can thus be studied under steady state conditions, because the lifetime of the metastable state ($\sim 10\,ms$) is much higher than the pulse width of high-speed digital signals ($< 1\,ns$ for systems in which the bit rate is more than $1Gbit/s$) or of analog signals ($< 100\,ns$). In fact, although the amplifier has a response time which can be more than 100 times faster than the lifetime of a metastable state, we can use steady state equations to calculate the population of the relevant energy levels by considering the average powers calculated on intervals of a few ms.

In order to determine the differential equations which form the EDFA model we need to consider a cylindrical coordinate system (r, ϕ, z) with the z-axis as the fiber axis. $N_T(r, \phi)$ is the erbium ions concentration in the transverse field, generally speaking a point function. At first, we ignore the ASE and we consider a single photon beam with photons of frequency $\nu_s = c/\lambda_s$ and intensity I_s. The propagation equation of I_s, as it was studied in (A.16), is:

$$\frac{dI_s(z)}{dz} = [\sigma_{es} N_2 - \sigma_{as} N_1]\, I_s(z), \tag{11.1}$$

where $\sigma_{as} = \sigma_a(\nu_s)$, $\sigma_{es} = \sigma_e(\nu_s)$ and I_p is the intensity of the pump beam with $\nu_p = c/\lambda_p$ frequency. The intensity I_s of the transverse field corresponds to the magnitude of the Poynting vector of the fundamental mode of the fiber at wavelength λ_s. The optical power of the beam travelling for a length z along the fiber will thus be:

$$P_s(z) = \int_0^{2\pi} \int_0^{\infty} I_s(r, \phi, z) r \, dr \, d\phi. \tag{11.2}$$

Since the functions of the modes do not depend on z, the intensity can be written as:

$$I_s(r, \phi, z) = P_s(z) \, i_s(r, \phi),$$

where i_s is the *normalised optical intensity*, corresponding to:

$$\int_0^{2\pi} \int_0^{\infty} i_s(r, \phi) r \, dr \, d\phi = 1. \tag{11.3}$$

The power propagation equation is thus obtained by integrating the equation in (11.1) over the transverse coordinate:

$$\begin{aligned} \frac{dP_s(z)}{dz} &= \sigma_{es} P_s(z) \int_0^{2\pi} \int_0^{\infty} i_s(r, \phi) N_2(r, \phi, z) \, r \, dr \, d\phi \\ &\quad - \sigma_{as} P_s(z) \int_0^{2\pi} \int_0^{\infty} i_s(r, \phi) N_1(r, \phi, z) \, r \, dr \, d\phi. \end{aligned} \tag{11.4}$$

The same process can be followed for the propagation equations of the pump and ASE powers, thus obtaining equations similar to (11.4). If then we associate these equations to those indicating the populations N_1 and N_2 as functions of the beam powers, we are able to calculate the amplifier parameters. Before doing so, however, some aspects need to be further clarified.

Figure 11.10 relates the pump and signal normalised intensity profiles to the size of the doped region. The erbium ions are uniformly distributed in a disk of radius b concentric to the core radius a. This allows simplification of the integrals in (11.4). Furthermore, since the variations of the normalised intensities i_s and i_p in the doped region are very low, the population inversion is generally uniform along the doped fiber. Consequently, N_1 and N_2 only depend on z and can be taken out of the integrals in (11.4). If we ignore the background loss, we thus have:

$$\begin{aligned} \frac{dP_s(z)}{dz} = \Bigg[&\sigma_{es} N_2(z) \int_0^{2\pi} \int_0^{b} i_s(r, \phi) \, r \, dr \, d\phi \\ &- \sigma_{as} N_1(z) \int_0^{2\pi} \int_0^{b} i_s(r, \phi) \, r \, dr \, d\phi \Bigg] P_s(z). \end{aligned} \tag{11.5}$$

The overlap integral, Γ_s, between the dopant and the mode at ν_s frequency is:

$$\Gamma_s = \int_0^{2\pi} \int_0^{b} i_s(r, \phi) \, r \, dr \, d\phi; \tag{11.6}$$

Γ_s is usually between 0.4 and 0.7. The equation in (11.5) can finally be written as:

$$\frac{dP_s(z)}{dz} = \left[\sigma_{es} N_2(z) - \sigma_{as} N_1(z)\right] \Gamma_s P_s(z). \tag{11.7}$$

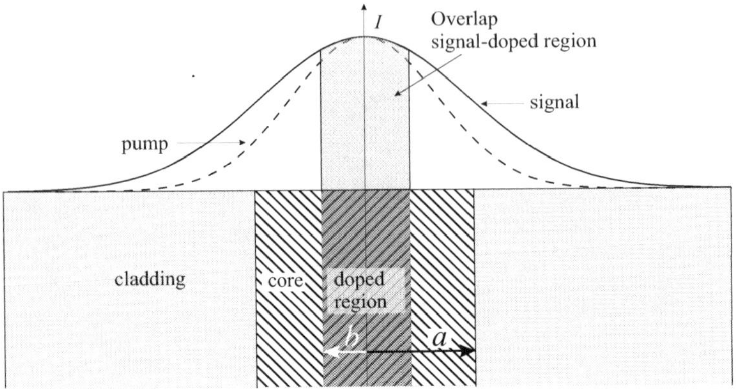

Fig. 11.10: Intensity profile of the fundamental mode at signal and pump wavelengths. The grey area indicates the overlap between the signal beam and the doped region of radius b.

It is now useful to introduce two further parameters, the absorption coefficient α_s and the gain coefficient g_s:

$$\begin{aligned} \alpha_s &= \Gamma_s \sigma_{as} N_T; \\ g_s &= \Gamma_s \sigma_{es} N_T. \end{aligned}$$

The two parameters are functions of the wavelength, their dimension is inverse length (m^{-1}), and they describe the spectral characteristics of a doped fiber. They can be calculated directly, or after finding the overlap integral, the absorption and emission cross sections, and the doping concentration. The equation in (11.7) thus becomes:

$$\frac{dP_s(z)}{dz} = \left(g_s \frac{N_2(z)}{N_T} - \alpha_s \frac{N_1(z)}{N_T} \right) P_s(z). \tag{11.8}$$

In WDM systems, the input to the EDFA comprises M signal beams, each with a different frequency ν_{s_k} and a different power P_{s_k} with $k = 1, ..., M$. We will thus obtain M equations of the type given in (11.8):

$$\frac{dP_{s_k}(z)}{dz} = \left(g_{s_k} \frac{N_2(z)}{N_T} - \alpha_{s_k} \frac{N_1(z)}{N_T} \right) P_{s_k}(z) \quad \text{con} \quad k = 1, ..., M. \tag{11.9}$$

Starting from the equation in (A.16) we can similarly obtain the pump propagation equation, considering that the stimulated emission cross-section σ_{31} is negligible in erbium. Therefore:

$$\frac{dP_p^{\pm}(z)}{dz} = \mp \alpha_p \frac{N_1(z)}{N_T} P_p^{\pm}(z), \tag{11.10}$$

where the superscript "+" refers to the copropagating pump beam, the superscript "-" to the counterpropagating pump beam, and $\alpha_p = \Gamma_p \sigma_{ap} N_T$, where:

$$\Gamma_p = \int_0^{2\pi} \int_0^b i_p(r, \phi)\, r\, dr\, d\phi,$$

is the overlap integral of the pump defined according to its normalised intensity $i_p = I_p/P_p$. Figure 11.10 shows that the fundamental mode is mainly confined at the pump wavelength, which is lower than the signal wavelength, so that $\Gamma_p > \Gamma_s$.

Finding the ASE equation is a bit more complicated. To start with, ASE is not described by an intensity, but by a power spectral density of intensity $p(\nu)$, expressed in $[W/(Hz\ m^2)]$. Secondly, it is no longer possible to ignore the spontaneous emission, because without the spontaneous emission the ASE would not exist.

We can overcome the first matter by dividing the spectrum into smaller frequency intervals $\Delta\nu$, each centred at a different frequency ν_j. We then substitute the spectral density with a sequence of intensities I_j of the type:

$$I_j = \int_{\nu_j - \Delta\nu/2}^{\nu_j + \Delta\nu/2} p(\nu)\, d\nu,$$

each centred in ν_j. The situation is thus similar to several signal beams propagating at the same time. The only difference is that for each j there is both a copropagating and a counterpropagating ASE beam. As for the second aspect, spontaneous emissions, it will not be treated in the present discussion because it would require the equations of quantum electrodynamics. We will thus only include the final result, which consists in adding to the propagation equations a term proportional to $m\, h\nu_j \Delta\nu_j$, where m indicates the number of modes which can propagate through the guide. In a single-mode fiber $m = 2$ because its fundamental mode shows two polarizations. The ASE will thus be:

$$\frac{dP_j^{\pm}(z)}{dz} = \pm \left(g_j \frac{N_2(z)}{N_T} - \alpha_j \frac{N_1(z)}{N_T} \right) P_j(z) \pm g_j \frac{N_2(z)}{N_T} 2h\nu_j \Delta\nu_j. \qquad (11.11)$$

The equations in (11.9), (11.10) and (11.11) must be associated with the boundary conditions for signal power, pump power and ASE power. The powers of the co-propagating beams are assigned at $z = 0$, whereas the powers of the counterpropagating beams at $z = L$, where L is the length of the doped fiber. Notice that the copropagating ASE is zero at $z = 0$, while the counterpropagating ASE is zero at $z = L$.

In order to complete the EDFA model it is necessary to formulate the equations describing the populations N_1, N_2 and N_3 as functions of the power of the signal, pump and ASE beams propagating through the fiber. These equations are known as *population equations.* At the beginning of this section, we observed that the lifetime of an excited state is so different from the lifetime of a metastable state that the excited state is empty, which means that $N_3 = 0$. The population density

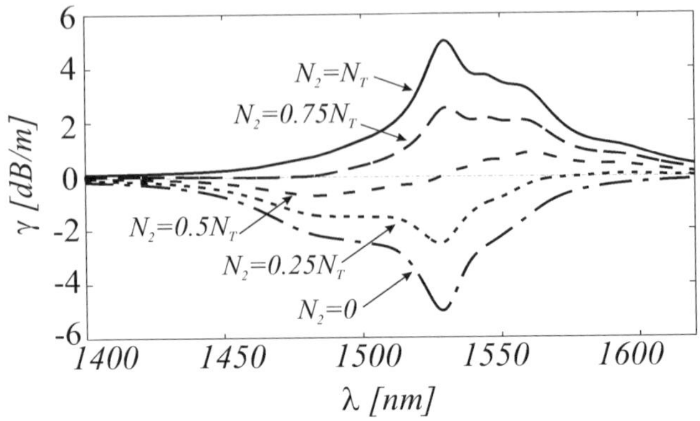

Fig. 11.11: Spectral variation of gain per unit length for different excited state population densities.

of the metastable state, which has been found in (A.18), is:

$$N_2(r, \phi, z) = N_T(r, \phi) \frac{\tau \left(\sigma_{as} \frac{I_s}{h\nu_s} + \sigma_p \frac{I_p}{h\nu_p}\right)}{1 + \tau \left[(\sigma_{as} + \sigma_{es}) \frac{I_s}{h\nu_s} + \sigma_p \frac{I_p}{h\nu_p}\right]}, \tag{11.12}$$

where $\tau = 1/A_{21}$ is the lifetime of the excited state and corresponds to the inverse of the spontaneous emission rate. Finally, the population in the ground state can be found through the ion conservation equation:

$$N_1(r, \phi, z) = N_T(r, \phi) - N_2(r, \phi, z).$$

By means of integration (11.12) over the transverse plane and thanks to the approximations introduced for propagation equations we finally obtain:

$$\frac{N_2(z)}{N_T} = \frac{\sum_k \tau \alpha_k \frac{P_k(z)}{h\nu_k \, \pi \, b^2 \, N_T}}{1 + \sum_k \tau(\alpha_k + g_k) \frac{P_k(z)}{h\nu_k \, \pi \, b^2 \, N_T}} \tag{11.13}$$

where the summations include all the co- and counterpropagating signal, pump, and ASE beams. The final model of EDFAs is thus obtained by considering the propagation equations in (11.9), (11.10) and (11.11), together with the corresponding boundary conditions and the population equations in (11.13). Apart from some exceptions, these equations cannot be solved analytically, so that it is necessary to solve them numerically.

11.3.3 Gain and Optimum Length

Consider (11.9) in the ideal cases of $N_2 = 0$ and $N_2 = N_T$ along the whole fiber. In the first case, all the Er^{3+} are in their ground state, $N_1 = N_T$, so that neither stimulated emission nor spontaneous emission occur. In particular, if we set $N_2 = 0$ in (11.9) and consider only the copropagating signals, we obtain:

$$\frac{dP_{s_k}(z)}{dz} = -\,\alpha_{s_k} P_{s_k}(z). \tag{11.14}$$

By integrating this differential equation from $z = 0$ to $z = L$, we obtain:

$$P_{s_k}(z = L) = P_{s_k}(z = 0)\, e^{-\alpha_{s_k} L}. \tag{11.15}$$

In this case, there is an exponential attenuation of the signals with α_{s_k} absorption coefficient. Therefore, if no population inversion occurs (e.g. when the pump is turned off) the doped fiber is strongly absorbent.

On the other hand, in case of complete population inversion, i.e. for $N_2 = N_T$ along the whole fiber, the equation in (11.9) becomes:

$$\frac{dP_{s_k}(z)}{dz} = +g_{s_k} P_{s_k}(z). \tag{11.16}$$

so that:

$$P_{s_k}(z = L) = P_{s_k}(z = 0)\, e^{g_{s_k} L}. \tag{11.17}$$

By definition, g_{s_k} is always positive, so that we obtain an exponential gain with coefficient g_{s_k}. We can calculate the value of g_{s_k} by using the equation in (11.8). Considering a $1550\,nm$ wavelength, a doping concentration of $N_T = 3 \cdot 10^{24}$ ions$/m^3$, an overlap $\Gamma_s = 0.5$, and $\sigma_s = 6 \cdot 10^{-25}\,m^{-2}$, we obtain $g_{s_k} = \Gamma_s \sigma_s N_T = 0.9\,m^{-1}$. Therefore, the gain in dB is:

$$g_{dB} = 4.343 g_{s_k} = 3.9\,dB/m.$$

This implies that in order to obtain gains around $20 - 30\,dB$, as required for optical systems, the length of the doped fibers must be of a maximum of a few tens of meters.

The particular cases so far described never occur in EDFAs. In fact, the *gain coefficient* $\gamma(\lambda)$ varies along the fiber due to the pump absorption and the variation of the signal power. $\gamma(\lambda)$ is thus:

$$\gamma(\lambda) = g(\lambda)\frac{N_2(z)}{N_T} - \alpha(\lambda)\frac{N_1(z)}{N_T}.$$

Figure 11.11 shows the spectral variation of $\gamma(\lambda)$ for different population density levels. The gain coefficient is wavelength dependent mainly due to the spectral variation of the emission and absorption cross-sections. The gain is positive, which means that the power is an increasing function of z, only when the metastable state is sufficiently populated. The population density, however, varies considerably along the fiber due to the effect of gradual pump absorption. In EFDAs, therefore, fibers may be strongly absorbent or they can be active media. Consequently, it is possible to identify an optimum length of the doped fiber in order to maximise the gain.

In order to understand, consider the case of a single signal, a copropagating pump and no ASE. The typical signal power evolution is shown in figure 11.12. On the input side of the amplifier, the high pump power allows an almost total population inversion, so that the signal power increases

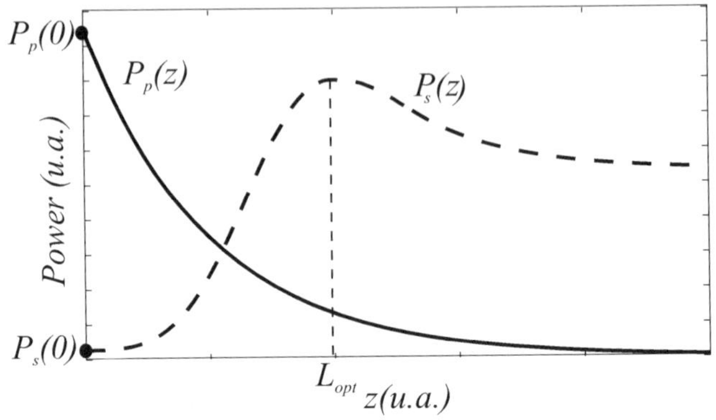

Fig. 11.12: Pump power and signal power along an EDFA.

exponentially. As the power transfers from the pump to the signal, P_p decreases, while P_s increases, thus determining a gradual reduction of the population inversion and, consequently, of P_s growth rate. It follows that there is a section of the amplifier where $g(\lambda)N_2(z)/N_T - \alpha(\lambda)N_1(z)/N_T = 0$. Beyond this section, the absence of population inversion makes the material absorbent for the signal wavelengths. The distance between the input port and the fiber section at which the population inversion ceases is called *optimum length*, L_{opt}, of the amplifier. If $L = L_{opt}$ and the other parameters are constant, the amplifier gain:

$$G = \frac{P_s(L)}{P_s(0)}$$

is maximised. Both the optimum length and G increase if the input pump power $P_p(0)$ increases. The higher $P_p(0)$ is, the longer the length of fiber in which the pump is able to create a population inversion.

According to the approximations so far described, L_{opt} does not depend on L. The situation changes, however, if we consider the ASE. In fact, if the fiber length increases, the counterpropagating ASE at the input part of the amplifier increases the reduction of the population inversion. L_{opt} slightly decreases if L increases. L_{opt} thus requires an iterative method to be calculated.

The optimum length depends on the pumping scheme. According to (A.20), the pump level required for population inversion depends on the signal level. Considering all the other parameters constant, the counterpropagating scheme has a higher optimum length than the copropagating scheme. This is because in the counterpropagating scheme both the pump and the signal power are low on the input side of the doped fiber. In the copropagating scheme, conversely, the pump power is low on the output side of the fiber, where the signal power is at its maximum. In the counterpropagating scheme, therefore, the gain is higher because it is possible to exploit a longer part of the fiber. It is thus clear that both the optimum length and the gain are maximised in a dual pump scheme where P_P is high along the whole fiber. These devices, however, are also more expensive.

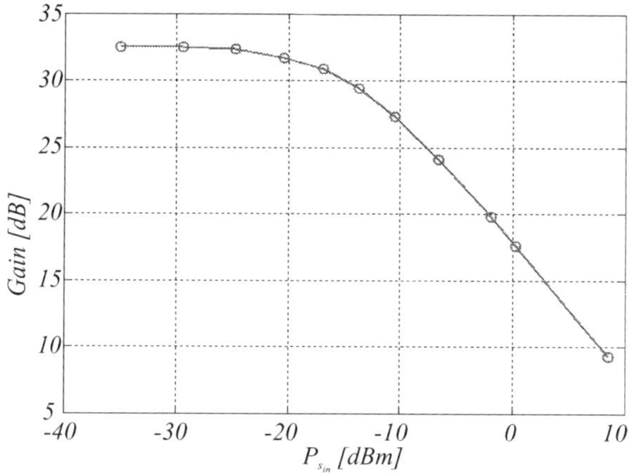

Fig. 11.13: EDFA gain saturation.

The fact that the minimum P_p required for population inversion depends on P_s causes another important effect known as gain saturation. According to (A.20), L_{opt} remains constant for low values of P_s, thus basically depending on the spontaneous emission. When P_s increases, however, both L_{opt} and the gain tend to decrease. Figure 11.13 shows the gain as a function of P_s.

11.3.4 Gain Equalisation

Since the ratio between the cross sections varies according to λ, some wavelengths are more efficient than others. Figure 11.11 shows that the maximum gain is around $1530\,nm$, that is, where the difference between the emission and absorption cross sections is at its peak. Consequently, there are significant gain variations which can have a negative impact on the performance of WDM systems. There are several approaches for gain equalisation in fiber amplifiers. To start with, we can use notch filters centred at the gain peak or filters based on long-period Bragg gratings. The filters are inserted in the middle stage of EDFAs and their loss is matched to the reciprocal of the gain variation. Therefore, as it is shown in figure 11.14, the EDFA gain spectrum is flattened. Gain flattening, however, causes a gain reduction, too.

11.3.5 EDFA Noise

All optical amplifiers generate an unavoidable noise resulting from spontaneous emission. The noise at the output of the amplifier is given by the amplified spontaneous emission (ASE). ASE is a broadband noise and it can be considered a white noise on the signal bandwidth for all the wavelengths within the band of the EDFA. Its power spectral density is expressed in W/Hz and

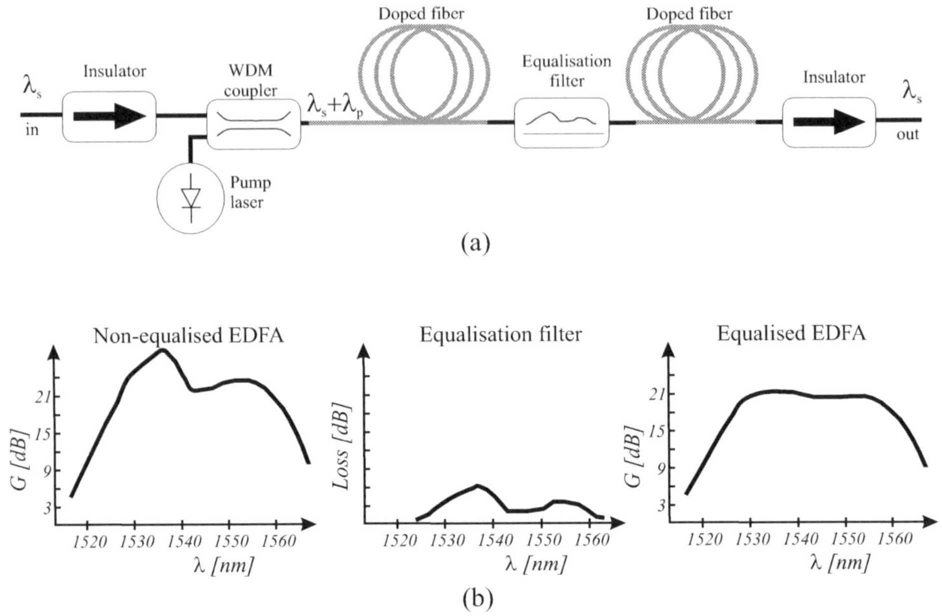

Fig. 11.14: (a) Schematic diagram of an EDFA gain equalisation filter. (b) EDFA gain spectrum. Left: Non-equalised gain spectrum. Center: Loss spectrum of a gain equalisation filter. Right: Equalised gain spectrum.

corresponds to:

$$p_{ASE} = 2n_{sp}h\nu_s(G-1) \qquad [W/Hz]$$

where n_{sp} is the *population inversion factor*, which approximately corresponds to:

$$n_{sp} = \frac{\overline{N}_2}{\overline{N}_2 - \overline{N}_1} = \frac{\overline{N}_2}{2\overline{N}_2 - \overline{N}_T}$$

where $\overline{N}_i$ is the average population of level i over the whole fiber length:

$$\overline{N}_i = \frac{1}{L}\int_0^L N_i(z)dz.$$

The equation for n_{sp} is physically possible only for $N_2 > N_1$, that is, for $N_2 > N_T/2$. Therefore, it is possible only in a medium with a gain greater than unity. Figure 11.15 shows the variation of the population inversion factor. Its minimum value of 1 coincides with total population inversion, whereas it tends toward infinity as it approaches the limit condition $N_2 = N_1 = N_T/2$.
The noise figure, NF, corresponds to:

$$NF = \frac{1}{G} + 2n_{sp}\frac{G-1}{G}.$$

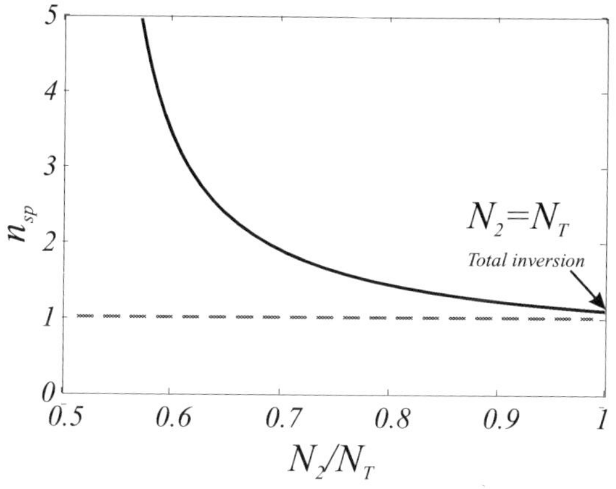

Fig. 11.15: Variation of the population inversion factor, n_{sp}, for different values of population density of the metastable state N_2.

Thus, in an ideal EDFA, in which $G >> 1$, it is:

$$NF \simeq 2n_{sp}.$$

According to the equations so far described, the higher the inversion population along the amplifier, the lower the noise. In the ideal case of total population inversion over the whole doped fiber length, we thus find the minimum noise figure, usually referred to as quantum limit, which is:

$$NF = 2 = 3dB.$$

A low noise figure, however, has a negative impact on the amplifier efficiency. A medium-high level of population inversion can only be obtained if L is much lower than L_{opt}, so that the pump power injected into the fiber cannot be fully exploited. The pumping scheme, however, can partially solve the situation. As far as noise is concerned, the best configuration is the dual pump scheme, because it keeps the pump power high along the whole amplifier. It is so because when the power of one pump is low, the power of the other pump is high. The copropagating scheme is the second most efficient configuration, followed by the counterpropagating scheme. In fact, although the counterpropagating scheme has the highest gain, the advantages of the copropagating scheme can be understood if we consider an amplifier formed by a series of amplifiers, that is, a cascade amplifier. In cascade amplifiers the performance of the whole system is basically determined by the characteristics of the amplifiers at the beginning of the cascade. In the counterpropagating scheme, in these amplifiers the noise is high because the pump level and, consequently, the population inversion, is low. In the copropagating scheme, conversely, the amplifiers at the beginning of the cascade show high pump power, so that the noise is low. Figure 11.16 shows the noise figure of the two pumping schemes for

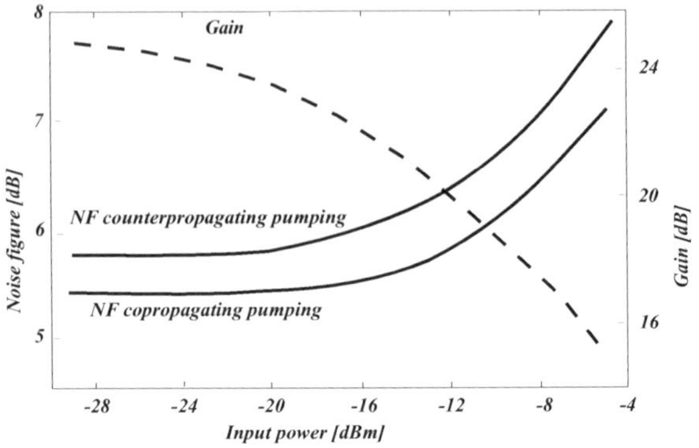

Fig. 11.16: Gain and noise figure as functions of the input signal power in two different pumping schemes.

different input signal powers (the gain is the same). The population inversion decreases at increasing input signal powers. This, in turn, causes a lower gain and a higher noise figure.

High gains ($40\,dB$ and over) can be obtained with high pump powers and long fiber lengths. In these cases, the ASE power is so high that it can reduce the population inversion. In particular, the backward-travelling ASE at the output end reduces the population inversion at the input end, so that also the noise figure is worsened. This effect can be solved by inserting an optical isolator approximately in the middle of the doped fiber.

It is also the pump wavelength that influences the noise figure of the amplifier. The $980nm$ pumping, for instance, generates a higher population inversion, and thus a lower noise figure, than the $1480nm$ pumping. In conclusion, the lowest noise figure in EDFAs is obtained with the copropagating scheme at $980nm$.

11.3.6 Typical configurations

EDFA configurations depend on the type of application considered, which are essentially of three types: booster amplifers, in-line amplifiers and preamplifiers. A booster amplifier is a power amplifier with high output power, up to $25\,dBm$. It is a dual stage amplifier in which:

$$\begin{aligned} NF &= NF_1 + \frac{NF_2}{G_1}; \\ G &= G_1 + G_2; \end{aligned}$$

where NF_1 and NF_2 are the noise figures of the first and the second stage respectively, and G_1 and G_2 their power gains. As figure 11.17 shows, the first stage consists of an EDFA with copropagating

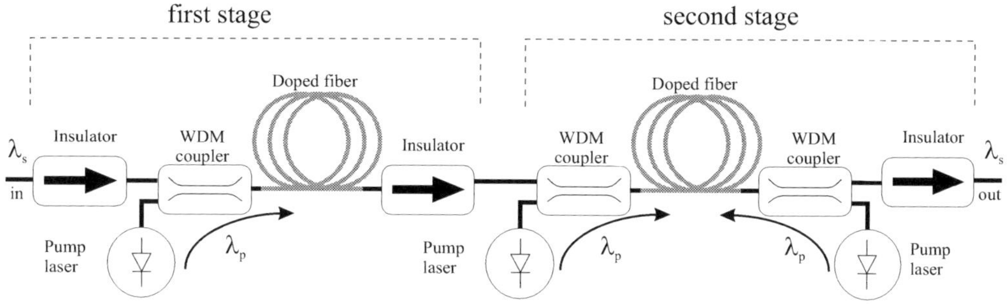

Fig. 11.17: Booster amplifier.

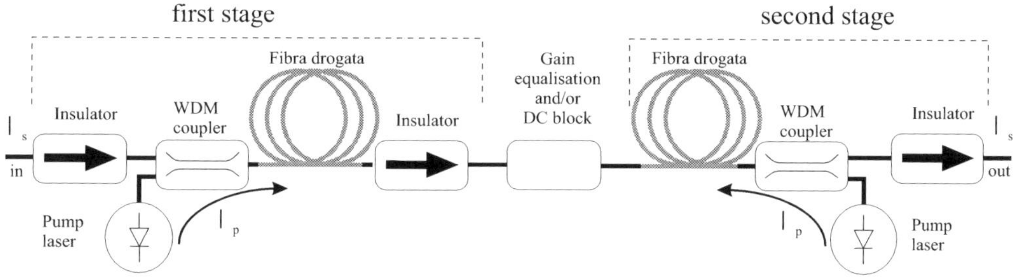

Fig. 11.18: In-line amplifier.

pumping scheme, so that the noise figure and the gain are low. A dual pumping scheme is introduced in the second stage in order to increase the saturation power and, consequently, the gain.

In-line amplifiers must achieve medium gain, around $20\,dB$, low noise figures in order to maximise the distance between the regenerators, and low costs, because there are several of them along the fiber. Figure 11.18 shows the two-stage scheme of the in-line amplifier. The first stage consists of an EDFA with copropagating pumping scheme which minimises the noise figure, while in the second stage there is a counterpropagating pump which maximises the gain. Between the two stages, an equalization block and a dispersion compensation (DC) block are often inserted. Since the block causes a loss, it is usually inserted after the first stage, so that the signal power is not too low. At the same time, it is usually placed before the second stage so that it does not reduce the signal power at which the signal is launched in the following stage.

Preamplifiers are used at the front end of optical receivers to maximise the signal power, which is not affected by the noise caused by the photodiode and by the front-end electronics. These devices, therefore, show low noise figures and use copropagating pumping schemes.

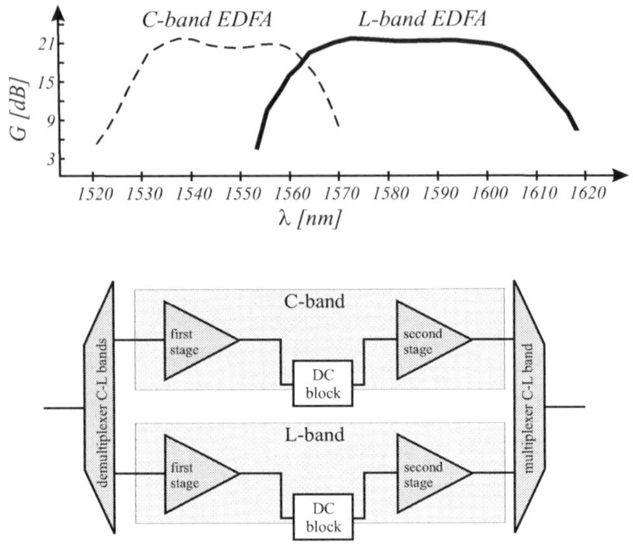

Fig. 11.19: Above: Spectrum of a C- and L- band EDFA. Below: Diagram showing a C+L band EDFA.

11.3.7 L-band EDFA

Theoretically, as it is shown in figure 11.11, it is possible to produce amplifiers working in the L-band, that is, between 1570 nm and 1610 nm. The gain per unit length, however, is extremely low, so that the doped fibers must be very long, up to 100 meters, and the pump power must be high. By doing so, the gain over the whole fiber is acceptable, as shown in figure 11.19 (above). Nevertheless, optimising the L-band amplification implies reducing the C-band gain, so that the most widely used amplifiers are C+L band amplifiers, shown in figure 11.19 (below). At first, the C-band and L-band channels are separated by a demultiplexer, then they are sent to two different amplifiers, each optimised for its band. Once the signals have been amplified, they can be multiplexed to a single fiber.

11.4 Semiconductor Optical Amplifiers

Semiconductor optical amplifiers (SOAs) are based on forward-biased p-n junctions, as shown in figure 11.20. Population inversion is thus achieved in the junction region. If the semiconductor is a direct band gap material, the population inversion strongly increases the probability of stimulated emission. The equations used for SOAs are more complex than those for EDFAs, and we will thus consider only the final propagation equation:

$$\frac{d\,I_s}{d\,z} = g(\nu) I_s \tag{11.18}$$

where g is the gain coefficient per unit length:

$$g(\nu) = \frac{c^2}{2h^2} \frac{(2m_r)^{3/2}}{\tau_s} \frac{\sqrt{h\nu - E_g}}{\nu^2} \left[f_c(\nu) - f_v(\nu)\right] \tag{11.19}$$

E_g is the energy gap of the semiconductor, τ_s is the average lifetime of carriers before they recombine, m_r is the geometric mean of the effective masses of electrons and holes, f_c and f_v the Fermi functions of the conduction and valence band respectively (see equation in (A.21)). If we integrate (11.18) over the transverse direction, considering g constant in the active region, we obtain:

$$\frac{d\,P_s}{d\,z} = g_s(\nu)\, P_s; \tag{11.20}$$

where g_s is the gain per unit length and corresponds to:

$$g_s(\nu) = \int\int g(\nu) i(x,y)\, dS = g(\nu) \int\int_A i(x,y)\, dS = g(\nu)\Gamma \tag{11.21}$$

A is the area of the active region, Γ the overlap integral between the signal and the active region. Following (11.19) and (11.21), the gain per unit length is non-zero only if $\nu \geq E_g/h$, because otherwise the photon energy is too low to allow band transitions. In addition to this, g is constant only if $f_c - f_v > 0$, that is, according to (A.21), only when $\nu < (E_{F_c} - E_{F_v})/h$, being E_{F_c} e E_{F_v} the quasi Fermi levels in the conduction and valence bands. For high energy photons, however, the conduction and valence bands are not sufficiently populated to achieve population inversion. All

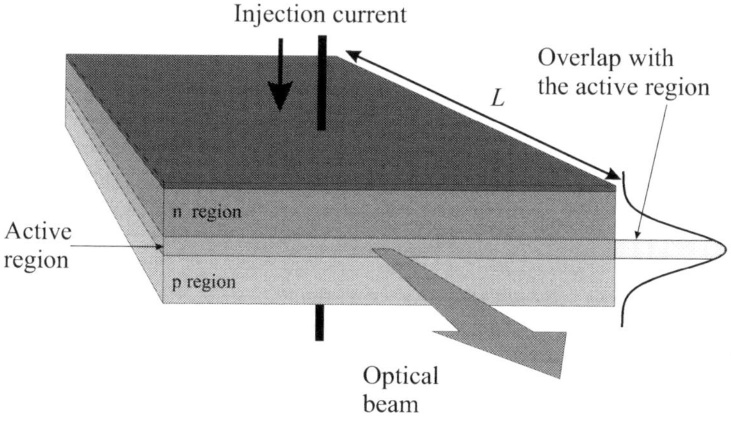

Fig. 11.20: Semiconductor optical amplifier (SOA).

things considered, the gain per unit length occurs only for frequencies v satisfying:

$$\frac{E_g}{h} < \nu < \frac{E_{F_c} - E_{F_v}}{h}.$$

The behaviour of amplifiers, therefore, depends on the carrier concentration, N, described by the two quasi Fermi levels E_{F_c} and E_{F_v}, which depend, in turn, on the electron and hole concentrations. The electron-hole recombination effected by stimulated emission is supported by an injection current I. Figure 11.21 (left) shows the spectral variation of g_s for different values of N. For low values of N the gain is negative. As the concentration increases, the gain increases until it reaches zero. N_0 is the concentration required, at least at one frequency, in order to achieve the transparency of the medium, and it is thus called *transparency carrier concentration.* The gain is positive above this concentration, thus extending the spectral bandwidth where the medium is active. Approximately, the peak gain g_{peak} linearly increases for increasing values of carrier concentration, following the relation:

$$g_{peak} = g_0 \Gamma \left(N - N_0\right),$$

where g_0 is referred to as *differential gain coefficient*, despite having the dimensions of an area. The variation of peak gain is shown in figure 11.21 (right).
Furthermore, if we observe the figure we notice that the value of g_s is approximately 3 times more than the gain of an EDFA. Consequently, the distances along which the signal travels in order to be amplified are reduced to a few hundred microns or a few millimeters, instead of meters or tens of meters. These values are well compatible with integrated optics.

11.4.1 Power Gain

In SOAs, population inversion is determined by the bias current density of the p-n junction. For the sake of simplicity, current density can be at first considered constant along the propagation

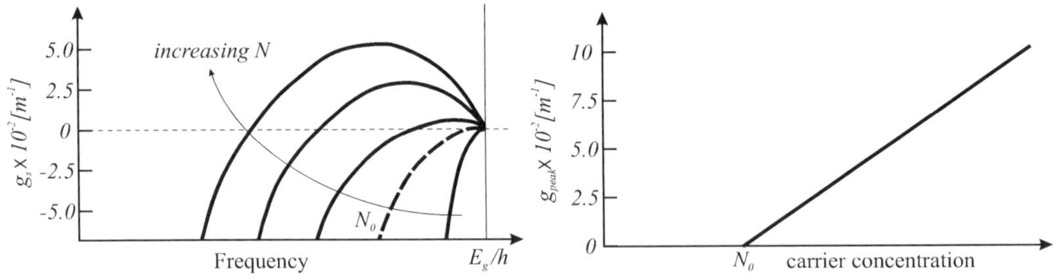

Fig. 11.21: Left: Spectral variation of the gain per unit length for different values of the carrier concentration. Right: Peak value of the gain per unit length as a function of the carrier concentration.

direction, so that also the gain per unit length is constant, and we can integrate the equation in (11.20):

$$P_s(z) = P_s(0)e^{g_s L}$$

where L is the amplifier length. The power gain will thus be:

$$G = e^{g_s L}, \qquad G_{dB} = 4.343 g_s L = 4.343 g \Gamma L. \tag{11.22}$$

An example of gain spectrum is shown in figure 11.22. The spectrum forms approximately a parabolic curve, the typical $3\,dB$ bandwidth spans about $70 - 80\,nm$. The type of polarisation needs to be specified because, due to the high birefringence of guides in integrated optics, the overlap integral Γ is not the same for the two types of polarisation. The result is a polarization dependent gain (PDG), which can cause serious problems. In fact, the gain can vary from a few to some tens of dB, so that it is necessary to face the phenomenon. In order to obtain a $< 1\,dB$ polarization dependent loss (PDL) several techniques have been developed. The most obvious is to produce guides in which the confinement factor is the same for the two polarisations. The required dimensions of the guides, however, are so small that this technique is not a commercially suitable solution. Another possibility is to exploit the birefringence of specific structures, such as quantum wells, in order to compensate for the birefringence of the waveguide. Finally, it is possible to introduce anisotropy in the material by applying small mechanical strains to the layers of the waveguide during the growth phase. By employing one of these techniques, it is possible to reduce PDL to a few tenths of a dB.

The equation in (11.22) considers g_s constant along the propagation direction. The gain per unit length along the z axis, however, actually depends on the signal power along the same axis. In particular, the increase of signal power along the z axis leads to a considerable depletion of the carriers due to stimulated emission. In fact, when the signal power becomes excessively high, the carrier density is reduced, thus reducing also g_s. The relation can be expressed as follows:

$$g_s(z) = \frac{g_{s0}}{1 + \frac{P_s(z)}{P_{s_{sat}}}}$$

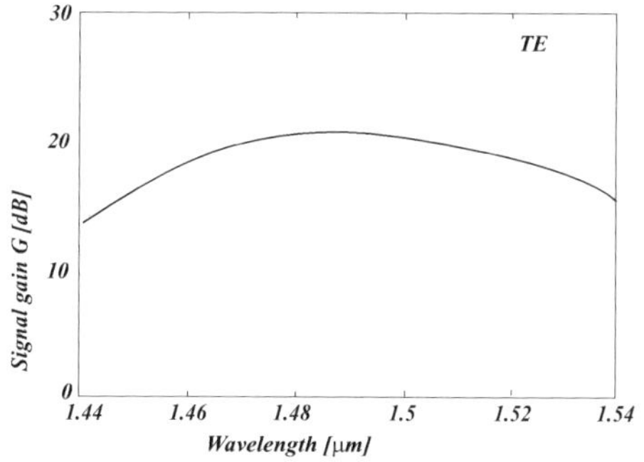

Fig. 11.22: Gain spectrum of a SOA for the TE polarisation.

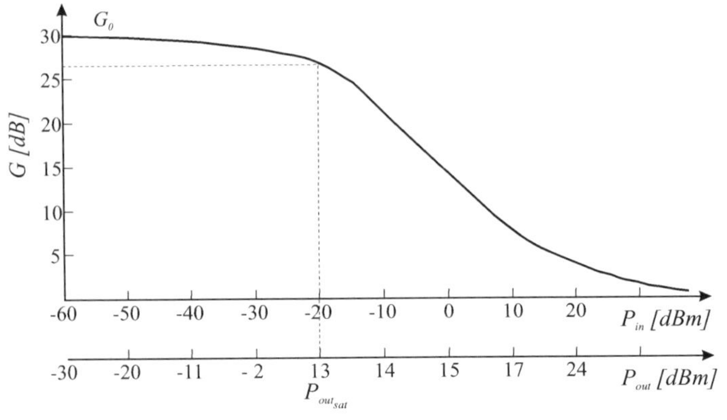

Fig. 11.23: SOA gain versus signal input power.

where g_{s_0} is the gain per unit length in the absence of signal, and $P_{s_{sat}}$ is the intrinsic saturation power of the amplifier, which can be defined as the internal power $P_s(z)$ at which the gain per unit length halves its value.

The power gain can thus be expressed as:

$$G = 1 + \frac{P_{s_{sat}}}{P_{in}} \ln\left(\frac{G_0}{G}\right)$$

where $G_0 = e^{g_{s_0} L}$ is the power gain in the absence of signal, also known as *small signal gain*. Figure 11.23 illustrates the dependence in a SOA with $G_0 = 30\,dB$. Furthermore, the figure shows the saturated output power, i.e. the value of the output power where gain is reduced by $3\,dB$.

11.4.2 Choice of Semiconductor Materials

A fundamental aspect of amplifiers and semiconductor lasers (see chapter 12) is that amplification can be obtained at any wavelength, provided that the suitable semiconductor is chosen. Figure 11.24 shows the relation between the lattice constant, the energy gap E_g and the corresponding wavelength of various binary, ternary and quaternary compound semiconductors used in telecommunications. In particular, if E_g is expressed in electronvolts (eV) and the wavelength λ in μm, the equation obtained is:

$$\lambda_0 \simeq \frac{1.24}{E_g}.$$

The dots correspond to binary compounds, the lines to ternary compounds, and the area within the lines represents quaternary compounds. In ternary compounds, E_g and the lattice constant may be varied for different mole fractions of the elements, whereas in quaternary compounds it is possible to vary E_g and keep the same lattice constant. The horizontal dashed line, for example, indicates the range of values for the band gap of $In_{1-x}Ga_xAs_yP_{1-y}$ which can be obtained by varying the x and y mole fractions of gallium and arsenic in order to create a quaternary compound with the same lattice constant of indium phosphide. This is particularly relevant because it allows to produce materials with different E_g without introducing crystal lattice defects. The relation between the mole fractions x and y is not arbitrary, but it must be $x/y = 0.45$. The compound can then be written only as a function of y. The energy gap thus becomes:

$$E_g = 1.35 - 0.72\,y + 0.12\,y^2.$$

The minimum energy gap is for $y = 1$ and corresponds to approximately $1.65\,\mu m$. For $y = 0$ the energy gap is $0.92\,\mu m$. $In_{1-x}Ga_xAs_{1-y}P_y$ SOA on InP substrates, therefore, it is possible to cover all the wavelengths between $1.0\,\mu m$ and $1.65\,\mu m$, that is, all the wavelengths used in modern optical communication systems.

11.4.3 Homojunction and Heterojunction

In order to maximise the efficiency of SOAs the semiconductors used for $p-n$ junctions should have different band gap. This type of interface is called *heterojunction*. As a matter of fact, by means of homojunction, which uses the same semiconductors but differently doped, the carriers are not well

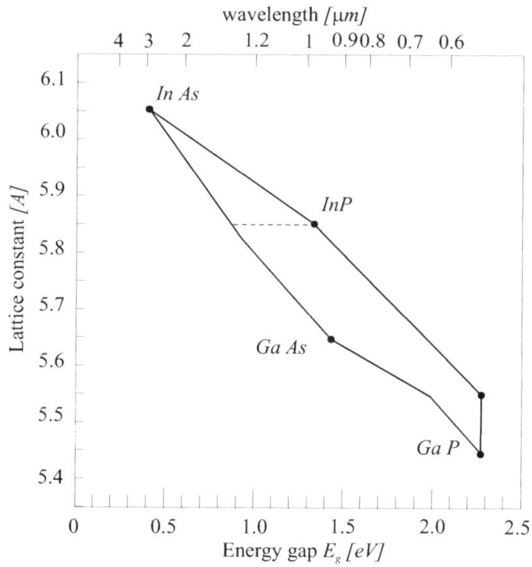

Fig. 11.24: Lattice constants, energy gaps, and corresponding wavelengths for different III-V compound semiconductors.

confined in the active region, thus reducing the SOA efficiency. With heterojunctions, on the other hand, there are a series of advantages, briefly described in figure 11.25. The central layer is almost as thick as the depletion layer and it is made of the semiconductor with lower band gap. Due to the potential barriers generated by the different bandgaps and doping concentrations, the injected carriers are basically confined to the active layer. The carrier density and uniformity of distribution in the active region reduce the quantity of electrons and holes required to reach transparency.
Heterojunctions have also another important function, due to the fact that in semiconductors the increase in energy gap leads to a reduction of the refractive index. Consequently, the refractive index is higher in the active layer than in the adjacent regions, thus creating a waveguide and improving the overlap integral. To conclude, since the band gap in the outer layers is higher than the amplified photons energy in the active region, it reduces the absorption coefficient, which should be theoretically zero, but which has in fact a very low value (although not zero) due to impurity assisted band-to-band transitions or other phenomena.

11.4.4 Structure

Figure 11.26 shows two different structures of SOAs. The first one is a ridge waveguide, in which lateral confinement is possible due to the different refractive indices of InP and SiO_2, whereas

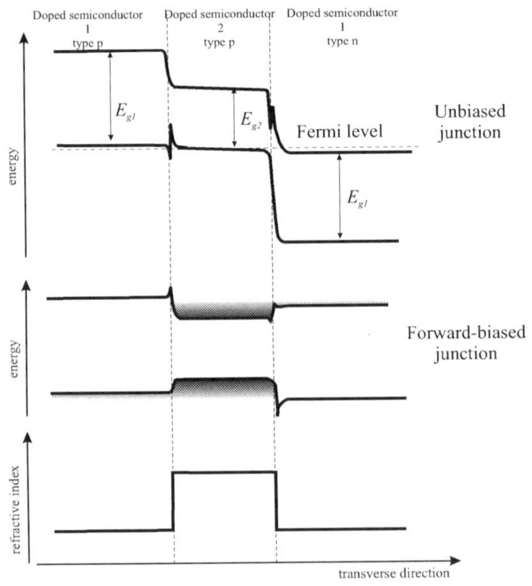

Fig. 11.25: Energy bands in a double heterojunction. Above: unbiased junction; center: forward-biased junction; below: diagram of the refractive index

vertical confinement is achieved thanks to the different indices of InGaAsP and InP. The second structure is a buried waveguide in which the active region in InGaAsP forms the core, which is surrounded by an InP cladding. Both the waveguides achieve a good confinement of the field, so that there is a significant overlap between the field and the active region. So far, we have implicitly assumed that the propagation direction of the signal beam along the longitudinal direction was only one. Nevertheless, the discontinuities at the semiconductor-air interface at the SOA ends cause a partial reflection of the beam. This phenomenon is illustrated in figure 11.27. The reflected signals travel backwards through the SOA until they reach the end of the component. These multiply reflected beams generated at the interface determine a gain fluctuation. At some wavelengths, the waveforms are in phase and create a resultant wave from the sum of the multiple waves (constructive interference). In this case, the gain is higher than for destructive interference. Given G as the gain of the SOA and R_1, R_2 as the facet reflectivities, the peak-to-peak gain fluctuations, also known as *gain ripple*, will be:

$$Ripple = \left(\frac{1 + G\sqrt{R_1 R_2}}{1 - G\sqrt{R_1 R_2}}\right)^2 .$$

The facets thus create an effect of cavity resonance which selects only some wavelengths. This phenomenon is useful in order to obtain high gains in a narrow range of wavelengths, as it occurs in lasers. It is, however, extremely negative for broadband amplification. In fact, a reflectivity below $1.5 \cdot 10^{-5}$ is required in order to obtain lower than $0.2\,dB$ fluctuations with a $G = 30\,dB$ gain.

SOAs can be classified in *Fabry-Perot* (SOA-FP) and *travelling wave* (SOA-TW) according to the facet reflectivity and the influence of multiple reflections on the gain value and gain spectrum. *Fabry-Perot SOAs* (SOA-FP) show high reflectivities, so that the signal travels repeatedly across the active medium. SOA-TW, on the other hand, have low reflectivities, so that the signal travels across the active medium only once. The gain of SOA-TW is lower, but it is compensated by a

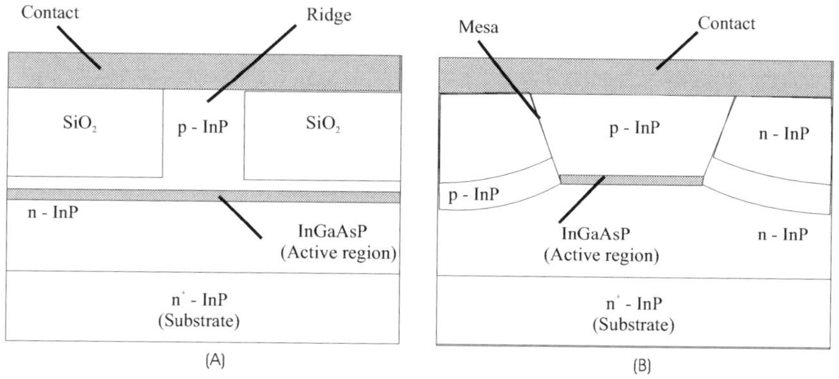

Fig. 11.26: Schematic structure of different SOAs. (a) ridge waveguide; (b) buried waveguide.

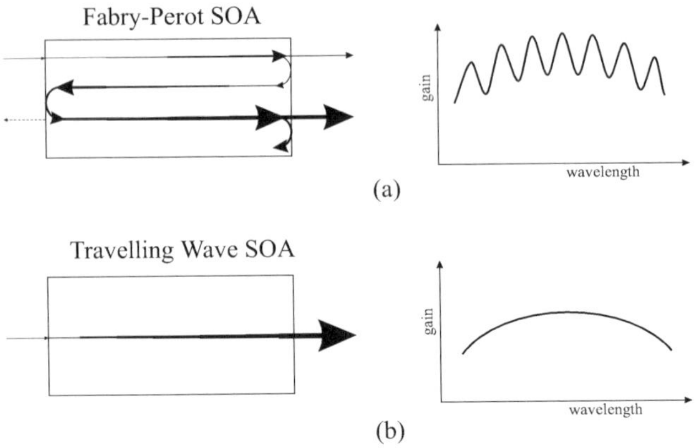

Fig. 11.27: (a) Fabry-Perot SOA. (b) Travelling wave SOA.

wider bandwidth.

Antireflection coatings can be applied to the facets to reduce the reflectivity of TW-SOAs, but this solution is not sufficient to reach a suitably low reflectivity. The techniques used to further reduce reflectivity are basically two. On the one hand, the waveguide can be produced so that the direction of the beam propagation has a tilt with respect to the normal to the interface. On the other hand, a homogeneous layer can be introduced between the facets and the antireflection coating. Figure 11.28 shows both the solutions, which are called respectively *tilted-stripe* and *window region*. The aim is to reduce the light which travels in the backward direction after being reflected by the antireflection coating. In the first case, by tilting the waveguide the beam is reflected at a non-zero angle with respect to the non-tilted position. As for the homogeneous medium, without a guiding mode the beam is diffracted both before and after reflection. The power which is recoupled to the waveguide is extremely low. Tilted-stripe obtains reflectivities of about 10^{-3}, or 10^{-4} at the maximum, whereas the window-region achieves a reflectivity of about 10^{-4}. By combining the two techniques, therefore, it is possible to achieve a reflectivity of 10^{-6}.

11.4.5 Gain Dynamics

As mentioned in the previous sections, saturation in SOAs depends on the carrier depletion of the active region which occurs for high power signals. The gain reduction caused by the excessive increase of signal power is also known as gain compression. Figure 11.29 shows what happens to the SOA gain when an input pulse of high power and short period is injected. When the pulse arrives, the gain is reduced by a few dB, and then it begins to recover when the pulse finishes. Nevertheless, the gain recovery time, i.e. the time required for the carrier concentration to go back to its initial value, is determined by the average carrier lifetime τ_s, which typically corresponds to about a hundred

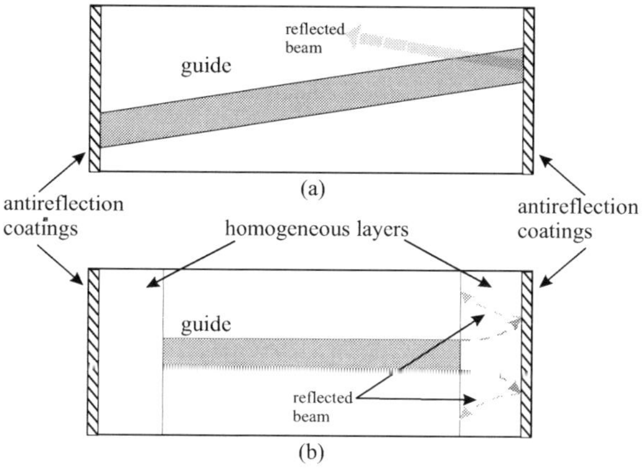

Fig. 11.28: Techniques for the reduction of facet reflectivity: (a) tilted-stripe, (b) window region.

ps. These values can be compared with the bit rate of modern optical systems, so that the gain dependence on the instant value of power causes the distortion of the transmitted bits or crosstalk phenomena in WDM systems.

The same gain dynamics is also typical of EDFAs, although their recovery time is similar to the lifetime at the metastable state (10 *ms*), which is significantly higher than the bit rate. Consequently, the population and the gain of EDFAs depend on the average power of the signal, which changes on a *ms* time scale. Therefore, there is no distortion or crosstalk for single bits. In WDM systems,

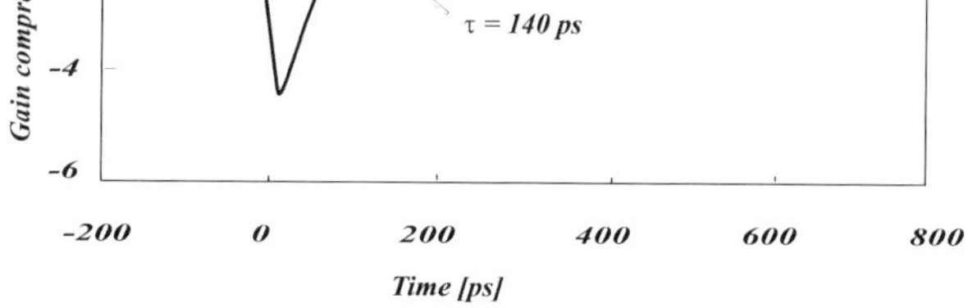

Fig. 11.29: Change in gain compression due to the injection of a high power pulse.

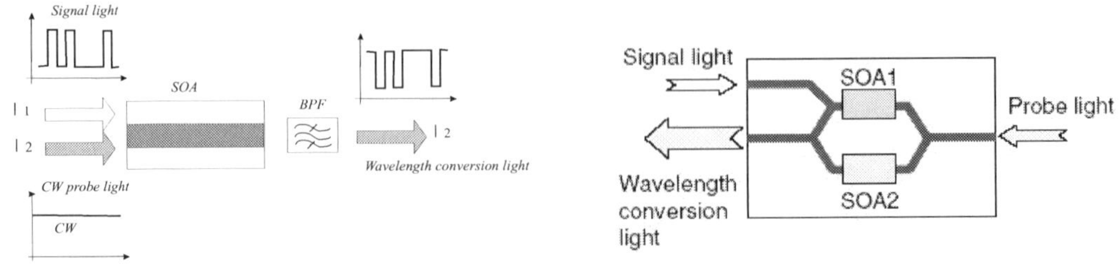

Fig. 11.30: Application of SOA as a wavelength converter. Left: XGM wavelength converter. Right: XPM wavelength converter.

however, the power variation depends on channel adding or dropping. Since WDM networks allow packet-mode and burst-mode transmission, their lifetime can be compared to the metastable lifetime, causing gain fluctuations which reduce the performance of the system.

Before concluding, it is important to highlight that the gain dynamics of SOAs reduces the amplifier performance, but it is useful in all-optical signal processing, such as, for example, in wavelength conversion and signal regeneration (see the following section).

11.4.6 All-optical Signal Processing Applications

SOA carrier dynamics is well suited for all-optical signal processing. Their first application has been for SOA-based wavelength conversion, which is fundamental for network traffic control.

In fact, wavelength converters reduce the blocking probabilities in WDM networks. In traditional networks, this result is achieved through an optical to electrical conversion, and the signal is then retransmitted at a different wavelength. However, if we want to obtain the same effect with an all-optical signal processing, we need to exploit the cross gain modulation (XGM) of SOAs.

Figure 11.30 shows how the process works. Two beams are injected into the SOA: the modulated signal at wavelength λ_1 (which needs to be converted), and the CW laser beam at the new wavelength λ_2. The signal beam reduces the carrier density at high signal power, thus also reducing the gain of the probe beam. Conversely, the probe beam can be amplified at low signal power values. This implies that the gain variation of the probe beam is inversely proportional to the signal at λ_1. The time fluctuations of the gain will then be present at the SOA output for the beam at λ_2. The SOA works under a saturation condition. The input power must be sufficiently high in order to reduce as much as possible the SOA recovery time thanks to the stimulated recombination of the carriers. Recently developed wavelength converters are able to operate at a bit rate of up to $40Gbit/s$.

The disadvantages of this kind of applications are basically two. The first one is the extinction ratio degradation (ratio between 1 and 0) of the converter when converting from shorter to longer wavelength. This effect is caused by the changing population in the different energy bands, which

makes it more difficult to compress the gain at longer wavelengths. The second problem is that the refractive index of a semiconductor depends on the carrier density. The gain modulation is thus associated with a change in the refractive index, which in turn causes an undesired phase modulation of the signal at wavelength λ_2 (Cross Phase Modulation - XPM).
Nevertheless, the wavelength converters can also exploit the refractive index dependence on optical power. In particular, the phase modulation caused by the variation of carrier concentration can be converted into amplitude modulation by inserting the SOA in an interferometric structure, such as a Mach-Zehnder interferometer. A schematic example is shown in figure 11.30. By means of these devices it was possible to obtain wavelength conversion up to $100\,Gbit/s$.
The nonlinear properties of SOAs can be exploited for signal regeneration when the signals are too seriously degraded by the amplifier noise and by the nonlinear distortions of the optical fiber. Figure 11.31 shows a 3R regenerator (Reamplifying, Reshaping, Retiming) based on XPM in a SOA. By means of Y-junctions and a 3dB coupler the degraded signal is injected into the two branches of the Mach-Zehnder interferometer together with the clock signal required for retiming. The interferometer is transparent to the clock only when the signal power is sufficiently high. The output signal thus has the clock wavelength but carries the information of the signal which has to be regenerated. Finally, figure 11.32 shows a SOA applied to an ADD-DROP multiplexer for OTDM systems. In this case, the Mach-Zehnder interferometer is unbalanced by using the XPM of SOAs.

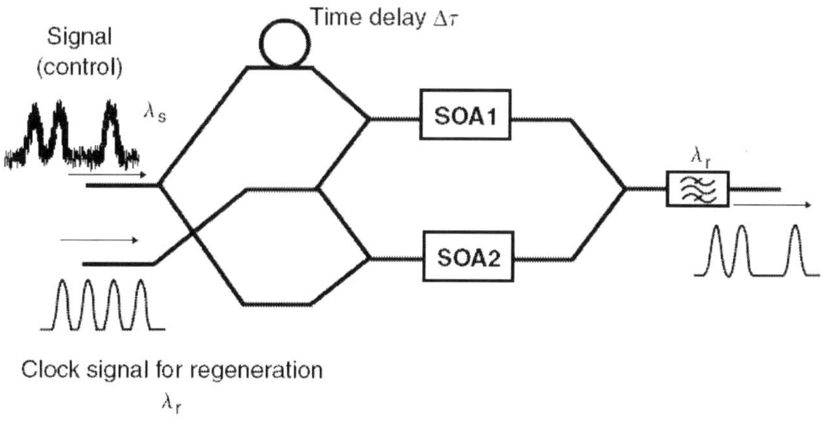

Fig. 11.31: 3R regenerator based on SOAs and Mach-Zehnder interferometers.

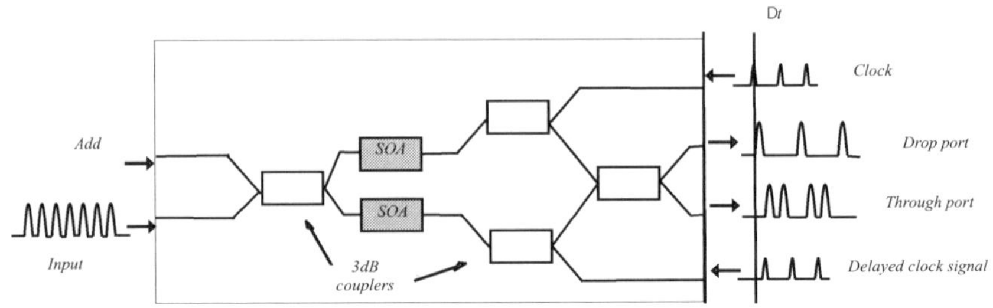

Fig. 11.32: ADD-DROP multiplexer for OTDM systems based on SOAs and Mach-Zehnder interferometers.

11.5 Raman Amplifiers

Raman amplifiers (RA) are based on the Raman effect. The Raman effect is the inelastic scattering occurring to some photons of the beam when they interact with the optical phonons which describe the vibrational modes of the molecules constituting the scattering medium. The interaction between the electromagnetic field and the vibrational modes creates scattered photons of energy (and frequency) either slightly lower or higher than that of the incident photons. The Raman effect was discovered in 1928 by the Indian physicist C. V. Raman, who was awarded the Nobel prize in 1930. It was then in the 1960s that the possibility of obtaining stimulated Raman scattering (SRS) was experimentally proved, thus leading to its application in optical amplification. Figure 11.33 shows a diagram of the SRS mechanism. Quantum dynamics uses the perturbation theory to describe the scattering. According to this theory, the Raman effect produces the absorption of a photon of $h\nu_p$ energy and the subsequent emission of a photon of $h\nu_s$ energy. The interaction is mediated by an electron, which is excited to a virtual energy state and then decays back to a lower energy state.

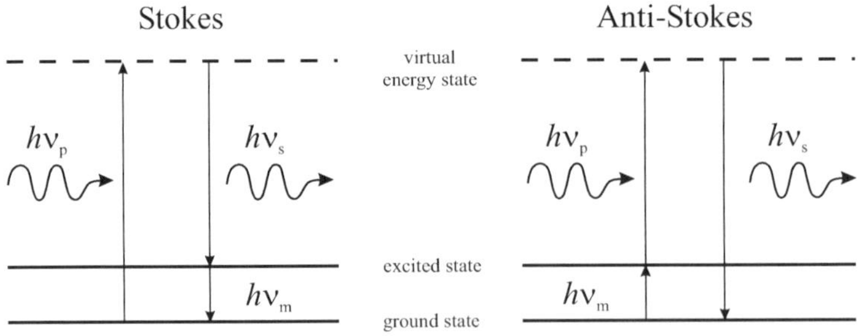

Fig. 11.33: Energy transitions involved in Raman scattering. Left: Stokes scattering. Right: anti-Stokes scattering.

According to the law of conservation of energy, if in the process there is an optical phonon emission of ν_m frequency, the energy of the absorbed photon will be:

$$h\nu_p = h\nu_s + h\nu_m,$$

and the scattered photon has a lower energy. In this case, the process is called Stokes scattering. On the other hand, if the scattering is characterised by the absorption of a phonon, the energy of the absorbed photon is:

$$h\nu_p + h\nu_m = h\nu_s,$$

and the scattered photon has a higher energy. The process is called anti-Stokes scattering. The decay can be either spontaneous or stimulated, thus causing, respectively, noise or amplification. By means of SRS it is thus possible to transfer power from a high power pump beam to a low power signal beam. If Stokes scattering is considered, the pump beam will have higher frequency and wavelength than the signal beam. The opposite result will be achieved through anti-Stokes scattering. The relevant aspect of this process is that it depends exclusively on the difference between the two wavelengths, and not by the wavelength in itself. Therefore, provided that the pump laser source has a suitable wavelength, it is possible to amplify on every band while using the same active medium.
Since anti-Stokes scattering requires the presence of a phonon, it is less probable than the Stokes process, at least at room temperature. Raman amplifiers only rely on Stokes scattering. The transition probability per unit time, W_R, from the ground state to the vibrational level (Stokes transition) by means of Stokes scattering can be approximately expressed by:

$$W_R \propto n_s \, n_p + n_p(n_m + 1),$$

where n_s is the number of signal photons, n_p is the number of pump photons, and n_m the number of optical phonons per unit volume. The first term of the equation represents the stimulated Raman scattering and it is proportional to the product of the pump and signal photons, whereas it does not depend on the number of optical phonons. The second term represents the noise generated during amplification, which depends on two values. The first one is the spontaneous emission, while the second one is the spontaneous emission plus the initial number of optical phonons. In thermal equilibrium, the average number of optical phonons obeys the Bose-Einstein statistics, so that:

$$n_m = C \frac{1}{e^{\frac{h\nu_m}{kT}} - 1}$$

where ν_m is the frequency of the optical phonon and C a multiplying constant which depends on the material. If we momentarily ignore the effects of spontaneous emission and attenuation in the optical medium, the propagation equations become:

$$\frac{d\,I_s}{dz} = K_{pol} g_{sr}(\nu_p - \nu_s) I_s \, I_p \,, \qquad (11.23)$$

$$\frac{d\,I_p}{dz} = -K_{pol} \frac{\lambda_s}{\lambda_p} g_{sr}(\nu_p - \nu_s) I_s \, I_p \,.$$

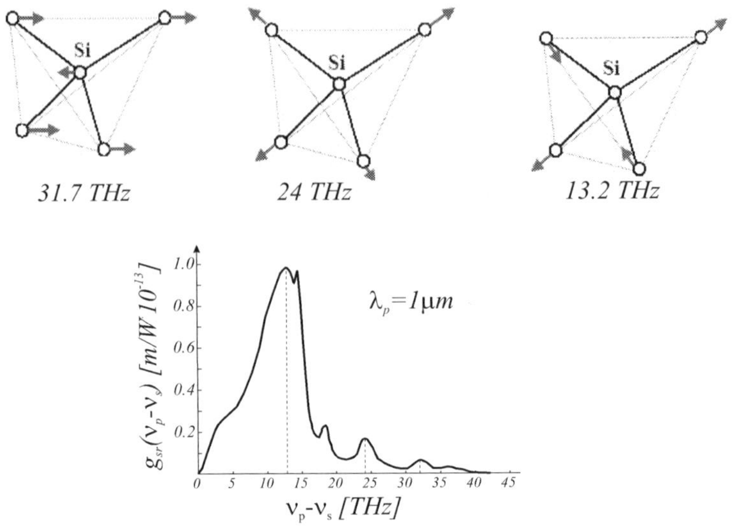

Fig. 11.34: Above: main vibrational modes of silica molecules involved in Raman scattering. Below: Raman gain efficiency spectrum of silica for $\lambda_{p_{ref}} = 1\,\mu m$.

where g_{sr} is the *Raman gain efficiency* and it is a function of $\nu_s - \nu_p$, whereas K_{pol} is a polarisation coefficient between 0 and 1 which accounts for the relationship between the polarizations of the two beams. In particular, it is 0 if the polarizations are orthogonal, 1 if they are coincident, 1/2 if the beams are depolarised.

In optical fibers, the polarisations at the ground state tend to degenerate, so that the beams can be considered depolarised. The value of K_{pol} will thus be $K_{pol} = 1/2$. If the energy levels of the vibrational modes were perfectly discrete, the spectrum of g_{sr} would be described by a zero function for each value except for $\nu_m = \nu_p - \nu_s$. Nevertheless, the interaction between molecules in solid materials splits the energy levels in a similar way to EDFAs, so that g_{sr} is a non-zero continuous function within a frequency interval depending on the type of material.

11.5.1 Fiber Raman Amplifiers

Fiber Raman amplifiers exploit the vibrational modes of the silica molecules constituting the fiber. The oxygen atoms of silica form a tetrahedron, whereas the atom of sylicon occupies the centre. Figure 11.34 shows the vibrational modes involved in Raman scattering with the corresponding resonance frequencies. The modes involved are those which do not alter the dipole moment of the molecule. Figure 11.34 also shows the Raman gain efficiency spectrum of pure silica. It is clear how each mode influences the spectrum and the spectral broadening related to the interaction between molecules. The spectrum has been calculated at $\lambda_{p_{ref}} = 1.0\,\mu m$ pump wavelength. The value for different wavelengths is expressed by the equation:

$$g_{sr}(\lambda_p) = g_{sr}(\lambda_{p_{ref}})\frac{\lambda_{p_{ref}}}{\lambda_p}.$$

Apart from silica, in both the core and the cladding of optical fibers there are other compounds which modify the refractive index. These compounds can increase the Raman gain efficiency. In particular, germanium oxide (GeO_2), which is used to increase the refractive index of the core, has an 8 times higher efficiency than silica. Since the dopant concentration varies according to the radius, in optical fibers the Raman gain efficiency is a function of the transverse coordinate: $g_{sr}(x, y)$.
The propagation equations (expressed in terms of power) are obtained by integrating along the transverse direction the equations in (11.23):

$$\frac{d\,P_s}{d\,z} = -\alpha_s\,P_s + \frac{1}{2}g_R\,P_p\,P_s + h\nu_s g_R P_p \Delta\nu\,, \tag{11.24}$$

$$\frac{d\,P_p}{d\,z} = \mp\alpha_p\,P_s \mp \frac{1}{2}\frac{\lambda_s}{\lambda_p}g_R\,P_s\,P_p + \frac{\lambda_s}{\lambda_p}h\nu_p g_R P_p \Delta\nu; \tag{11.25}$$

to which the values of spontaneous emission and attenuation of the fiber must be added. P_s and P_p are the signal power and the pump power, α_s and α_p are the values of attenuation introduced by the fiber at the signal and pump wavelengths, $\Delta\nu$ is the bandwidth in which the noise power is calculated, g_R is the *Raman gain coefficient*:

$$g_R = \int_\infty g_{sr}(x, y) i_s(x, y)\, i_p(x, y)\, dx\, dy\,,$$

where i_s and i_p are the normalised intensities of the two beams. The Raman gain coefficient can be substituted by the Raman effective area:

$$A^R_{eff} = \frac{1}{\displaystyle\iint_S i_s(x, y) i_p(x, y)\, dx\, dy},$$

where the double integral is calculated on S, the transverse area of the fiber. Approximately, the ratio between the Raman gain coefficient and the effective area is the following:

$$g_R \simeq \frac{\bar{g}_{sr}}{A^R_{eff}}\,,$$

where $\bar{g}_{sr}$ takes into account the weighed factors due to the materials composing the fiber.
The value of g_R varies according to the different fibers used in transmission systems. The peak, however, is always around $13\,THz$ which, at $1550\,nm$, corresponds to a wavelength shift of about $105\,nm$. The dependence on fiber is related to both the different concentration of germanium oxide and to the field confinement. In optical fibers, therefore, the dispersion compensation is approximately of one order of magnitude higher than in standard fibers. No specific material is

required for the fibers of a fiber Raman amplifier, because the same fiber used for long-distance transmission is a suitable active medium.
If we consider:

$$g_R = 3 \cdot 10^{-4}\, 1/(W\, m),$$

a $P_p = 2\,W$ pump power is required to obtain a gain per unit length in the order of:

$$g = g_R\, P_p = 6 \cdot 10^{-4}\, 1/m,$$

which is three orders of magnitude lower than the gain in EDFAs. This implies that in order to obtain gains similar to EDFAs the signal must propagate for several tenths of km. Another element to consider is that the fiber itself can be used as an active medium. The two aspects explain the typical counterpropagating pump scheme of fiber Raman amplifiers, which is shown in figure 11.35. At a certain point along the fiber, the pump light is injected by means of a WDM system. The amplification occurs as the signals propagate along the fiber, and since the fibers is several km long, it is called *distributed amplification*, as opposed to discrete amplification, which is not used in long-distance communication and which occurs in a specific point of the fiber.

11.5.2 Small Signal Gain

Due to the low gain coefficient of RAs, the pump power tends to be significantly higher than the signal power. In these cases, the second element of the right side in (11.25) can be ignored. Nevertheless, since the propagation distance is long, the pump power is not constant, but varies due to the fiber attenuation, α_p. In RAs, in fact, the reduction of pump power is basically caused by the fiber attenuation, rather than by the amplification process. Therefore, all the values related to ASE in (11.24) and (11.25) can be ignored when compared to the pump and signal power. The two equations (11.24) and (11.25) can thus be solved analytically, so that the signal power along the fiber becomes:

$$P_s(z) = P_s(0)e^{\frac{g_R}{2} P_p(0) L_{eff}(z) - \alpha_s z} \tag{11.26}$$

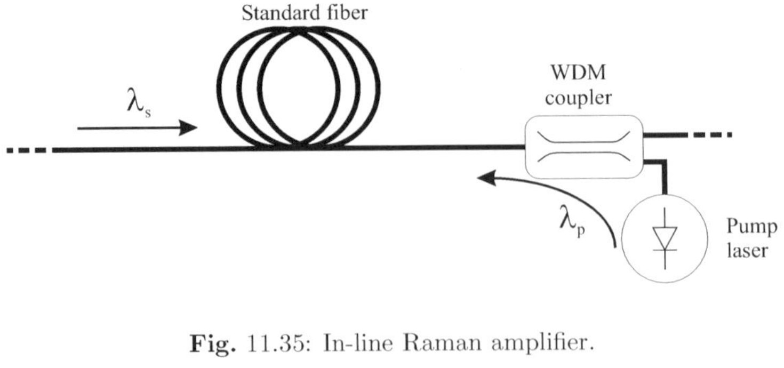

Fig. 11.35: In-line Raman amplifier.

where L_{eff} is the *effective interaction length*, which corresponds to:

$$L_{eff} = \int_0^z \frac{P_p(\zeta)}{P_p(0)} d\zeta = \frac{1 - e^{-\alpha_p z}}{\alpha_p}.$$

The gain of Raman amplifiers usually refers to the on-off gain, that is, the gain relative to the output power in presence or absence of the pump power:

$$G_R^{on-off} = \frac{P_s(L)|_{P_p=\text{on}} - P_{ASE}(L)|_{P_p=\text{on}}}{P_s(L)|_{P_p=\text{off}}} = e^{g_R P_p(0) L_{eff}(L)}$$

where L is the fiber length and P_{ASE} is the output ASE power calculated at $P_s(0) = 0$. G_R^{on-off} differs from the net gain:

$$G = \frac{P_s(L)}{P_s(0)} = e^{g_R P_p(0) L_{eff}(L) - \alpha_s L}$$

because it ignores the α_s fiber losses and measures the increase of power related to the introduction of a Raman pump.

Consider, for instance, a single-mode fiber of $g_R = 0.3\, 1/(W\, km)$ gain, $L = 10 km$ length, $P_p(0) = 400\, mW$ pump power, $\alpha_p = 0.25\, dB/km$ attenuation at $\lambda_p = 1450\, nm$ wavelength. The effective interaction length is:

$$L_{eff} = \frac{1 - e^{-\frac{0.25 \cdot 10}{4.343}}}{0.25\, 10/4.343} = 7.6\, km$$

and the on-off gain:

$$G_R^{on-off} = e^{0.3 \cdot 400 \cdot 10^{-3} 7.6} = 2.48$$

which approximately corresponds to a $4\, dB$ gain. By increasing the fiber length at $80\, km$ and the pump power at $P_p(0) = 800\, mW$ we obtain:

$$L_{eff} = 17.2\, km \quad \text{e} \quad G_R^{on-off} = 62 \cong 18\, dB.$$

11.5.3 Noise

If we consider the ASE as the only source of noise, the noise figure becomes:

$$NF = \frac{P_{ASE}}{G h \nu_s B}$$

where G is the net gain and P_{ASE} is the ASE power measured within the bandwidth B. It follows that in RAs the population inversion is always total, because the phonon lifetime is so low that also the noise figure is extremely low.

There is, however, another source of noise, i.e. the photons which are at first backreflected due to the Rayleigh scattering, and then propagate along the direction from which they came, always due to Rayleigh scattering. The phenomenon is called *Double Rayleigh BackScattering* (DRBS), and causes noise because the backscattered photons have an arbitrary delay with respect to the signal.

11.5.4 Discrete and Distributed Amplifiers

Raman amplifiers cannot substitute EDFAs and SOAs, but are complementary to both of them. In particular, combined RA/EDFA amplifiers improve the transmission distances and/or reduce the nonlinear distortions of the fiber. Furthermore, they can be used for high capacity systems in bands where EDFAs do not work, especially in the S-band.

The advantages of an amplifier consisting of an EDFA and a distributed Raman amplifier are shown in figure 11.36. By using only discrete amplification, the transmission distance is limited by the impossibility of increasing or reducing the signal power too much. The minimum P_s is determined by the effects of noise, whereas the maximum power is determined by the necessity of limiting the nonlinear distortions generated by the fiber, which are proportional to the signal power and are extremely damaging for the amplifier because they cannot be compensated. The transmission distance is thus determined by these two phenomena and by the fiber attenuation. If we want to increase the transmission distance, thus increasing also the distance between regenerators, we can introduce a counterpropagating Raman pump which amplifies in the same band as the EDFA. The result of the process is graphically shown in figure 11.36. Raman pumping has no effect when the transmission starts, because it is a counterpropagating pumping. Furthermore, the signal power decreases at a fixed rate determined by the fiber attenuation. As the signal propagates, however, the pump power increases until the Raman gain per unit length is higher than the fiber attenuation. From

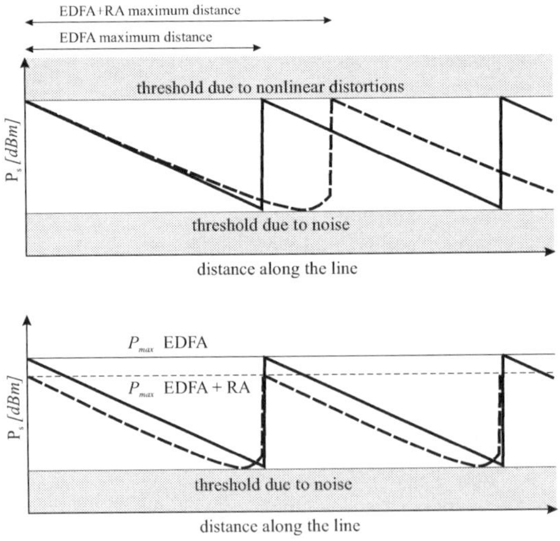

Fig. 11.36: Power variation along fibers relying on EDFAs (Continuous line) and along fibers relying on RA/EDFAs (dashed line). Above: increase in transmission distance for the same maximum and minimum power. Below: reduction of maximum power for the same transmission distance.

this point onwards, the signal power increases again. Similarly, if we fix the transmission distance, the effect of Raman counterpropagating pumping is to reduce the signal power and, consequently, the nonlinear distortions.

Discrete Raman amplifiers have different applications. The fibers showing higher Raman gain are dispersion compensating fibers (DCF). Dispersion compensating fibers are obtained by winding the fibers (which are some km long) into small coils, which are then placed in fixed points along the transmission line and show high losses. In order to compensate for the losses of the compensating fibers, a Raman pump with an adequate gain coefficient can be introduced in the fiber.

Consider, for example, a DCF fiber of $g_R = 3\,1/(W\,km)$ gain, $L = 10km$ length and $P_p(0) = 400\,mW$ pump power. At $\lambda_p = 1450\,nm$ pump wavelength, the DCF fiber attenuation is higher than the SMF fiber attenuation, which is typically $\alpha_p = 0.6\,dB/km$. The higher attenuation causes a lower effective interaction length, which corresponds to:

$$L_{eff} = 5.4\,km$$

Nevertheless, the higher gain coefficient leads to an on-off gain:

$$G_R^{on-off} = 652 = 28\,dB$$

which is higher than the on-off gain of SMF fibers.

11.5.5 Gain Equalisation

The gain spectrum of a Raman amplifier is quite similar to the gain coefficient spectrum, so that the $3\,dB$ bandwidths range between 5 and $8\,THz$, which corresponds approximately to $60\,nm$ in C-band. In order to increase the bandwidth and, above all, to equalise the band gain, several pump lasers with different wavelengths are used at the same time. Figure 11.37 illustrates this technique. By using different wavelengths and by calculating carefully the power of each pump, the gain can be equalised in both the C-band and the L-band at the same time. The total gain, however, is not the mere sum of the gains obtained with the single pumps, because the wavelength difference between the pumps is comparable to the Raman shift, so that there is a partial exchange of power between the pumps which makes the design more complex.

11.6 Amplifier Type Comparison

Table 11.1 compares the main characteristics of the optical amplifiers described in this chapter. As it has already been mentioned, the amplifiers have complementary aspects which make them suited for different applications. EDFAs are well-established and reliable devices compatible with the fibers used in long distance communication. They achieve high efficiency and do not cause distortion or crosstalk. On the other hand, however, they allow amplification only around $1550\,nm$, they can

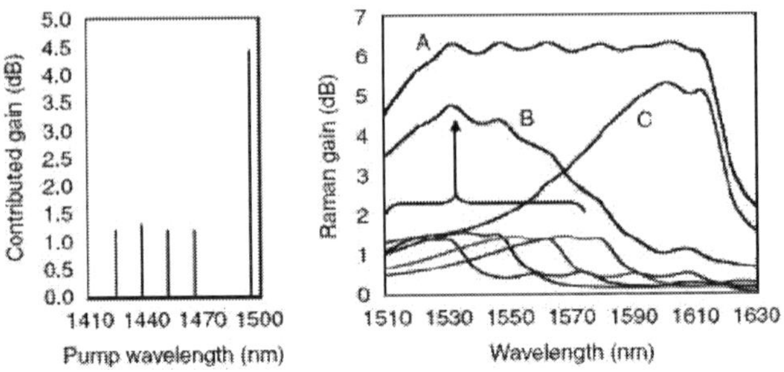

Fig. 11.37: Bandwidth broadening and gain equalisation in multipump Raman amplifiers.

only be made of special fibers, and they employ pump lasers and equalisers, so that they are quite expensive. Therefore, they are ideal devices for high capacity transmission systems operating in C-band.

SOAs are compact devices which can be easily integrated with other components. They have a high bandwidth, which may vary according to the band gap, and they are cheap because they do not require additional components. Nevertheless, they cause distortion, crosstalk and polarisation sensitivity. They are thus suited for local or metropolitan networks (LAN and MAN), and for low cost applications which do not require high bit rates. Finally, Raman amplifiers are quite expensive

Parameter	SOA	EDFA	Raman	Unit of measurement
Gain	> 30	> 40	> 25	dB
Wavelength λ	1280 - 1650	1530 - 1560+	1280 - 1650	nm
Bandwidth (3 dB)	60	30 - 60	pump dependent	nm
Saturation power	15	20	0.50∗pump	dBm
Max saturation power	18	22	0.75∗pump	dBm
Polarisation	< 0.5	0	0	dB
Noise figure	8	5	5	dB
Pump power	$< 400\,mW$	25	30	dBm
Time constant	0.2 ns	10 ms	$\approx fs$	-
Size	compact	rack mounted	bulk module	-
Price	low	medium	high	-

Tabella 11.1: Comparison of optical amplifiers.

and not very efficient, because they require several high power pump lasers and provide limited gains. Nevertheless, they do not require specific materials and can amplify in every band, so that they can be used in high capacity systems in S, C, and L band. Furthermore, Raman amplification is the only one which can be applied to pure silicon, which is an indirect band gap semiconductor. In the future, this characteristics may be exploited in *silicon photonics*, which aims at obtaining all the optical functionalities (sources, modulation, switching, detection) by employing CMOS technology, thus aiming at high integration with microelectronics.

Chapter 12

Lasers

Laser is the acronym for Light Amplification by Stimulated Emission of Radiation. The term is now commonly used to refer to an optical oscillator with two basic functions. The first one is light amplification by means of a gain medium which becomes active due to the effect of pumping; the second one is the optical feedback by means of two-mirror optical cavities.
This chapter deals with the different types of laser devices, focusing in particular on their application in optical fiber communication. To begin with, in order to understand some basic concepts about light emission by means of semiconductor materials, we will consider LEDs (Light Emitting Diodes), simpler semiconductor light sources without a proper resonant cavity. Lasers will be treated afterwards, focusing on their main characteristics, such as, for example, the types of material, the different resonant cavities and mirrors, or the spatial and spectral emission properties.

We now briefly introduce some of the aspects which will be analysed in detail in the following sections.
The characteristics of laser radiations highly depend on the type of material used as an active medium. Due to stimulated emission, single molecules or semiconductor crystals emit light at a wavelength which is inversely proportional to the difference in energy between the excited state and the ground state. The wavelength varies according to the active medium. Amplification based on stimulated emission practically occurs within a limited bandwidth centred around one wavelength. In gas lasers, the bandwidth is typically around $0.5\,nm$, whereas in solid-state lasers it ranges from $1\,nm$ to $10\,nm$. In semiconductors, due to the structures of bands, the bandwidth is significantly higher and can reach $40\,nm$. The emission wavelength may be selected by using compound semiconductors with the suitable molar fractions, typically III-V ternary and quaternary compound semiconductors such as Ga, In and As, P respectively.
The usual optical resonator is a simple Fabry-Perot cavity, consisting of two opposite flat mirrors. Due to the cavity configuration, light is continuously reflected back and forth, thus generating a

standing wave. Typically, there are two Bragg gratings acting as mirrors, so that the amount of reflected light is accurately controlled and high reflectivities are achieved. Gratings are also useful devices for the selection of cavity modes in the cavity, so that lasers with extremely narrow linewidths are obtained.

Consider $g\,(cm^{-1})$ as the optical gain produced by stimulated emission when light propagates along 1 cm of the active medium (gain per unit length), $\alpha\,(cm^{-1})$ the losses per unit length, R_1 and R_2 the mirror reflectivities, and L the length of the laser cavity. If the intensity of light inside the cavity is constant, the gain equals the losses, thus giving the *threshold gain* g_{th}, which is:

$$g_{th} = \alpha + \frac{1}{2L} ln\left(\frac{1}{R_1 R_2}\right).$$

Oscillation occurs only if the optical gain is higher than the total losses (i.e. the sum of the absorption and the reflection losses) and if the resonance wavelength is within the gain bandwidth of the active medium. The laser beam thus created has a spectral linewidth considerably narrower than that of the luminescence or of the spontaneous emission associated with the corresponding stimulated emission transition.

In a Fabry-Perot cavity, the oscillation occurs at a wavelength λ such that the waves reflected by the mirrors are in phase (constructive interference), thus satysfying the condition:

$$L = m\frac{\lambda_0}{2n} \tag{12.1}$$

where n is the refractive index of the medium inside the cavity and m an integer. Since usually $\lambda << L$, there are several modes that lie within the gain bandwidth of the active medium. If the value of m changes by one, the wavelength in the cavity changes by $\Delta\lambda$, which, according to (12.1), corresponds to:

$$\Delta\lambda \simeq \frac{\lambda_0^2}{2nL}.$$

If the gain g at $\lambda + \Delta\lambda$ wavelength reaches the threshold, oscillation occurs also at this second wavelength. Therefore, it is possible to select the oscillation frequencies dynamically, thus enabling the production of wavelength-tunable lasers. Apart from lasers characterised by a very small size, the bandwidth of the gain spectrum includes a wide range of modes, called *longitudinal modes.* On the other hand, the field distributions in the direction perpendicular to the direction of propagation between the two cavity mirrors are called *transverse modes.* Their characteristics and fabrication techniques will be analysed in detail in the following sections.

One last remark must be made about the characteristics of the laser beam, which should have high spatial and temporal coherence. Furthermore, the beam spot size should be as similar as possible to the spot size of the fiber mode or of the waveguide mode excited by the laser itself.

12.1 LED

LEDs are optical sources based on the spontaneous emission of electromagnetic radiation occurring mainly in the visible or infrared spectrum. They consist of p-n junctions which emit light when directly polarised. By applying a direct polarisation voltage V to a $p-n$ junction, the electrons are able to overcome the potential barrier of the depletion region, so that they can recombine with the holes. More specifically, the process is:

- The mobile electrons in the n region (attracted to the p region due to the bias voltage applied) enter the depletion region, i.e. they enter the conduction band. At the same time, the positive holes propagate in the opposite direction, thus creating more holes in the valence band;
- In the depletion region, the electrons and holes tend to recombine, i.e. electrons tend to fall back to the valence band, thus neutralizing the holes;
- Due to recombination, the LED emits light.

This process actually occurs in every diode, not only in LEDs. The difference, however, is that in traditional diodes the energy released due to recombination is thermal (or non-radiative), whereas in LEDs the recombination generates electromagnetic radiation. In LEDs, therefore, the radiative recombination fraction (i.e. the recombination that results in the release of a photon) must be higher than the non-radiative one. We define as *internal quantum efficiency*, η_{int}, the ratio between the radiative recombination rate R_r and the total recombination rate, which is the sum of R_r and R_{nr} (non-radiative recombination rate). We thus obtain:

$$\eta_{int} = \frac{R_r}{R_r + R_{nr}}.$$

The internal quantum efficiency can also be expressed as a function of the carrier recombination lifetime, that is, the average time an electron spends in an excited state before it recombines. We thus obtain:

$$\eta_{int} = \frac{\tau_{nr}}{\tau_r + \tau_{nr}}, \tag{12.2}$$

where $\tau_r = 1/R_r$ and $\tau_{nr} = 1/R_{nr}$ are, respectively, the radiative and non-radiative recombination lifetimes. Therefore, the internal quantum efficiency is high if the radiative lifetime is low.

The total optical power emitted by the LED, P, corresponds to the total energy per second, which is the product between the total number of photons and the energy of a single photon per second, E_p. According to the definition of radiative recombination, the number of photons corresponds to the product between the number of excited electrons, N, and the internal quantum efficiency, η_{int}. It follows that:

$$P = (N\eta_{int}E_p) / t, \tag{12.3}$$

whereas the bias current corresponds to:

$$I = Ne\,/\,t, \tag{12.4}$$

where e is the charge of the electron. Following (12.3) and (12.4) we obtain:

$$P = (It/e)(\eta_{int}E_p)/t = \frac{\eta_{int}E_p}{e}\,I. \tag{12.5}$$

It is thus clear that the optical power is proportional to the bias current of the junction.

So far, we have described LEDs based on p and n semiconductors with the same *band gap*, that is, homojunction LEDs. These devices can be further classified into *surface emitting* and *edge emitting* LEDs. There are two main disadvantages for both. To start with, the active region, which consists of the depletion region and its surrounding area, is too diffuse. It follows that the efficiency is very low, because high current density is needed to obtain the required recombinations. Furthermore, the light beam is quite wide, and approximately half of its power is concentrated within a visibility cone of angular width $\theta = 120°$ in the plane parallel to the edge, and a much narrower angle in the plane perpendicular to the edge.
This drawbacks can be overcome by using heterojunction devices, which are made of two different types of semiconductors, each one having its own *energy gap*. Recombinations are thus confined in a restricted active region, and it is possible to guide the light radiation in a single direction. If we place, for example, a semiconductor with a small energy gap, e.g. gallium arsenide, GaAs (E_g=1.42 eV), between two layers of a material with a wider energy gap, e.g. aluminium gallium arsenide, AlGaAs (E_g=1.92 eV), we are able to reduce the active region and, at the same time, to guide the light in only one direction, as if the central region were a waveguide, since $n_{GaAs} = 3.66 > n_{AlGaAs} = 3.2$. Finally, there are also double heterostructure LEDs, which are produced by using three different types of semiconductor.

LEDs and Lasers

LEDs differ from lasers because they have no optical cavities or reflective surfaces, so that they can be considered an *incoherent* light source. Furthermore, their spectral linewidth is quite wide, up to 100 times wider than the linewidth of lasers. In GaAs devices, for instance, the linewidth at room temperature ranges between 30 and 40 nm. Consequently, LEDs support different modes and are often used as *multimode optical sources*.
Compared to lasers, LEDs have lower modulation bandwidths and lower fiber coupling efficiency. This last defect basically depends on the low focusing of the emitted beam, and it is usually avoided by using *direct coupling*, which consists in inserting a multimode fiber directly in a semiconductor structure in order to place it as close as possible to the active region. The light power coupled into a fiber also depends on the numerical amplitude, NA, of the fiber. In step-index fibers, the coupled

power corresponds to:

$$P_{in} = P_0 \, NA^2, \tag{12.6}$$

where P_0 is determined by (12.3). In a multimode step-index fiber, for example, where $n_1 = 1.48$, $n_2 = 1.46$ and $P_0 = 100\ \mu W$, we obtain $NA = (n_1^2 - n_2^2)^{1/2} = 0.2425$ and P_{in}=5.88 μW. Theoretically, by applying the same formula to a *graded-index* fiber with NA=0.275 we would obtain P_{in}=7.56 μW, which means that the power launched into the fiber is 10% lower than the power P_0 emitted by the LED.

Nevertheless, LEDs can be exploited in short-distance and narrow-bandwidth applications due to other positive features. To start with, their *simplicity of manufacturing* and *low costs*, which are linked to the absence of reflective surfaces, and their *reliability*, since they have a longer lifetime due to their lower tendency to degrade. In addition to this, LEDs have a *reduced temperature dependence* (because they are not threshold devices like lasers), and they are formed by *simpler circuits*, due to the absence of temperature stabilising devices and to the lower current. Finally, their *linearity* is useful for analog modulation.

12.2 Semiconductor Materials for Lasers

In optical systems, the most interesting semiconductor materials for light-emitting devices are based on $GaAs$ and InP. Nevertheless, many different types of semiconductors can be used to produce laser devices such as optical memories, printers, CD players and molecular spectroscopes.

The most important feature to consider when selecting the material for lasers (usually heterostructure lasers) is the quality of the heterojunction interface between the two semiconductors. In fact, it is essential to minimise the crystal defects, so that the lattice constants of the two materials must be the same, within a 0.1 % maximum lattice mismatch. The relation between the energy gap, E_g, the lattice constant and the corresponding wavelength for some binary, ternary and quaternary compounds used in telecommunication have already been shown in figure 11.24. In particular, the figure shows that by varying the mole fractions of the compounds it is possible to obtain the same lattice constants, thus being able to produce lasers with a selected output wavelength. GaAs, for instance, which has a 564 pm lattice period, is used as the active layer of double heterostructure lasers. The outer layers are usually made of AlGaAs, which has a comparable 566 pm lattice period. Furthermore, as it has already been shown, the relation between the energy gap and the emission wavelength of the photon involved in the emission or absorption processes is:

$$\lambda_0 \simeq \frac{1.24}{E_g},$$

where E_g is expressed in eV and λ in μm. By accurately selecting the composition of $InGaAsP$ lasers, so as to obtain the same lattice constant of the InP substrate, it is thus possible to cover the

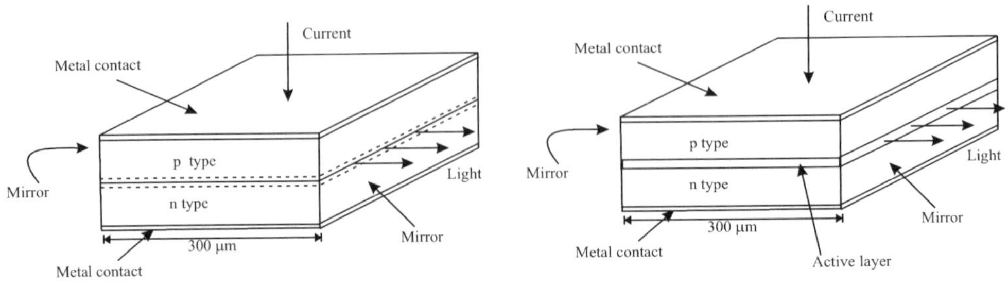

Fig. 12.1: $p - n$ junction (left) and heterojunction (right).

bandwidths between $1.1\,\mu m and 1.65\,\mu m$, that is, all the bandwidths used in modern optical fiber communication systems.

Homojunction and Heterojunction

The core of a semiconductor laser is formed by a junction between two semiconductor materials, p and n, with different doping levels. An example is shown in figure 12.1 (left). The concept of $p - n$ junction lasers was first discovered in 1961, and the first laser was then produced the following year. Semiconductor $p - n$ junction lasers work very simply. When the semiconductors with different doping levels are joined, an electron diffusion from the n-region to the p-region occurs, whereas the holes travel in the opposite direction. The process continues until a steady state is achieved, i.e. when the electric field generated by the acceptors (negatively charged) in the n-region and the donors (positively charged) in the p region opposes to a further transfer of carriers. If direct polarisation (forward bias) is applied, however, an external voltage is applied to the junction, so that its internal electric field decreases, thus reducing also the barriers to the diffusion of carriers across the junction. A small region where electrons and holes recombine is thus created. Recombination can be radiative, when it results in the emission of a photon whose energy equals the energy gap, or non-radiative. The inverse process, called absorption, can also occur. Furthermore, if the forward voltage is increased beyond a certain value, the carriers injected into the device cause the so-called *population inversion*, so that stimulated emission of photons becomes more probable than their absorption. The injected current required to obtain population inversion is called the *threshold current* of the laser. By further increasing the injected current, the emitted power rapidly increases and the typical current-voltage characteristics of lasers are obtained.

When population inversion is achieved, the $p - n$ junction is able to amplify the electromagnetic radiation, thus obtaining an *optical gain.* Since in homojunction lasers the carriers are not easily confined, the gain region, i.e. the region in which both electrons and holes are present, is very small, approximately $0.01\,\mu m$, and the threshold current is thus very high.

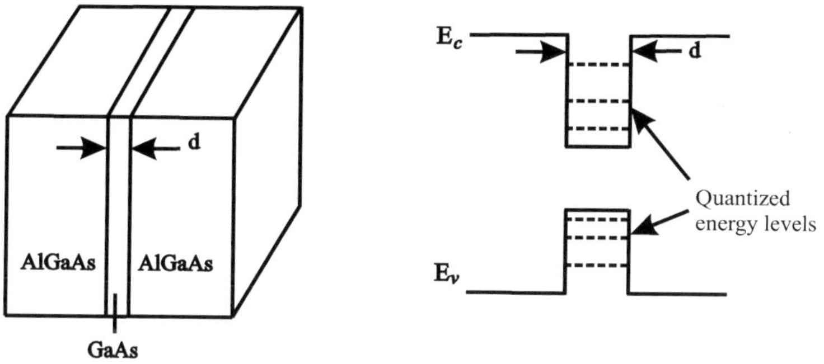

Fig. 12.2: Schematic representation of a QW structure in which a thin $GaAs$ layer is incorporated between two semiconductor $AlGaAs$ regions with higher band gap (left). The thickness d of the $GaAs$ layer in which the conduction electrons are confined is so tight that the energy levels can be quantized (right)

Thus, in order to increase the laser efficiency, $p-n$ homojunctions have been substituted by heterojunctions, structures in which a layer of smaller band gap material is placed between two layers of semiconductor materials with higher band gap and different doping level, as shown in figure 12.1 (right). Heterostructures show considerable advantages. In homojunction there are significant gain losses because the threshold currents are very high due to the absence of confinement for the injected current and for the optical field propagating in the device (see previous section). In heterojunctions, on the other hand, the injected carriers are confined into the central layer, called *active layer*, by means of the potential barriers which are created thanks to both the different doping concentration and lower band gap of the central layer with respect to the outer layers. The situation has already been illustrated in figure 11.25. The uniformity and the carrier density in the active layer reduce the number of electrons and holes required to reach the laser threshold.
As for in SOAs, the increase of energy also determines the reduction of the refractive index. It follows that the active layer has a higher refractive index than the outer layers, so that it creates a waveguide. This is an important aspect because the good performance of lasers also relies on the possibility of confining the optical field into the gain region. In fact, this also results in the reduction of the absorption losses in the outer layers, which further reduces the threshold current.

Quantum Wells

An extension of heterostructures can be realised by reducing the thickness of the central active layer to the order of the De Broglie wavelength of electrons and holes, which ranges between $1\,nm$ and $50\,nm$; the layer thus created is known as *quantum well* (QW). The reduction of thickness alters

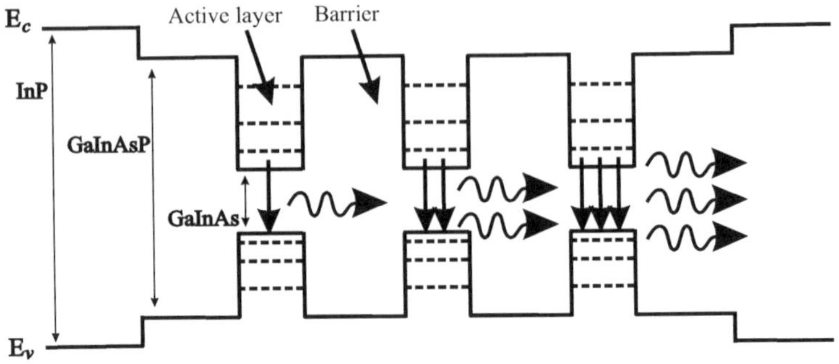

Fig. 12.3: Schematic representation of an $InGaAs/InGaAsP$ MQW laser structure.

the energy levels of the carriers in the semiconductor, which can be quantized within the potential barriers of the heterostructure, as shown in figure 12.2. The performance of the semiconductor source is thus enhanced because the electrons and holes tend to move to the first energy level where the charge density is higher and the threshold current thus decreases (up to 50% lower). If high optical gain is required, in order to compensate for the reduced thickness of the active layer, *multiple quantum wells* (MQW) can be exploited. As shown in figure 12.3, MQW structures consist of a small number of periodically arranged quantum wells separated by a significant distance, which is necessary to avoid the overlap between the wave functions of the electrons.

Quantum well lasers thus show higher gain and lower threshold currents than traditional lasers, in which quantum efficiency is limited only by the optical absorption of the electrons and holes. Consequently, they are nowadays the most widely used devices to produce semiconductor lasers.

12.3 Optical Resonators

Optical resonators (or resonating cavities) have two main functions. On the one hand, they enhance the action of the active medium because the electromagnetic field is able to cross the medium repeatedly due to the multiple reflections of the mirrors. On the other hand, they define the resonant frequency. The most simple type of optical resonator is known as *Fabry-Perot cavity* and consists of two opposing mirrors with the active medium inbetween. In semiconductor lasers, the easiest way to realise the mirrors is to place two flat surfaces at the ends of the active layer. Nevertheless, if no other treatment is applied, their reflectivity R is quite low, as it can be understood by the formula $R = (n-1)^2/(n+1)^2$, where n is the refractive index. In fact, if we consider the typical refractive index of a semiconductor active medium, $n = 3.5$, and the refractive index in air, $n = 1$, we obtain $R = 0.3$. The reflectivity of one of the mirrors is usually made close to 100% at the laser wavelength, whereas the transmittivity of the output mirror is non-zero, so that light can be transmitted outside the cavity, thus creating the laser beam. In order to understand how the cavity works, figure 12.4 shows the one-dimensional cavity model. In the cavity length L, a wave of phase constant β propagates through an active medium of refractive index n, gain per unit length (usually assumed to be 1 cm) $g\,(cm^{-1})$, and power loss $\alpha\,(cm^{-1})$. The medium is between two opposing mirrors, which are placed in $z = 0$ and $z = L$ positions along the cavity length L. R_1 and R_2 are the reflectivities of the mirrors. At least one of the reflectivities (in the case considered, R_2) must be below 1, so that the radiation can propagate outside the cavity. Consider E_0 as the electric field of the wave in $z = 0^+$ propagating along the positive direction of the z axis. The wave will propagate through the active medium for a length L and will hit the mirror in $z = L$ with an incident electric field $E_2 = E_0 e^{-\frac{\alpha}{2}L} e^{\frac{g}{2}L} e^{-j\beta L}$, since α and g are parameters which refer to the wave power. The wave will thus be partially reflected by the mirror, it will travel back in the opposite direction, hitting the mirror in $z = 0$. Finally, it will hit once more the mirror in $z = L$ with an electric field:

$$E_0' = E_0 \sqrt{R_1} \sqrt{R_2} e^{-\alpha L} e^{g L} e^{-j\beta 2L}.$$

As the field inside the cavity oscillates at constant amplitude, the incident and the back-reflected electric field must be the same ($E_0 = E_0'$). By substituting to E_0' the equation just found, we thus obtain:

$$\sqrt{R_1} \sqrt{R_2} e^{-\alpha L} e^{g L} e^{-j\beta 2L} = 1.$$

The result is thus a relation between complex numbers, which needs to be verified in terms of magnitude and phase. The minimum gain which allows oscillation is obtained by equating the

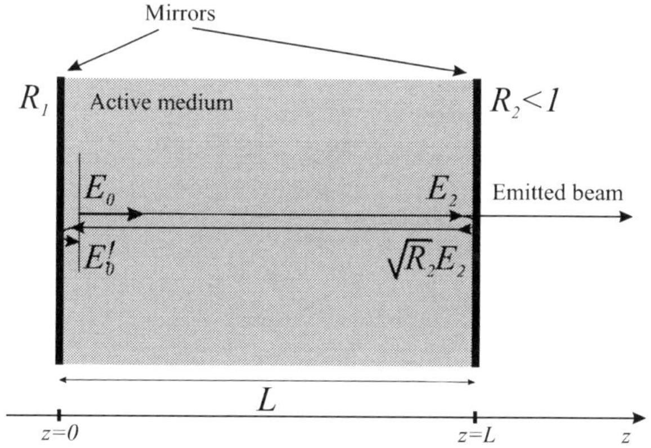

Fig. 12.4: One-dimensional Fabry-Perot cavity. The active medium is placed between two mirrors, one of which must necessarily be semireflective so that the field can be transmitted outside the cavity. Inside the cavity, the wave reflects back and forth between the two resonator mirrors (multiple reflection).

magnitudes, and it is called *threshold gain*, which corresponds to: g_{th}:

$$g_{th} = \alpha + \frac{1}{2L}\, ln\left(\frac{1}{R_1 R_2}\right); \tag{12.7}$$

Consider the phase:

$$arg(R_1) + arg(R_2) + 2\,\beta\, L = m\, 2\pi, \tag{12.8}$$

where m is an integer($m = 0, 1, 2, ...$). This equation allows to determine the wavelengths at which resonance may occur:

$$\lambda_0 = \frac{2\, n\, L}{m - \frac{\phi}{2\pi}}, \tag{12.9}$$

where $\beta = n\, 2\pi/\lambda_0$, n is the refractive index of the active medium, and $\phi = arg(R_1) + arg(R_2)$. Frequency can thus be expressed as:

$$\nu = \left(m - \frac{\phi}{2\pi}\right)\frac{c}{2nL}. \tag{12.10}$$

Note that there are several wavelengths (or frequencies) at which resonance may occur. The resonance frequencies are usually called *cavity modes*. The minimum frequency spacing between them is:

$$\Delta\nu = \frac{c}{2nL} \tag{12.11}$$

which in terms of wavelengths becomes:

$$\Delta\lambda \simeq \frac{\lambda_0^2}{2nL}. \tag{12.12}$$

According to the equations so far described, the threshold gain decreases for increasing values of the reflectivities and of the cavity length. Reflectivities can be increased only up to a certain level, whereas the cavity length can be increased indefinitely, at least theoretically, until the threshold gain is reduced to a lower limit determined by the absorption of the medium. The length of the active region, however, has an upper limit due to (12.11) and (12.12). If we consider typical values for semiconductor lasers, such as $L = 300\,\mu m$ and $n = 3.5$, the spacing will be $\Delta\nu \simeq 143\, GHz$, which corresponds to $\Delta\lambda \simeq 1.1\, nm$ at $\lambda_0 = 1.5\,\mu m$. In a Fabry-Perot cavity, then, there are several resonance frequencies that can be amplified and contribute to the laser beam, in particular those within the spectral gain curve of the active medium. In semiconductors, the bandwidth can be a few tens of nm. An example of gain curve is shown in figure 12.5. As we will see in the following sections, there are different configurations of gratings which can be used to select the laser lines even more accurately than with Fabry-Perot resonators, thus enhancing the properties of the output beam. This is useful also because it is not possible to reduce too much the cavity length just to select a single or a few wavelength, because the threshold gain would then increase too much. Finally, note that for each resonant frequency the spectral width must be greater than zero, due to the finite value

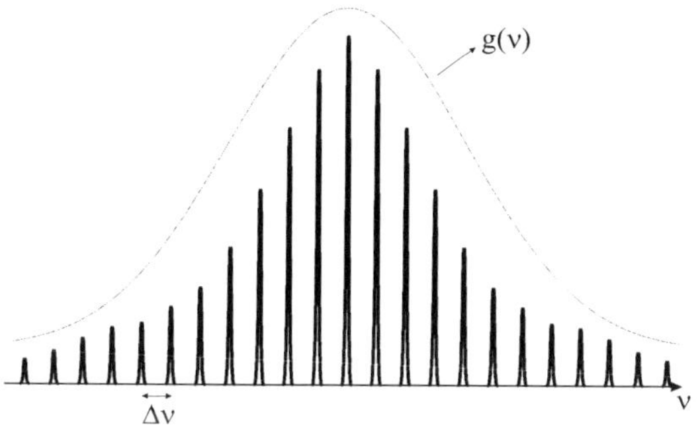

Fig. 12.5: Spectrum of the signal emitted by an active medium with gain $g(\nu)$ in a typical Fabry-Perot resonator with frequency spacing $\Delta\nu$.

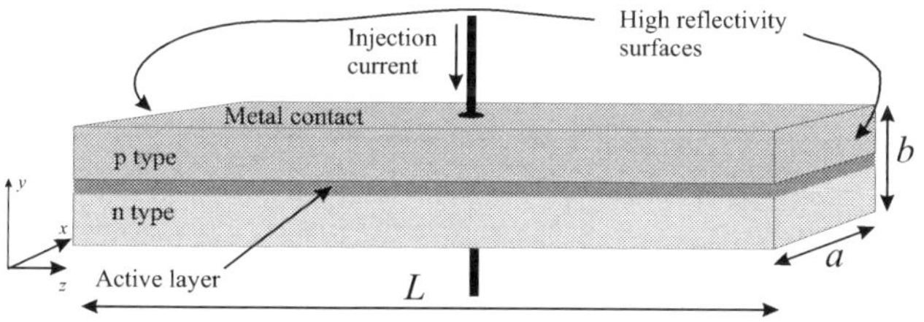

Fig. 12.6: Fabry-Perot cavity. The regions with different colours have a different doping level, but the same refractive index.

of the quality factor of the resonator. We can calculate the spectral width in wavelength, $FWHM_\lambda$, as it has been done in chapter 5 for the Fabry-Perot interferometer, thus obtaining:

$$FWHM_\lambda = \Delta\lambda \frac{1 - |\sqrt{R_1}||\sqrt{R_2}|}{\sqrt{\sqrt{|R_1||R_2|}}}.$$

For example, given $R_1 = 1$ and $R_2 = 0.95$, we obtain $FWHM_\lambda = 0.05\Delta\lambda$.

So far, we have considered a simplified one-dimensional model. Nevertheless, we cannot ignore the transverse direction. Consider, for example, a Fabry-Perot resonator of homogeneous material in which the cavity is a parallelepiped of length L, width a and height b (see figure 12.6). The dark area corresponds to the active region, where g is non-zero, whereas in the outer layers $g = 0$ and $\alpha \neq 0$. In semiconductor lasers, the active region extends along the longer side of the cavity, i.e. along the z axis. We thus assume that amplification and resonance will mainly occur for waves travelling along the z direction. Therefore, we approximate the waves propagating inside the cavity with the modes of a rectangular waveguide of side lengths $2a$ and $2b$. Since only part of the wave propagates in the active medium, if we suppose that the mirrors do not introduce any phase shift, i.e. the reflectivities are real, the equation in (12.7) becomes:

$$g_{th} = \frac{1}{\Gamma}\left[\alpha + \frac{1}{2L}\, ln\left(\frac{1}{R_1 R_2}\right)\right], \tag{12.13}$$

where Γ is the overlap integral between the mode and the active medium. The cavity modes, that is the wavelengths at which resonance may occur, are those for which the phase constants satisfy the resonance condition expressed in (12.8), which according to our hypotheses becomes:

$$\beta = m\frac{\pi}{L}.$$

In order to determine the wavelengths (or frequencies) of the cavity modes it is necessary to solve the equation for the phase constant β. In a dielectric guide with a rectangular core, however, the equation cannot be solved analytically, so that we need a further hypothesis, consisting in the following approximation of β:

$$\beta = \sqrt{\left(2\pi\frac{\nu}{nc}\right)^2 - k_x^2 - k_y^2}, \tag{12.14}$$

where k_x and k_y are the constants of the transverse modes of two flat layers, the first one having a core of size a and the second one of size b. The core refractive index is n for both, while the cladding is the air surrounding the cavity. In order for this propagation to be possible, it is essential that the modes are highly confined to the core. In fact, the length of the cavity of hundreds μm and the step index between the core and the cladding (between 2 and 3) satisfy this condition. Furthermore, due to the size of the core and the high step index, the modes are reasonably far from cutoff, so that, according to the properties of the modes in flat layers:

$$k_x \simeq i\frac{\pi}{a};$$

$$k_y \simeq l\frac{\pi}{b};$$

where i and l are integers. By substituting these terms into (12.14) we obtain:

$$\beta = \sqrt{\left(2\pi\frac{\nu}{nc}\right)^2 - \left(i\frac{\pi}{a}\right)^2 - \left(l\frac{\pi}{b}\right)^2}.$$

If we impose the resonance condition, ν becomes:

$$\nu = \frac{c}{2n}\frac{m}{L}\sqrt{1 + \left(\frac{L}{m}\frac{i}{a}\right)^2 + \left(\frac{L}{m}\frac{l}{b}\right)^2}.$$

Supposing that $i, l \ll m$ the frequencies of oscillation can be expressed as:

$$\nu \simeq \frac{c}{2n}\left\{\frac{m}{L} + \frac{1}{2}\frac{L}{m}\left[\left(\frac{i}{a}\right)^2 + \left(\frac{l}{b}\right)^2\right]\right\}. \tag{12.15}$$

The condition

$$\nu = m\frac{c}{2nL}, \tag{12.16}$$

previously found represents the limit for $i, l = 0$ of the equation (12.15).
By using the equation in (12.15) it is possible to state that for each sequence of i, l and m exists a well-defined resonance frequency and a well-defined mode. The difference of frequency between two modes having the same i and l and differing in m by 1 is:

$$\Delta\nu_m = \frac{c}{2nL}, \tag{12.17}$$

which corresponds to (12.11). The two modes have the same transverse configuration but differ in the field distribution along the longitudinal direction z. Consequently, $\Delta\nu_m$ is often called the spectral difference between two adjacent *longitudinal modes.* In semiconductor lasers, $\Delta\nu_m$ ranges from tens to hundreds of GHz. The spectral difference between two modes differing in i or l by 1 is, respectively:

$$\Delta\nu_i = \frac{cL}{2ma^2n}\left(l + \frac{1}{2}\right) \tag{12.18}$$

$$\Delta\nu_l = \frac{cL}{2mb^2n}\left(i + \frac{1}{2}\right) \tag{12.19}$$

and it is called the spectral difference between *transverse modes.* In fact, the modes have a different distribution on the transverse plane but the same distribution in the longitudinal direction due to the same frequency of spatial oscillation, β. Typical values of $\Delta\nu_{i,l}$ are of a few GHz.
Note that there are not two different types of modes, i.e. longitudinal and transverse modes, because all the modes oscillating inside the cavity have a spatial distribution in every direction. Nevertheless, in laser literature it is common to use the term "longitudinal modes" to refer to waves with different m, regardless of the transverse configuration. Therefore, the indices of two adjacent longitudinal

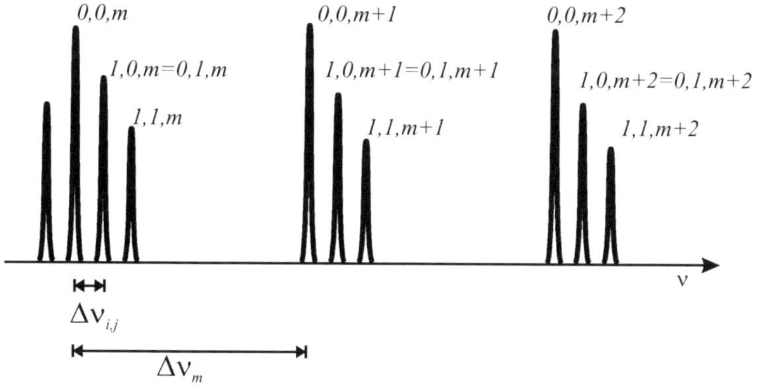

Fig. 12.7: Resonance frequency in a dielectric resonator using flat mirrors.

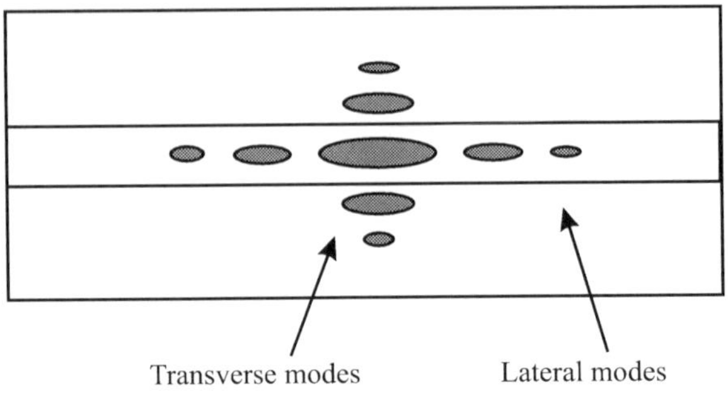

Fig. 12.8: Spatial modes of a laser. The so-called transverse and longitudinal modes are highlighted.

modes will be: m and $m+1$, or m and $m-1$. The same concept can be applied to transverse modes. To conclude, note that modes with the same m index and different i and l, but in which $i + l =$ *costant*, have the same frequency and are thus called degenerate modes (see figure 12.7). Although they cannot be distinguished spectrally, their field distribution in the transverse plane is different, because the spatial frequency along x and y is different. This concept will be further discussed in the following section.

12.4 Lateral Guiding

As we have seen in the previous section, even in a small interval of frequencies there are several modes with different field distributions in the plane orthogonal to the z axis. By combining the different modes we obtain the distribution of the laser beam, which is shown in figure 12.8. In laser literature, these modes are referred to as the *spatial modes* of the laser. We refer to *transverse modes* when the peaks of intensity are mainly distributed in the direction orthogonal to the junction plane. Conversely, when the peaks are in the parallel direction to the junction plane we speak of *lateral modes*. In single spatial mode lasers, light is emitted as a single spot. This is a favourable condition because it reduces the fiber coupling loss with optical fibers and considerably enhances the features of the emitted beam. As a matter of fact, higher-order modes are usually associated with nonlinear current-power characteristics and an increase in the intrinsic noise level.

In order to obtain a single spatial mode, however, it is not possible to use the whole semiconductor as a waveguide, but a single waveguide must be created inside it. By doing so, only the fundamental mode propagates in the cavity, which means that $i = l = 0$ in the equations previously found. Since the junction used in lasers is always a heterojunction, the waveguide in the transverse plane is obtained by exploiting the index difference between the active region and the outer layers.

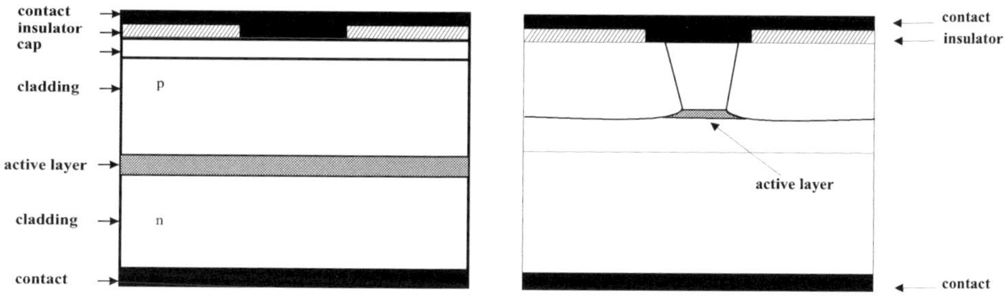

Fig. 12.9: Structure of a gain-guided laser (left) and of an index-guided laser (right).

On the other hand, *lateral guiding* can be obtained with different techniques which exploit either the optical features of the medium related to the refractive indices, or the electrical features related to current injection. The three main types of structures are:

- *gain-guided lasers*;
- *index-guided lasers*;
- *quasi-index-guided lasers.*

Figure 12.9 (left) shows the structure of a gain-guided laser. In these devices, the current is confined to the central region thanks to the deposition of an insulating material surrounding the contact layer. By doing so, the guiding effect along the lateral direction is obtained by increasing the refractive index. The increase of refractive index, in turn, is determined by the temperature increase caused by the current injection in the central region. These types of lasers are easy to produce, but cannot be exploited in sophisticated applications because their current-power characteristics are often nonlinear and their threshold current is relatively high, up to $100\,mA$.

Figure 12.9 (right) shows the cross section of an index-guided laser. In this case, lateral confinement is obtained by surrounding the active region with a layer of material with lower refractive index. Their fabrication is technically complex because they require multiple stage growth processes, but they achieve high performances, i.e. low threshold current (approximately $10\,mA$), linear current-power characteristics, and high output power.

A compromise between the two structures so far described are quasi-index-guided lasers, shown in figure 12.10. In this case, lateral confinement is created by fabricating a ridge on top of the layers of the junction. By selecting the width of the ridge, typically $2-3\,\mu m$, a single lateral mode emission can be achieved. This type of devices has linear current-power characteristics with a threshold current of about $30\,mA$. They are technically easy to produce because they do not require any regrowth.

12.5 Single Longitudinal Mode

Single spatial mode is not the only condition to obtain efficient lasers. In modern applications, such as when a laser is directly modulated at frequencies up to several gigahertz, it is important to produce lasers which are able to emit a single wavelength, i.e. single longitudinal mode lasers, also called *single frequency lasers*. As it has already been explained, the longitudinal modes can be spectrally defined by means of the equation in (12.15) for different values of the index m, which can be simplified to obtain (12.16). The spacing between the modes can thus be represented by (12.17). If no selection mechanism is applied, all the modes within the gain curve of the active medium can oscillate, so that several solutions have been developed to solve the problem. The most widespread and successful are:

- Distributed Bragg Reflector Lasers (DBR);
- Distributed Feed Back Lasers (DFB).

DBR lasers use distributed Bragg reflectors as mirrors at the opposite ends of the cavity instead of traditional discrete mirrors. Apart from reflecting the incident radiation, distributed Bragg reflectors are able to select inside the cavity the wavelengths which satisfy the Bragg condition:

$$\Lambda = m\frac{\lambda}{2n_{eff}} \qquad (12.20)$$

where Λ is the grating period, m an integer, and n_{eff} the effective index of the mode. Note that, in order to consider the guiding effects, the refractive index must be substituted by the effective index, n_{eff}, of the mode propagating in the waveguide. Figure 12.11 shows the structure of DBR lasers. Neither in DFB lasers the optical feedback is obtained by using the traditional discrete semireflective mirrors at the ends of a dielectric cavity. In fact, the optical feedback is provided by a grating obtained by chemical etching on one of the laser layers. The chemical etching does not usually involve the active medium because the result would be an increase in nonradiative recombination. This kind of devices is based on a principle already studied for gratings: each variation of the

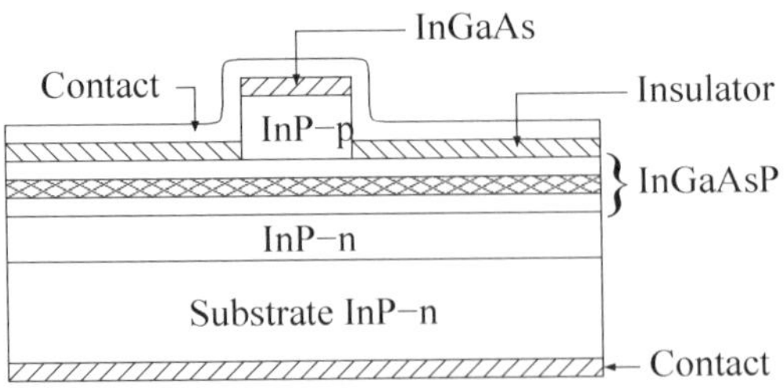

Fig. 12.10: Structure of a quasi-index-guided laser.

refractive index is associated with a small reflection of the electromagnetic field, which is added in phase to the field reflected at the previous index variation. The relation expressed in (12.20) is used to determine the grating period of DFB lasers once the operating wavelength has been selected. We thus obtain single frequency devices with side mode suppression ratio of more than 30 dB.
A DFB laser can be created with every type of lateral structure. Figure 12.12 shows a ridge-type DFB laser, i.e. a quasi-index-guided laser, which is efficient and easy to build.

Single frequency lasers can also be produced by employing short Fabry-Perot cavities. As a matter of fact, according to (12.17), for low values of L the spectral difference between possible frequencies can be widened up to the point where only one frequency falls within the gain curve of the active medium and can thus be amplified. In particular, if the spacing between the cavity modes, $\Delta\nu_m$, is higher than $\Delta\nu_0/2$, where $\Delta\nu_0$ is the width of the gain curve, that is:

$$\Delta\nu_m = \frac{c}{2L} \geq \frac{\Delta\nu_0}{2}, \tag{12.21}$$

there is only one longitudinal mode which can be amplified by the active medium. By considering (12.21) we can also define the maximum cavity size as a relation between the characteristics of the medium:

$$L \leq \frac{c}{\Delta\nu_0}. \tag{12.22}$$

This kind of device, however, is limited by the technological availability of ultrashort cavities. Furthermore, shorter cavity lengths also reduce the interaction between the wave and the active medium, thus decreasing the gain as well.

12.6 Optical Power

In order for the laser effect to be achieved, it is necessary that the gain compensates, and then overcomes, every loss which is generated inside the cavity, which implies at least to reach the

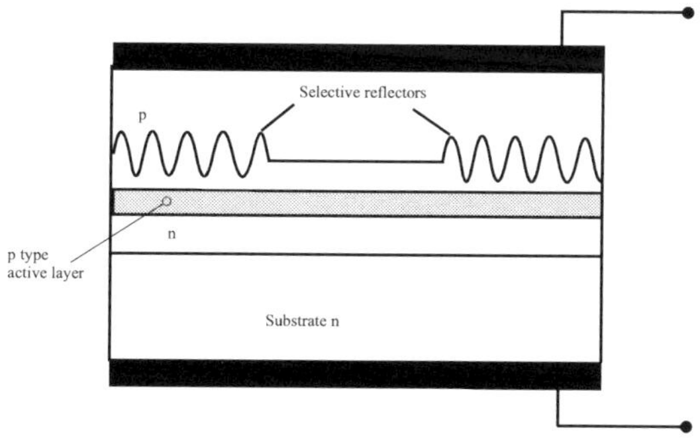

Fig. 12.11: Structure of a DBR laser.

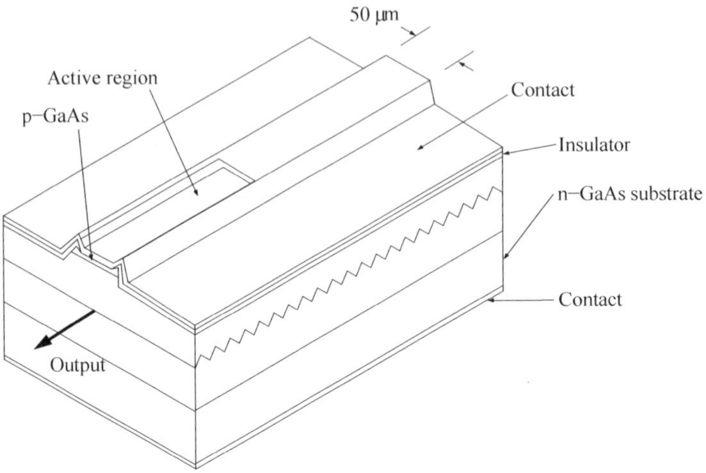

Fig. 12.12: Structure of a DFB laser.

threshold gain coefficient, reported in (12.13). The gain coefficient can also be expressed by means of the relation in (A.17), which can be reformulated so as to highlight that the cross-sections depend on the wavelength:

$$g(\lambda, z) = \sigma_e(\lambda)N_2(z) - \sigma_a(\lambda)N_1(z) .$$

It is thus clear that the pumping process of the active medium must achieve an adequate population inversion. In other words, the injection current I must compensate the effects of spontaneous recombination, as expressed by the following relation:

$$\frac{I}{qdWL} = \frac{N}{\tau_{sp}} + \frac{N_{ph}}{\tau_{ph}}, \tag{12.23}$$

where d, W and L are the linear dimensions of the active region, q is the electron charge, N and N_{ph} are, respectively, the concentrations of electrons and coherent photons in the cavity, τ_{sp} is the average lifetime of electrons before spontaneous recombination, and τ_{ph} is the average lifetime of photons. Consequently, in the equation (12.23), the left side equals the number of electrons in the cavity, the first term of the right side equals the number of electrons which spontaneously recombine, and the second term equals the photons generated due to spontaneous recombination.

It is possible to ignore N_{ph} at the threshold, because the coherent radiation will be generated only for higher injection currents. The equation in (12.23) thus allows the formulation of one of the most important parameters in semiconductor lasers, i.e. the threshold current I_{th}, as a function of N_{th}, the threshold concentration of electrons:

$$I_{th} = \frac{N_{th}}{\tau_{sp}} qdWL. \tag{12.24}$$

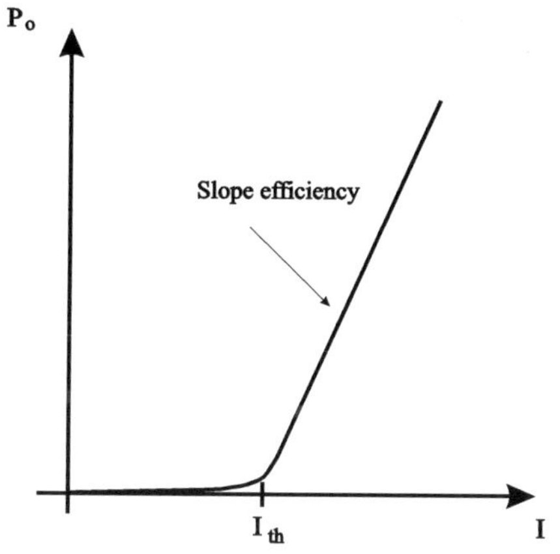

Fig. 12.13: Current-output power characteristics.

It is thus clear that I_{th} highly depends on the geometry of lasers as well.
On the other hand, the concentration of electrons beyond the threshold is practically constant and the same as N_{th}. Through the equations in (12.23) and (12.24) we find the concentration of photons inside the cavity:

$$\frac{I}{qdWL} = \frac{I_{th}}{qdWL} + \frac{N_{ph}}{\tau_{ph}}$$

so that:

$$N_{ph} = \frac{\tau_{ph}}{qdWL}(I - I_{th}). \qquad (12.25)$$

Consider now a single photon travelling across the optical cavity of refractive index n in a time interval $\Delta t = nL/c$. Only part of the photons, $N_{ph}/2$, reaches the output mirror, and an even smaller part, $(1 - R)$, is transmitted outside the cavity (R is the mirror reflectivity). The output power P_o can thus be expressed as:

$$P_o = \frac{N_{ph}(dWL)(1 - R)h\nu}{2\,\Delta t}.$$

If we substitute the equation in (12.25) and solve for Δt we obtain:

$$P_o = \frac{\tau_{ph}c^2 h(1 - R)}{2qn\lambda L}(I - I_{th}). \qquad (12.26)$$

The equation in (12.26) describes the current-output power characteristics, which is shown in figure 12.13. When the output power reaches the threshold current, it increases rapidly, with a gradient defining the *slope efficiency* of the laser, η_{slope}:

$$\eta_{slope} = \frac{P_o}{(I - I_{th})} \quad [W/mA].$$

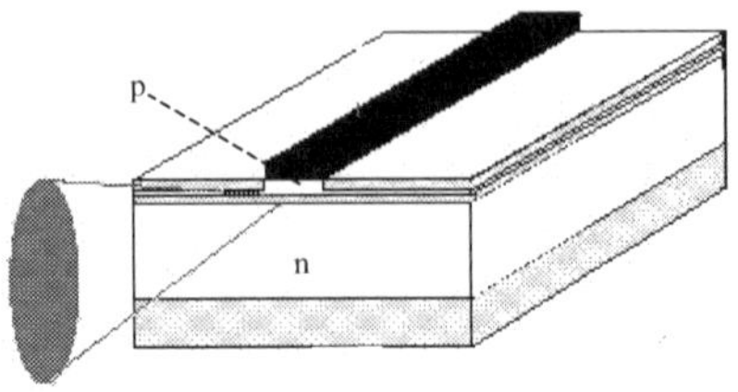

Fig. 12.14: Divergence of the laser beam while propagating in air.

The *power conversion efficiency*, on the other hand, is the ratio between the output optical power and the electrical input power (P_o/P_i). In the most efficient devices, power conversion efficiency can reach up to 30 – 40%.

12.7 Characteristics of the Emitted Laser Beam

As shown in figure 12.14 laser diodes do not emit a circular and collimated beam, but an elliptical and inherently divergent beam. The intensity distribution of the light beam near the laser output facet is called *near field*. When the laser beam propagates through air, the spot size increases due to the divergence of the beam. The angular intensity distribution of the light beam far from the laser facet is called *far field*.

In semiconductor lasers, divergence can be quite high, forming an angle of up to 30°. These results are a direct consequence of the extremely thin active layers. Since the output facet of the active layer is rectangular, the emitted laser beam will be approximately elliptical, with the major axis of the ellipse parallel to the junction. Due to diffraction, the radius diverges more rapidly in the direction perpendicular to the junction, so that in proximity of the output facet the elliptical cross section becomes circular. If the beam further propagates, it becomes once more elliptical, but the major axis will be perpendicular to the junction. If we consider the far field, we call *aspect ratio* the ratio between the divergence in the perpendicular direction to the junction and the divergence in the parallel direction. Gain-guided lasers usually have a high *aspect ratio* and are characterised by astigmatism, i.e. a shift along the z axis of the origin of divergence in the two transverse directions. In particular, astigmatism is higher than for index-guided lasers, where it is almost eliminated.

When laser diodes operate beyond the threshold, they emit a linearly polarised light. The electric field vector is parallel to the junction, whereas the magnetic field oscillates along the perpendicular direction. On the other hand, at low currents and below the threshold, the higher quantity of nonpolarised light causes spontaneous emission, and the polarization ratio is very low.

Another fundamental property of laser beams is their coherence. Waves are coherent if they have the same phase, i.e. their maxima and minima coincide over time and space. According to this definition, in order to obtain coherent light all the waves of the laser beam must be of the same wavelength. A light beam can have different temporal and spatial coherence. Temporal coherence is related to the monochromatic nature of light, whereas spatial coherence refers to the directionality of lasers. These properties can be calculated by using, for example, Michelson and Young interferometers, which measure the coherence time and distance. These are conventional quantities which indicate how long and how far the initial characteristics of the laser beam are kept unaltered, so that they can be calculated in a different point and at a different instant.

12.7.1 Gaussian Beams

Lasers often generate Gaussian beams, i.e. beams for which the electrical field distribution can be described by a Gaussian function:

$$E(r, z) \propto e^{-\frac{r^2}{W(z)^2}} e^{j\phi(r,z)}$$

where r is the transverse distance from the beam axis, z the coordinate along the propagation direction, $W(z)$ the so-called radius of the Gaussian beam, and $\phi(r, z)$ is the term which describes the phase evolution along the beam and the curvature of the wavefronts:

$$\phi(r, z) = kz - arctan\frac{z}{z_R} + \frac{kr^2}{2R(z)},$$

where $k = 2\pi/\lambda$ is the wave number, $R(z)$ the curvature of the wavefronts, and $z_R = \pi W_0^2/\lambda$ is the Rayleigh length, calculated at the beam focal point, where the beam radius W_0 reaches its minimum value. The beam radius evolves as:

$$W(z) = W_0\sqrt{1 + \left(\frac{z}{z_R}\right)^2}$$

whereas the curvature radius as:

$$R(z) = z\left[1 + \left(\frac{z_R}{z}\right)\right].$$

For $z >> z_R$ the radius evolves almost linearly, with a divergence angle corresponding to:

$$\theta = \frac{\partial W}{\partial z} = \frac{\lambda}{\pi W_0}. \qquad (12.27)$$

While propagating in a homogeneous medium, the Gaussian beams remain Gaussian and only their parameters change. This kind of beams is thus important in optics, and is also used to define the features of other types of beams, as we will see in the next section about the M^2 factor.

12.7.2 M^2 Factor

The fundamental parameters to determine the quality of a laser beam are the M^2 *factor* and the *Beam Parameter Product* (BPP), which are related as follows:

$$M^2 = BPP\frac{\pi}{\lambda}. \tag{12.28}$$

The M^2 factor is commonly used for gas lasers, such as CO_2 lasers, while the BPP is used for solid lasers, such as fiber lasers and $Nd:YAG$ lasers. The BPP is the product between the beam radius at its narrowest point (beam waist) and the beam divergence half-angle measured in the far field. Generally speaking, the higher the BPP, the lower the beam quality. The lowest BPP is achieved with a diffraction-limited gaussian beam and corresponds to λ/π. The M^2 factor, also referred to as *beam quality factor* or *beam propagation factor*, is related to the beam divergence. The beam divergence half-angle θ with respect to the beam axis corresponds to:

$$\theta = M^2\frac{\lambda}{\pi\, W_0}, \tag{12.29}$$

where W_0 is the minimum value of the beam radius. The total divergence angle of the beam is thus $2\,\theta$. The M^2 factor measures the divergence of a laser beam from *diffraction-limited* Gaussian beam, for which M^2 is 1, as can be derived from (12.27). Smaller values of M^2 are physically not possible and laser beams are often said to be M^2 times diffraction-limited. The M^2 factor also limits the degree to which the beam can be focused. For non circularly symmetric beams, the M^2 factor might differ in the two directions orthogonal to each other and to the beam axis. The M^2 factor can thus be calculated by measuring the evolution of the beam radius along the propagation direction. Similarly, knowing the M^2 factor helps predicting the evolution of the beam radius.
Apart from optical power, the beam quality factor also determines the *brightness*, B, of a laser beam. Brightness is the ratio between the total power and the product between the modal area at the beam waist and the solid angle in the far field:

$$B = \frac{P}{\pi W_0^2\,\pi\theta^2} = \frac{P}{\pi W_0^2\,\pi\,M^4\left(\frac{\lambda}{\pi W_0}\right)^2} = \frac{P}{\lambda^2}. \tag{12.30}$$

Brightness, therefore, also depends on the wavelength and is measured in $W\,sr^{-1}\,cm^{-2}$. For diffraction-limited beams, the brightness coincides with the optical power and it is generally used as a measure of the maximum optical intensity which can be generated at a given aperture and working distance.

An important property of optically pumped lasers is that they often generate laser beams with a much higher brightness than the one generated by the pump source, although the output power is lower. Consequently, optically pumped lasers are sometimes defined brightness converters.

Finally, note that brightness decreases due to deterioration of beam quality, thus implying that radiation is distributed from a few to several modes.

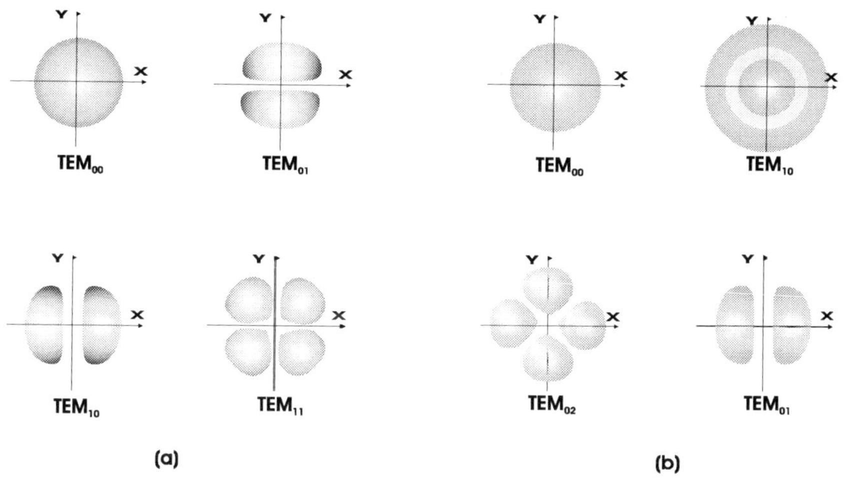

Fig. 12.15: Examples of TEM_{mn} modes generated in a cavity laser. (a) Modes with rectangular symmetry; (b) Modes with circular symmetry.

12.7.3 Modes

The power density of the laser beam is not constant along the transverse section, but varies at different positions mainly due to the number of oscillating modes and according to the size and geometry of the cavity. By varying the mirror geometries, for instance, it is possible to concentrate the field in proximity of the axis or to distribute it in a more homogeneous way. The beam intensity distribution can be described by a Gaussian function:

$$I(r) = I_0 \, e^{-r^2/r_0^2} \tag{12.31}$$

where I_0 is the on-axis intensity, r the radial distance from the centre of the beam, and r_0 is a typical parameter related to the beam radius. This type of distribution is widely used in industrial applications because, for the same power, it reaches extremely high peak values. Semiconductor lasers and optical fiber lasers have a field distribution which cannot be described by means of a Gaussian function, because all the oscillating modes within the cavity are considered. The modes are usually referred to as TEM_{mn}, where m and n are the indices of the modes. In modes with rectangular symmetry, m and n indicate the number of minima along the two directions orthogonal to the transverse plane, whereas in modes with circular symmetry they indicate the radial and angular mode orders. Figure 12.15 shows some field distributions of TEM modes. The TEM_{00} mode, for instance, has no minima, but only a central maximum, so that its distribution is very similar to a Gaussian distribution. The TEM_{10} mode, on the other hand, has a minimum along the x axis and none along y if we consider a rectangular symmetry, whereas it has a central maximum surrounded by a depressed area and then by a higher external ring if we consider a circular symmetry.

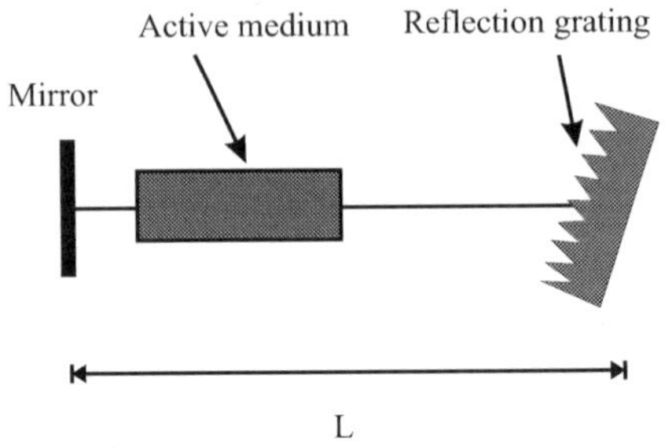

Fig. 12.16: Structure of a wavelength tunable laser with a diffraction grating used for the selection of the output wavelength.

12.8 Tunable Lasers

An inportant property of lasers is their ability to *tune*, i.e. their ability to adjust the output frequency. Some devices use filters external to the semiconductor laser, such as diffraction gratings or Fabry-Perot cavities, which guarantee the emission of a single wavelength. An example is shown in figure 12.16. The beam emitted by the diode laser is collimated by a lens, it travels through a diffraction grating and it is then reflected on a mirror. The diode laser facet facing the inside of the cavity has an anti-reflection coating, so that the diode only operates as a gain medium. The emission wavelength can be tuned by rotating the mirror, which in turn tunes the wavelength back-reflected by the grating inside the cavity. By doing so, the amplitudes of non-lasing modes are reduced by more than 40 dB below the amplitude of the mode selected at the output.

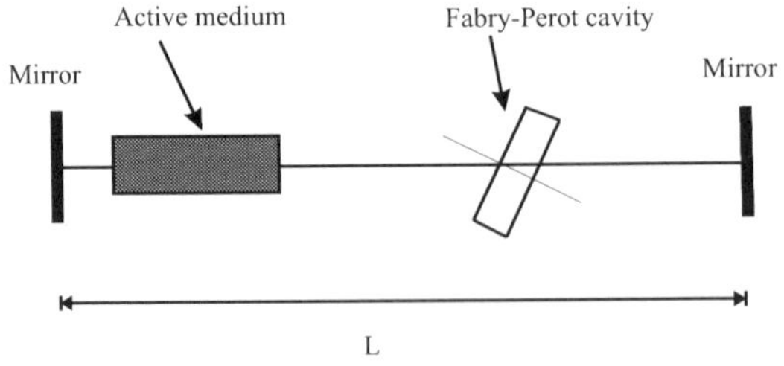

Fig. 12.17: Structure of a wavelength tunable laser with a Fabry-Perot cavity used for the selection of the output wavelength.

Figure 12.17 graphically shows another type of tunable laser. In this case, the inclination of the Fabry-Perot laser diode with respect to the axis of the structure can be varied to allow the wavelength selection. The process can be understood by looking at figure 12.18. As it is known, the spacing between the modes inside a cavity of length L is: $\Delta\nu_m = c/2L$. A Fabry-Perot placed inside the cavity introduces transmission peaks of spectral width $\Delta\nu_c$ which, in turn, are spaced by the so-called *free spectral range*, $\Delta\nu_{fsr}$. The ratio between these quantities defines the Fabry-Perot cavity finesse:

$$\mathcal{F} = \frac{\Delta\nu_{fsr}}{\Delta\nu_c}. \tag{12.32}$$

Consider also the $3 - dB$ bandwidth of the gain curve of the active medium, $\Delta\nu_0$. Two conditions must be satisfied for a single-mode emission. First of all, the $\Delta\nu_m$ between the cavity modes must be wider than half width of the Fabry-Perot transmission lobes. Secondly, the Fabry-Perot $\Delta\nu_{fsr}$ must be wider than half width of the gain curve. The first condition:

$$\Delta\nu_m \geq \frac{\Delta\nu_c}{2}, \tag{12.33}$$

allows the transmission by the Fabry-Perot resonator of only a single longitudinal mode of the cavity. On the other hand, the second condition:

$$\Delta\nu_{fsr} \geq \frac{\Delta\nu_0}{2}, \tag{12.34}$$

allows the amplification of only that specific mode. By combining (12.33), (12.34) and (12.32) we obtain:

$$\frac{\Delta\nu_0}{2} \leq \Delta\nu_{fsr} = \Delta\nu_c\,\mathcal{F} \leq 2\Delta\nu_m\,\mathcal{F} = \frac{c}{L}\,\mathcal{F}, \tag{12.35}$$

and thus the relation:

$$L \leq \frac{c}{\Delta\nu_0}\,2\,\mathcal{F}. \tag{12.36}$$

We can compare (12.36) to (12.22) in order to highlight that thanks to the Fabry-Perot resonator it is possible to obtain a single-mode emission with cavity lengths increased by a factor $2\,\mathcal{F}$. By

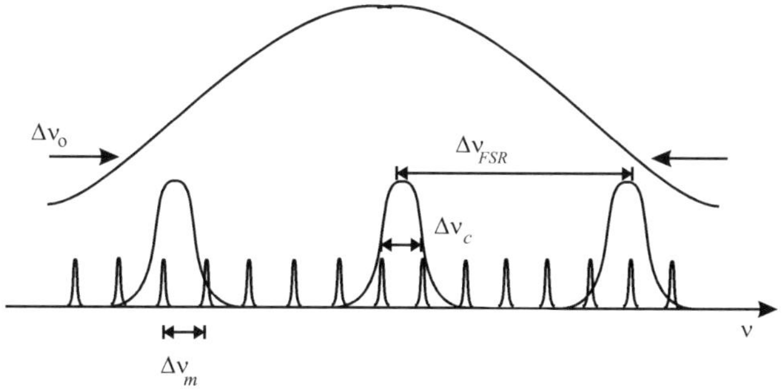

Fig. 12.18: Selection of the laser line.

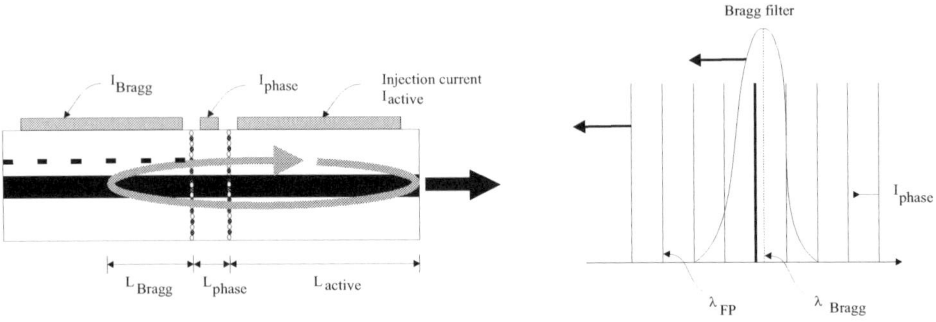

Fig. 12.19: Structure of a tunable semiconductor laser.

selecting the value $\mathcal{F} = 40$, for example, the cavity length L is increased by a factor of 80. If it is necessary, more than one Fabry-Perot resonator can be added, either of the same or of different type. If the Fabry-Perot resonators are of the same type, the following relation must be satisfied:

$$L \leq \frac{c}{\Delta\nu_0}(2\mathcal{F})^2 \tag{12.37}$$

which relaxes the previous condition.

The advantage of lengthening the cavity combines with that of tuning the laser emission. By rotating the Fabry-Perot resonator with respect to the z axis the angle of incidence of the rays with respect to the Fabry-Perot cavity itself is modified, thus changing also the oscillation frequency inside the cavity.

Nowadays, it is necessary to manufacture very compact tunable lasers and technologically advanced solutions have been developed thanks to the integration of planar technology. It is thus possible to obtain compact single-longitudinal-mode lasers which are wavelength tunable in a range of wavelengths of a few tens of nanometres. A schematic representation is shown in figure 12.19. The optical power depends on the current I_{active} injected in the active layer, usually made of a multiple quantum well material. In a traditional DFB laser, the grating would be fabricated directly on top of the active layer, but in this case a different section is used, the so-called *Bragg section*, which is made of a passive quaternary material. By injecting a I_{Bragg} current the carrier density N increases, thus causing a variation of the refractive index of the quaternary material, dn_Q/dN. Consequently, the effective index of the propagation mode is modified according to the following formula:

$$\Delta n_{eff} = \Gamma \frac{dn_Q}{dN} \Delta N \tag{12.38}$$

where Γ is the optical confinement factor of the field in the quaternary medium and ΔN the carrier density variation caused by I_{Bragg}. Since in gratings:

$$\lambda_{Bragg} = 2n_{eff}\Lambda,$$

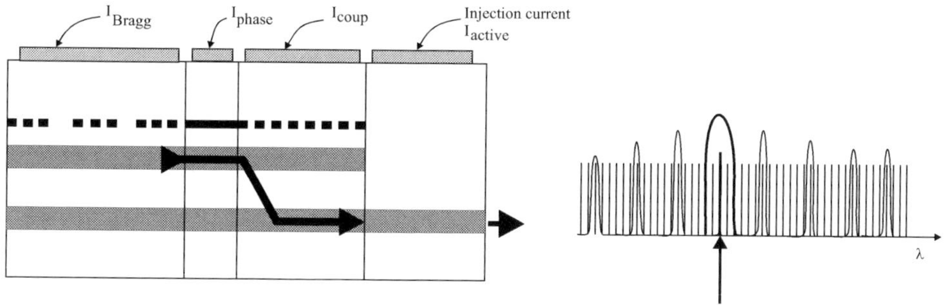

Fig. 12.20: Structure of a tunable semiconductor laser with two gratings used for the selection of the emission wavelength.

where Λ is the grating period, the effective index variation Δn_{eff} reduces the length of the wavelength reflected by the grating in the Bragg section, λ_{Bragg}.
As already seen, the wavelength of the laser oscillation can be calculated from:

$$L = m\frac{\lambda}{2n},$$

so that

$$\lambda = 2\frac{nL}{m},$$

where nL is the optical length of the Fabry-Perot cavity between the distributed mirror formed by the grating and the output facet of the active medium. In this structure, the optical length nL is:

$$nL = n_{active}L_{active} + n_{phase}L_{phase} + n_{Bragg}L_{Bragg}$$

where L_{Bragg} is the penetration length of the grating. The laser, therefore, emits on the Fabry-Perot mode at λ_{FP} closest to λ_{Bragg}. Furthermore, by current injection in the so-called phase section, the Fabry-Perot modes are further tuned due to the variation of n_{phase}. To conclude, by controlling the currents I_{Bragg} and I_{phase}, it is possible to tune the emission spectrum. Note that I_{Bragg} causes higher variations than I_{phase}, which thus provides a fine control of tunability.
The tunability interval only depends on the bandwidth of the gain curve of the active medium and on the desired side mode suppression ratio (SMSR). Generally speaking, the tunability interval ranges between 15 and 40 nm. The maximum SMSR, up to 40 dB, is obtained when the wavelength of the Fabry-Perot mode is exactly the same as the maximum reflection wavelength of the grating, i.e. $\lambda_{FP} = \lambda_{Bragg}$. On the other hand, the so-called *mode hop* occurs when two Fabry-Perot modes have the same spacing from the Bragg wavelength. In this case, the SMSR is at its minimum.
In order to achieve better performance in terms of tunability and emitted power, the structure can be further complicated. Figure 12.20, for example, shows tunable lasers using two gratings to control

the emission wavelength. More specifically, the second grating, by means of the injected current I_{coup}, allows to control the coupling wavelength with the waveguide beneath, which faces the active medium. Fine control of tunability, on the other hand, relies on the super structure grating with injection current I_{Bragg}, and finally on the current I_{phase}, which has a similar function to the case previously described.

12.9 Direct Modulation of Lasers

Lasers can be *internally* modulated by acting directly on the injection current. The threshold current in the output characteristics indicates that in order to cause the laser effect it is necessary that the injected carriers rise beyond a certain value. Consequently, if we inject an input current step, the signal arrives at the output with a delay t_r, which is given by the sum of the time t_s required by the device to arrive at the threshold, and the subsequent time required in order for the stimulated emission to produce an efficient optical beam. The first term influences the final delay more, due to the fact that the production of photons which are summed in phase occurs due to an avalanche ionization mechanism, which is described by the relation:

$$t_s = \tau_{sp} \ln \left(\frac{I}{I - I_s} \right) \tag{12.39}$$

where I is the amplitude of the current step, I_s the threshold current, and τ_{sp} the time required for electron-hole recombination. Thus, in most practical cases, the overall rise time of the laser is given by:

$$t_r \simeq t_s. \tag{12.40}$$

An efficient way to reduce the delay is to polarise the diode by injecting a current value close to the threshold current, so that it achieves a faster response time to the current variations. The rise time delay thus becomes:

$$t_r \simeq \tau_{sp} \ln \left(\frac{I - I_b}{I - I_s} \right) \tag{12.41}$$

where I_b is the bias current, which is selected as slightly lower than I_s. In commercially available devices, such as the one shown in figure 12.21, the rise time delay is approximately $0.5\,ns$, which equals a modulation velocity of about $2\,GHz$.

The only disadvantage is that since lasers are close to the switching threshold, they emit power also when unnecessary, thus causing a small waste of power and consequently decreasing the extinction ratio, i.e. the ratio between the high and low values of the transferred power. In directly modulated lasers a compromise is required between the rise time delay, which limits the modulation frequency, and its operating condition at low signals, which influences the quality of transmission.

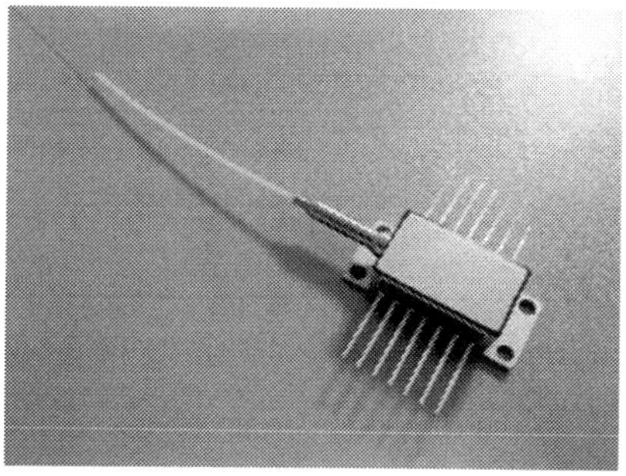

Fig. 12.21: *Example of a directly modulated laser up to 2.5Gb/s.*

Another problem is the so-called *frequency chirping*. The variation of injection current required to modulate the optical beam causes a continuous variation of the free carrier concentration inside the semiconductor used as an active medium, thus also varying the absorption coefficient and the refractive index. More specifically, the variation of the refractive index influences the position of the spectral lines, which are no longer stable in time, thus generating noise. The values of modulation velocity which can be achieved with direct modulation are not adequate to satisfy the current user bandwidth necessities, so that if a higher performance is required, *external* modulation of the beam can be achieved by means of optical modulators.

12.10 Noise

There are different sources of noise which affect semiconductor lasers. Nevertheless, many of them can be controlled, so as to limit the total noise of the device efficiently. The major sources of noise are *mode hopping* and *relative intensity noise.*

Mode hopping is a shift from one longitudinal mode to another. It occurs due to temperature changes in the active layer. If the temperature increases, the corresponding energy gap is reduced, thus enhancing the oscillation of modes with longer wavelengths. The shift of the output wavelength usually causes a brief noise transient. Once the temperature goes back to the initial value, the oscillation frequency will go back to its original value, too. Temperature changes are the most common source of noise and cause shifts of tens of nm. In fact, this phenomenon can also be exploited for laser frequency tuning within narrow ranges of frequencies. This result can be obtained by increasing the injection current, so that also the temperature of the active layer is raised. Temperature control is thus essential in semiconductor lasers, not only in order to avoid permanent damage to the devices, but also because it allows minimising or eliminating the shift between longitudinal modes.

Laser emissions are subjected to power fluctuations due to the statistics of photon generation in the active region. The power fluctuation causes relative intensity noise (RIN), which rapidly decreases when the injection current exceeds the threshold current.
In lasers, RIN can increase also due to uncontrolled light reflections, caused by a semiconductor active medium inserted into an external cavity, or generated at the ends of an optical fiber coupled with the laser itself. In this case, the noise is called *optical feedback noise*, thus indicating the phase and amplitude changes caused by the external cavity. Single-frequency lasers are very sensitive to optical feedback noise.

12.11 Vertical Cavity Surface Emitting Lasers

Vertical Cavity Surface Emitting Lasers (VCSELs) are the new generation of semiconductor lasers and have been key components in the development of photonics in recent years. They can be used for telecommunications, especially for local area networks, optical memories, sensors, laser printing or in spectroscopy and biomedics. VCSELs are semiconductor lasers which allow emission in the direction perpendicular to the substrate. They consist of an active medium, whose thickness is one laser wavelength, comprised between two DBR mirrors. The cavity axis is perpendicular to the active layer. The Bragg reflectors are produced using molecular beam epitaxy (MBE) or metal organic vapor phase epitaxy (MOVPE) and alternate layers of materials with low refractive index, e.g. high aluminium AlGaAs, to layers of materials with high refractive index, e.g. low aluminium AlGaAs. Each layer has a thickness of a quarter of the laser wavelength. In the active region there might be multiple quantum wells. Figure 12.22 shows a schematic representation of a VCSEL structure. The

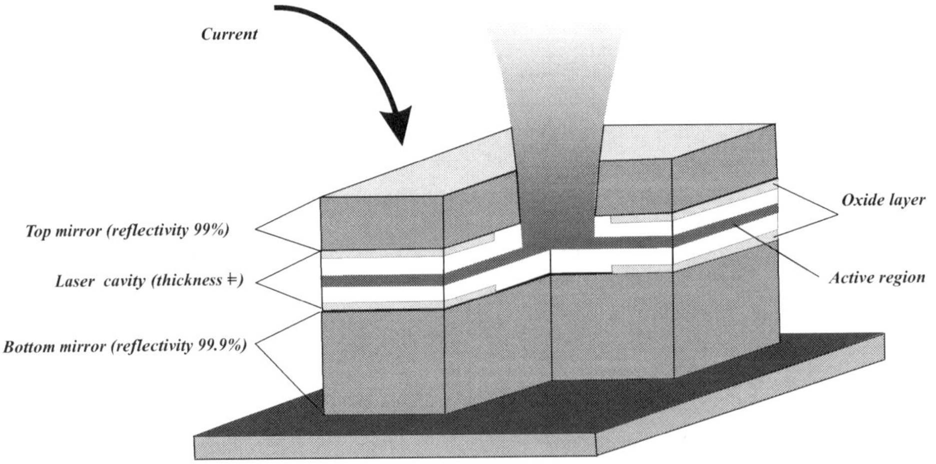

Fig. 12.22: Schematic representation of a VCSEL. The top and bottom mirrors consist of layers with alternating refractive index.

interest for this kind of devices started in the 1990s thanks to the enhancement of their performance. In fact, they are compatible with low-cost technologies, they enable on-wafer characterisation, and they can eliminate many of the drawbacks of traditional semiconductor lasers. Semiconductor lasers testing and characterisation, for instance, require an entire production cycle, because one must wait mirror formation, which is obtained by cutting the wafer vertically. Furthermore, since the two mirrors obtained have the same reflectivity, usually quite low (approximately 30%), it is necessary to modify the reflectivity of one mirror, so as to avoid that light is emitted on both ends, thus reducing the laser efficiency of 50%. Both the processes so far described are quite expensive. In addition to this, another major disadvantage of traditional lasers is the output beam divergence, which has an elliptical form.

Optical fiber coupling, therefore, requires efficient strategies to solve the problems so far highlighted. VCSELs are an adequate solution due to their peculiar structure. To start with, VCSEL testing and characterisation can be performed directly on wafer before cleaving the single chips. Secondly, the mirror reflectivity can be selected before production, so that *top-emitting* or *bottom-emitting* lasers can be created. Furthermore, by exploiting modern deposition techniques it is possible to control the mirror characteristics and achieve reflectivities up to 99.9%. The small size of VCSELs is another considerable advantage. In fact, it allows high density integration in two-dimensional (2D) arrays, it causes low power consumption and low parasitic capacitance, thus enabling high speed switching. VCSEL modulation can thus occur at $20\,Gb/s$, while in traditional DFB lasers it would be at $10\,Gb/s$. Moreover, by means of surface emission it is possible to control the output laser beam accurately, so as to obtain a low divergence beam with circular profile and no astigmatism. The beam can thus be easily coupled to an optical fiber, reaching an 80% coupling efficiency.

Finally, note that the extremely small size of the resonance cavity determines very low threshold currents, which can reach as low as $0.01\,mA$, and a spacing between two consecutive longitudinal modes of about $150\,THz$, which corresponds to a wavelength spacing of a few microns. This property makes the longitudinal modes of VCSELs intrinsically single mode. On the other hand, however, the thin active layer reduces amplification. Nevertheless, this aspect can be compensated by the high reflectivity of the mirrors, which are therefore constituted by long gratings of $4-5\,mm$.

Lateral Guiding

VCSEL transverse modes are linearly polarised (LP) due to the low index difference between the central region of the cavity and the outer layers. Figure 12.23 shows the intensity distributions of the first four LP modes. In order to obtain a single transverse mode, it is necessary to reduce the transverse size of the cavity or to modify the cavity so that the modes can be selected while the diameter of the cavity remains the same. Four basic structures of VCSEL have thus been created, each using different strategies for lateral waveguiding and lateral carrier confinement within the active region. Figure 12.24 shows these typical structures, which are: *mesa* structure, buried heterostructure, ion-implanted structure, and multi-oxide layer structure.

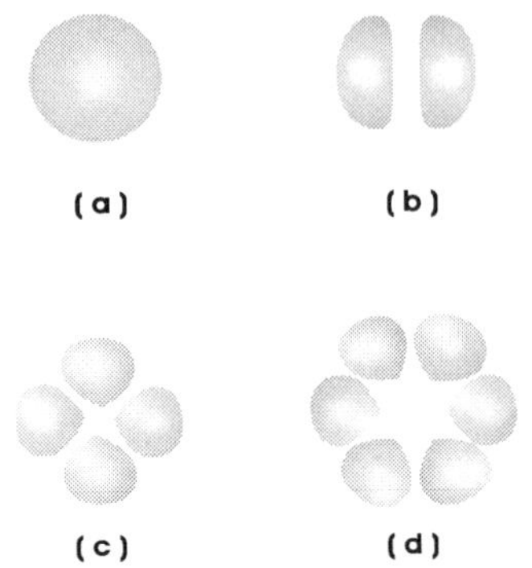

Fig. 12.23: First four LP modes of VCSELs: a) LP_{01}, b) LP_{11}, c) LP_{21}, d) LP_{31}.

Mesa structures and buried heterostructures use chemical etching of the epitaxial material, thus forming a circular mesa structure. Furthermore, in VCSEL buried heterostructures the confinement layer is implemented by epitaxial regrowth of a semi-insulating material with a wide energy gap. The process so far described enhances many aspects of the simple mesa structure because it avoids carrier loss which would occur during surface recombination at the borders of the mesa itself. Moreover, it enhances heat dissipation, which would reduce the output optical power. Finally, the process is used to *planarise* the surface, so that it can be integrated later on.

Both structures allow index-guiding because the semiconductor material is a circular waveguide with a higher refractive index than air (mesa structure) or than the semi-insulating material (buried heterostructure).

On the other hand, ion-implanted VCSELs are gain-guided, because the diameter of the active gain region is defined by the ion implantation region, which confines the current to the central region. Due to current confinement the central region is heated, so that the refractive index increases, thus causing also an index guiding. This process is known as *thermal lensing*. Since ion implantation is relatively simple, cheap, and creates planar structures, ion-implanted VCSELs are the most widely used devices.

Multi-layer oxide VCSELs, however, are the most advanced and promising structures. They consist of multiple high Al content AlGaAs layers placed above the active medium and obtained through aluminium oxidation. The layers act as confinement barriers for the injected current because the aluminium oxide is not conductive. Furthermore, they operate as optical waveguides, since Al_2O_3 has a lower refractive index than AlGaAs (1.6 versus 3). Multi-layer oxide VCSELs are more efficient, and they are being used to develop devices emitting in the visible window and in the second or third windows of telecommunications. Nevertheless, oxidation must be accurate because the oxidation time and quality are influenced by several parameters, so that reproducibility is lower than ion-implanted structures.

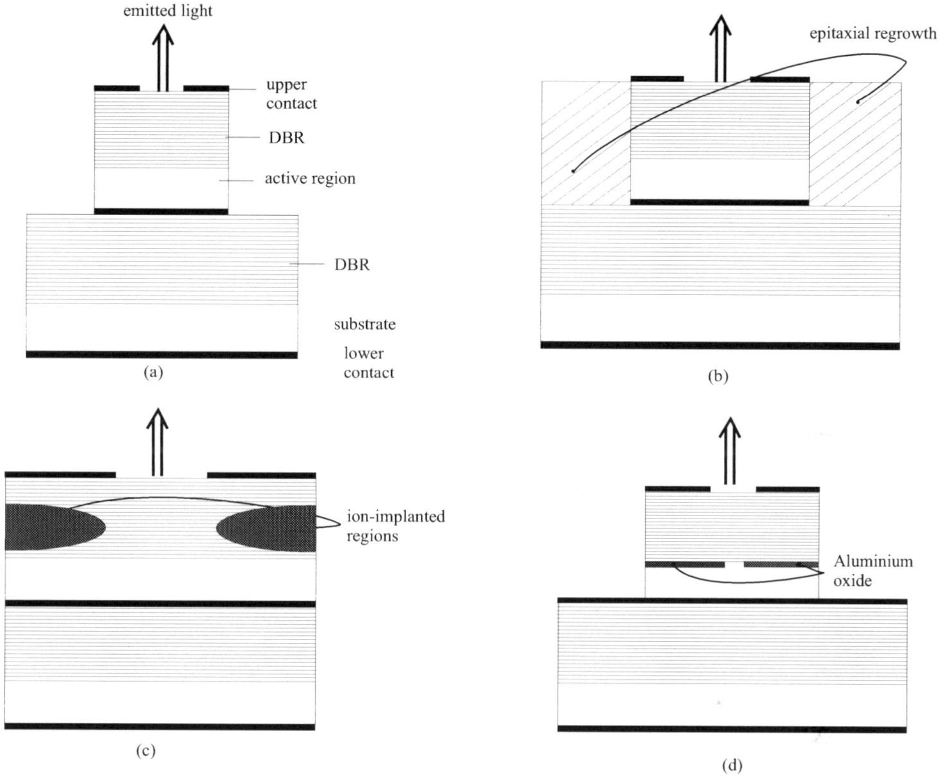

Fig. 12.24: VCSEL structures: a) mesa structure, b) buried heterostructure, c) ion-implanted structure d) multi-layer oxide structure.

Chapter 13

Photodetectors

A *photodetector* represents the dual component of a laser, since it converts electromagnetic energy carried by the propagating modes into electrical form. The electrical signal is then amplified and processed by the receiving circuit. Photodetectors work differently and can be classified according to the way in which different materials absorb the electromagnetic radiation. The most common photodetectors are *semiconductor detectors*, which exploit the *photovoltaic effect*, and *vacuum photodiodes*, based on the *photoelectric effect*. The properties and manufacturing process of these devices highly depend on the operating wavelength and the type of application desired.

The most widespread techniques to detect electromagnetic energy, and more specifically to detect light in optical communication systems, are based on the photovoltaic effect, i.e. the interaction between light and semiconducting materials. In simple terms, the effect can be described as follows. Each incident photon on an intrinsic semiconductor is absorbed, thus creating an electron-hole pair. When an electrical field is applied, the electron-hole pair contributes to the *photocurrent*, i.e. the current which can be measured at the ends of the device. In the absence of incident photons, the current of the external circuit is near zero, because the intrinsic semiconductor has a low conductivity. Conversely, if the incident radiation has enough optical power, the number of carriers produced through absorption becomes significant, thus producing an appreciable current which causes a voltage drop across the bias resistor at the output inside the external circuit.

In the photoelectric effect, on the other hand, electrons are ejected from the material if they are exposed to adequate photon energy. In particular, the photon energy must be higher than the potential barrier which electrons have to overcome in order to become free. Photodetectors based on the photoelectric effect are important devices in several fields, but they have the major disadvantage of low infrared sensitivity. Nevertheless, thanks to the recent introduction of specific alloys, their use has been extended to wavelengths of up to 2 μm. A commercially available photodetector is shown in figure 13.1.

13.1 Photodetection Process

Consider a photon of energy $h\nu$ propagating in a semiconductor with energy gap E_g. If $h\nu < E_g$, the photon energy is not able to excite the electron, so that the photon is not absorbed and the medium will be *transparent.* If $h\nu \geq E_g$, the photon may be absorbed by the semiconductor, thus generating an electron-hole pair, which makes the medium *opaque.* Note that photon absorption is considered a *stochastic* process, and that the higher the probability that photon absorption occurs, the more opaque the medium will be. Since $c = \lambda\nu$ and $h\nu \geq E_g$, so that $h\frac{c}{\lambda} \geq E_g$, in order for the absorption to occur we must have:

$$\lambda(\mu m) \leq \frac{hc}{E_g} \simeq \frac{1.24}{E_g\ (eV)}\ . \tag{13.1}$$

Therefore, the absorption process highly depends on the wavelength, so that semiconductors may be suitable for the first, second, or third window.
It is not only the energy gap, however, which determines the wavelength. Consider, for example, the GaAs band structure, in which the valence band maximum and conduction band minimum coincide at the same value of $\overline{k}$, the wave vector. GaAs is thus defined a direct band gap material, and the transition from the valence to the conduction band involves only one electron and one photon. On the other hand, if we consider the Ge or the Si band structure, the valence band maximum and conduction band minimum have different $\overline{k}$. The band gap is thus indirect, so that transition requires an electron, a photon, and a *phonon,* because during the transition $\overline{k}$ is also modified. Due to the need of one more particle, the phenomenon is thus less probable, and the absorption is lower. The material can become highly absorbent only if the photon energy is further increased, so that transition from the valence to the conduction band can occur without involving the phonon. Indirect

Fig. 13.1: Commercially available photoreceiver. On the right, the fiber optic input connector.

band gap materials can thus be successfully used to produce photodetectors with high energy incident beams which allow photon absorption without involving phonons. Due to vibrational relaxation, the excited electron then goes back to the absolute minimum in the conduction band. The result of the absorption process so far described is that in indirect band gap materials, transition from the transparent to the opaque region occurs gradually. Furthermore, since the transition probability is low, indirect band gap materials are never used for lasers.

In the following sections, we will consider some more basic parameters which are necessary to describe the interaction between the electromagnetic field and the medium in photodetectors, i.e. absorption coefficient, quantum efficiency, and responsivity.

13.1.1 Absorption Coefficient

Consider a field of power P_0 propagating along z and incident on a layer of material with thickness d, as shown in figure 13.2. If we ignore the reflection loss, after crossing the layer the power, P, is:

$$\frac{P}{P_0} = e^{-\alpha d} , \qquad (13.2)$$

where α is the *absorption coefficient* measured in cm^{-1}. The parameter α thus indicates the inverse of the distance which the field has to travel for its power to decrease by $1/e$. Figure 13.3 shows the absorption coefficient as a function of the photon energy in Ge, Si, GaAs, and other binary III-V semiconductors for $T = 300\ K$. Note that semiconductors are transparent for $h\nu < E_g$, whereas they are opaque at higher energies. The transition, as already mentioned, is sharper in direct band gap materials, whereas it is gradual in indirect gap materials.

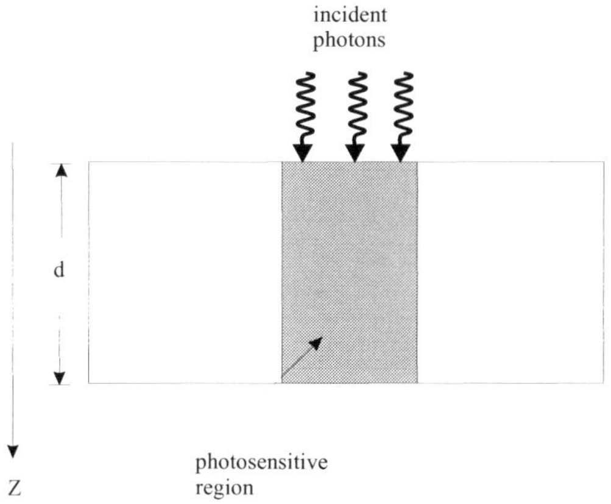

Fig. 13.2: Photons incident on a semiconductor material.

13.1.2 Quantum Efficiency

Quantum efficiency, η , is the probability that a single photon incident on the device generates an electron-hole pair which contributes to the photocurrent. As a matter of fact, the electron-hole pairs do not always contribute to the photocurrent because, in order to do so, the carriers must recombine after being separately diffused to regions with different doping levels.

Semiconductor photodetectors use quantum efficiency to describe the probability of quantum interaction occurrence, and the field is usually defined in terms of the *photon flux*, Φ. If we consider ν as the frequency of the incident field with power P, the flux Φ is formed by photons of energy $h\nu$, and in the time interval Δt the transmitted energy, E, is:

$$E = P\,\Delta t.$$

The number of received photons, N, in the same interval is:

$$N = \frac{E}{h\nu} = \frac{P}{h\nu}\Delta t.$$

The photon flux, Φ_{ph}, i.e. the number of photons per unit time, is thus:

$$\Phi_{ph} = \frac{P}{h\nu} \; .$$

For a given photon flux Φ_{ph}, the quantum efficiency η is defined as:

$$\eta = \frac{\Phi_e}{\Phi_{ph}} \; , \tag{13.3}$$

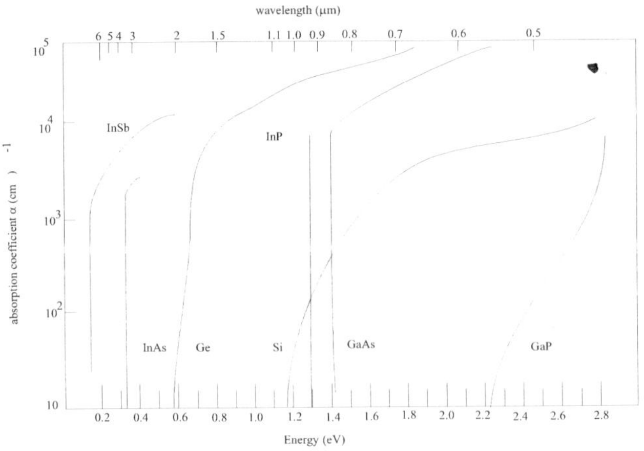

Fig. 13.3: Absorption coefficient as a function of photon energy in Ge, Si, GaAs, and other binary III-V semiconductors for $T = 300\ K$. Note that the transition in direct band gap materials is sharper than in indirect band gap materials.

where Φ_e is the flux of electron-hole pairs contributing to the current. Quantum efficiency also depends on the fraction of photons reflected by the surface, i.e. by the reflectivity, R. Therefore, it can be written as:

$$\eta = (1 - R)\xi(1 - e^{-\alpha d}) . \tag{13.4}$$

where $(1 - R)$ represents the transmitted power with respect to the incident power, and can thus be maximised by reducing the reflectivity R as much as possible, e.g. by applying antireflective coatings to the surface. The term $(1 - e^{-\alpha d})$, on the other hand, represents the absorbed power, i.e. the fraction of photons absorbed and, ultimately, the electron-hole pairs generated. According to the equation in (13.4), in order to achieve an adequate photon absorption the thickness d must be sufficiently high when compared to $1/\alpha$. Finally, the parameter ξ represents the fraction of electron-hole pairs which do not recombine before contributing to the photodetection current. The value of ξ is always lower than 1 and depends on the structure and the material of the device considered. Recombination usually occurs on the surface and can be increased by adding adequate layers. The spectral variation of photon flux is shown in figure 13.4. Figure 13.5, on the other hand, shows quantum efficiency as a function of the wavelength for some semiconductor materials. For high values of λ (i.e. when $\lambda > \lambda_g$, being λ_g the gap wavelength) the photons have not enough energy to cause the band-to-band transition and the quantum efficiency is approximately zero. As the wavelength decreases, η increases until it reaches a maximum and then decreases again if we further reduce λ. At extremely low wavelengths, the absorption is so high that photons are absorbed on the surface for the most part, so that they do not generate electron-hole pairs which can contribute to the photodetection current. In fact, the carriers tend to annihilate by recombining immediately. The problem, however, can be solved by using heterostructure photodetectors.
Note also that figure 13.5 highlights the possibility of operating on broad spectral bandwidths, especially on those used in communication systems.

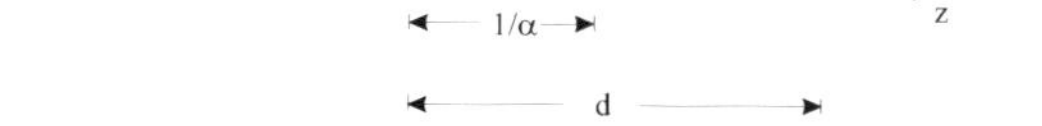

Fig. 13.4: Photons incident on a layer of material with thickness d are either reflected, absorbed, or transmitted by the medium.

13.1.3 Responsivity

Quantum efficiency refers to microscopic quantities. Nevertheless, the macroscopic quantities are more useful in order to characterise a device, especially the incident electromagnetic field and the photocurrent. The power associated to a given photon flux Φ_{ph} is:

$$P = h\nu\Phi_{ph} \; , \tag{13.5}$$

whereas the current associated to the electron flux Φ_e is:

$$I = q\Phi_e \tag{13.6}$$

where q is the electron charge. *Responsivity*, $\Re$, is the ratio between the photocurrent and the incident power:

$$\Re = \frac{I}{P} \quad [A/W] \; . \tag{13.7}$$

The ratio can also be written as:

$$I = \Re P \tag{13.8}$$

which highlights that responsivity is in fact the slope of the incident optical power-versus-photocurrent curve. Typical values of $\Re$ are around $1\,A/W$ and are supplied on the device data sheet. A photodiode with $0.85\,A/W$ responsivity which absorbs a light beam of $1\,mW$, for instance, releases a photocurrent which corresponds to: $I = \Re P = 0.85 \cdot 10^{-3} = 0.85\,mA$. Note, however, that the variation described in (13.8) is not indefinitely linear, because at high power the numbers of incident photons can be significantly higher than the electron-hole pairs available in the material. Therefore, the current saturates and reaches a maximum beyond which it cannot further increase.
By substituting (13.5) and (13.6) to (13.7), and by exploiting the definition of η in (13.3), we thus

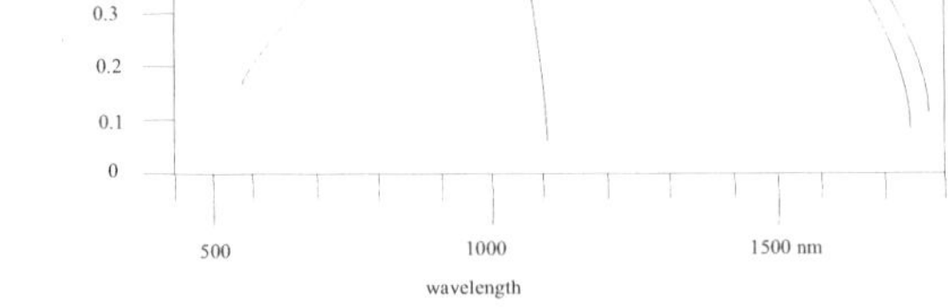

Fig. 13.5: Quantum efficiency as a function of wavelength.

obtain:

$$\Re = \frac{q\Phi_e}{h\nu\Phi_{ph}} = \frac{q}{hc/\lambda}\eta \simeq \eta\frac{\lambda\ (\mu m)}{1.24} \quad [A/W]\ . \tag{13.9}$$

Responsivity, therefore, relates the microscopic level, described by η, to the macroscopic one, defined by incident optical power and photocurrent, by means of the term $\frac{\lambda\ (\mu m)}{1.24}$. $\Re$ thus increases linearly with λ, but then tends to decrease due to quantum efficiency η. In fact, as described in the previous section, quantum efficiency abruptly falls at increasing values of λ, until it becomes zero when it reaches the gap wavelength λ_g.

13.2 Photodiodes

A *photodiode* is an inversely polarized $p-n$ junction. As shown in figure 13.6, when a beam of photons of sufficient energy strikes the junction, it is absorbed by the material at a rate described by the absorption coefficient α and generates electron-hole pairs due to the photovoltaic effect. Photon absorption occurs over a distance $1/\alpha\ cm$. The pairs contribute to the photocurrent only if the electrons and holes are separated and moved in opposite directions by the electric field. In the junction, this occurs only in the *depletion region*, where it is also better for the electron-hole pair to be generated.

Three circumstances may occur. In the first case, the electron-hole pairs are generated in the depletion region and are immediately separated by the electric field. When the electrons and holes arrive in the quasi-neutral region, they can recombine with opposite charges, which are provided, directly or indirectly, by the generator. The current generated thus contributes to the photocurrent.

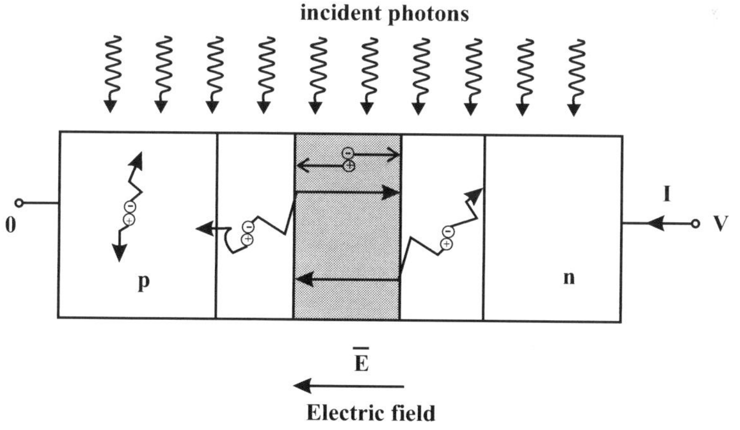

Fig. 13.6: Photons incident on a reverse biased $p-n$ junction photodiode. The grey area is the depletion region, which is placed between two diffusion regions.

In the second case, the electrons and holes are generated into the quasi-neutral regions (far from the depletion region) and randomly move around until they recombine. Since recombination occurs in the same neutral region where the carriers were generated, the process does not contribute to photocurrent. There is only an increase in the number of carriers, which increases conductivity. In the third and last case, the electrons and holes are generated in the quasi-neutral regions, but near the depletion region, so that they can cross the junction by diffusion and reach the depletion region before recombining. There, the electric field causes the carriers to cross the junction rapidly. The current thus generated contributes to photocurrent. The region in which this process can occur corresponds to the *minority carrier diffusion length*, i.e. the diffusion length through which minority carriers travel before recombining in the quasi-neutral regions. The diffusions lengths of silicon and gallium arsenide, for instance, are $L_{Si} = 500\ \mu m$ and $L_{GaAs} = 3\ \mu m$ respectively. All in all, the total contribution to photocurrent is generated within the depletion region and at either side of it, in the quasi-neutral regions, only within a width corresponding to minority carrier diffusion length. One last aspect to consider is that photon absorption also depends on the position of the $p-n$ junction. If the depth from the surface on which the photons strike is high, the photons enter the semiconductor and are absorbed before reaching the depletion region. Consequently, the depletion region ought to be near the surface, but not too much, in order to avoid a region with many defects.

13.2.1 Response Time and Bands

The response time of a photodiode has a lower limit which is set by three factors: the transit time through the depletion region, the diffusion time through the quasi-neutral region, and the junction parasitic capacitance.

The *transit time*, τ_t, is the time that the electrons and holes generated in the depletion region require in order to reach the quasi-neutral regions. It is extremely low (up to fractions of ns in silicon) because the electric field applied to the junction is very high.

The *diffusion time* is the time that minority carriers (electrons or holes) require to reach the depletion region, from where they are then carried to the other quasi-neutral region due to the action of an electric field. *Diffusion* processes are much slower than *drift* processes. The diffusion time is similar to the carrier lifetime and it is proportional to the minority carrier diffusion length. For example, the hole diffusion time through $10\,\mu m$ of silicon is approximately $40\,ns$, whereas the electron diffusion time over a similar distance is around $8\,ns$.

The *parasitic capacitance* is the junction capacitance and can be expressed as: $C_j = \varepsilon \frac{S}{d}$, where ε is the dielectric permittivity of the material, S is the photosensitive area and d is the depletion layer width. Consequently, a small depletion layer width, d, gives a high junction capacitance. Nevertheless, capacitance should be as small as possible in order to reduce the τ_{RC} time constant of the equivalent circuit of the photodiode. As it is shown in figure 13.7, apart from the junction capacitance, C_j, there is a current generator, I, which represents the photogenerated charge flux.

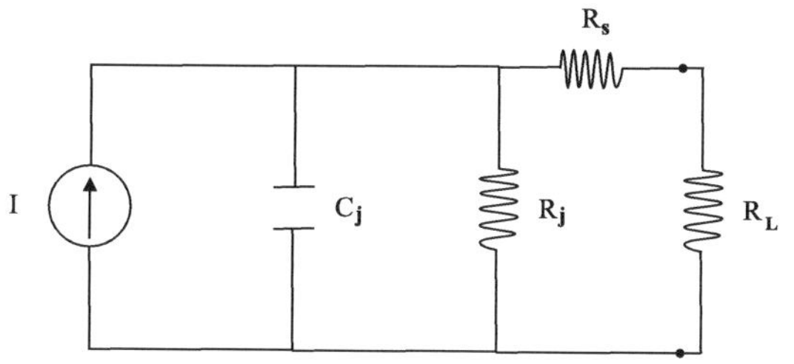

Fig. 13.7: Equivalent circuit of the $p-n$ junction diode. I is the photocurrent, C_j is the junction capacitance, R_j is the junction resistance, R_s is the series resistance, and R_L is the load resistance.

Furthermore, there is the parallel resistance, R_j, and the series resistance, R_s, which takes into account the contact resistance, the conductors, and the resistance in the p and n regions of the semiconductor. Finally, R_L is the load resistance. The parallel resistance, R_j, is intrinsically extremely high and can reach some tens of $M\Omega$, whereas the series resistance, R_s, is very low (only a few Ω). Consequently, it is the load resistance R_L which determines the τ_{RC} constant in the circuit:

$$\tau_{RC} = (R_s + R_L)\, C_j \approx R_L\, C_j .$$

Since the typical values of C_j are between 1 and $2\,pF$, if $R_L = 50\,\Omega$ the τ_{RC} time constant is between $50\,ps$ and $100\,ps$.

We can thus find B, the bandwidth of the photodiode, i.e. the maximum operating frequency or bit rate. If we ignore the diffusion time, B becomes:

$$B = \frac{1}{2\pi(\tau_t + \tau_{RC})} . \qquad (13.10)$$

Note once again that the concept of bandwidth, expressed in Hz, is applied to analog signals, whereas the term bit rate, expressed in bit/s, to digital signals. Nowadays, the general term bandwidth is used for both types of transmission. In a photodiode with $\tau_t = 100\,ps$ and $\tau_{RC} = 100\,ps$, for instance, the bandwidth B is approximately $0.8\,Gbit/s$.

13.2.2 Current-Voltage Characteristic

The current produced during photodetection travels from the n region to the p region, i.e. it is a reverse bias current with negative sign which is added to the current passing through the diode:

$$I = I_s \left[e^{\left(\frac{qV}{k_B T}\right)} - 1 \right] - I_p \ , \qquad (13.11)$$

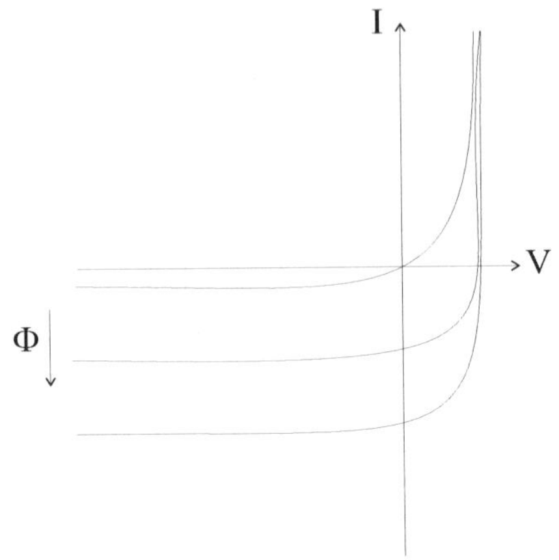

Fig. 13.8: Current-voltage characteristic of a photodiode for different values of photon flux.

where q is the electron charge, V is the voltage applied to the junction, T the absolute temperature (in K), k_B the Boltzmann constant, I_s the reverse saturation current, and I_p is the photocurrent proportional to the incident photon flux.

Figure 13.8 shows the current-voltage curve of a photodiode for different values of photon flux. When a reverse bias current is applied, in the third quadrant the absolute value of current increases proportionally to the increase of incident photon flux. The diode thus operates as a photodetector in the *photoconductive* mode. In an open circuit, however, the photodiode produces current also if no external bias is applied, i.e. $I = 0$. When the incident light is sufficient to create electron-hole pairs, the field of the depletion region separates the pairs, thus increasing the electric field, which in turn creates a voltage at the ends of the photodiode. The voltage increases if the incident photon flux increases. The diode, in this case, operates in the *photovoltaic* mode, which is typical of solar panels which convert sunlight into electricity.

The responsivity of a photodiode in photovoltaic mode is measured in V/W rather than in A/W (see, for example, (13.7)). In general, whether with or without an applied voltage, the photodiode becomes a source of electric signal. Nevertheless, the application of a reverse bias voltage considerably enhances the performance of a photodiode, because it increases the electric field, thus allowing a faster separation of the electron-hole pairs before they are able to recombine. Furthermore, it reduces the transit time and increases the size of the depletion region, i.e. the region in which absorption is more effective. Furthermore, the increase of the depletion region has a positive effect also on the response time of the photodiode, because the parasitic capacitance of the junction is reduced. Another advantage of reverse bias voltage is the reduction of the negative effect of the *dark current*, i.e. the thermally generated current flowing through the device even in the absence of an incident

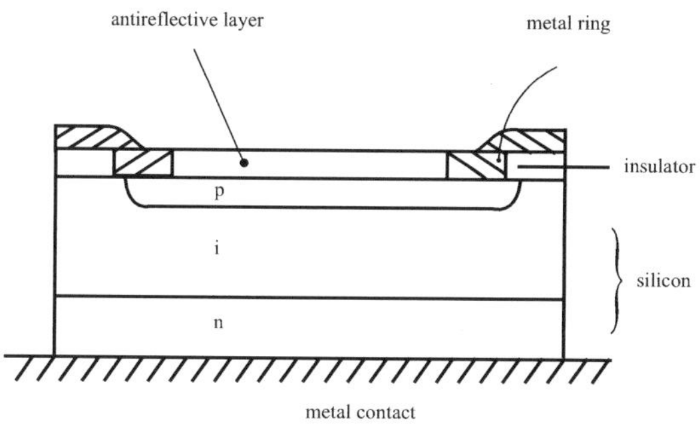

Fig. 13.9: Cross section of a silicon PIN photodiode.

optical power. As a matter of fact, heat is sufficient to produce the energy required to free the electron-hole pairs. Due to the electric field generated by the reverse voltage, then, the carriers are separated immediately and contribute to the photocurrent. All in all, the reverse bias voltage enhances the linearity and efficiency of a photodiode while at the same time reducing the response time and the dark current.

13.2.3 PIN Photodiode

The region in which the electron-hole pairs contribute to the photocurrent is formed by the depletion region and the quasi-neutral regions within the width corresponding to the minority carrier length. Nevertheless, photocurrent generation in the depletion region is much faster than in the quasi-neutral regions, where the carrier diffusion is random and slow, due to the fact that no electric field is applied. It is useful, therefore, to find an efficient way to extend the depletion region.
Figure 13.9 shows the structure of *PIN photodiodes*, in which the depletion region is extended by inserting a *quasi-neutral*, intrinsic (I) region between the p- and n-doped regions. It is thus possible to obtain an extended photosensitive region without reducing the doping concentration of the n and p regions, thus avoiding the increase of their resistance and, consequently, of the time constant. This occurs because the depletion layer width, d, is inversely proportional to the doping concentration of the p and n layers.

The *PIN* structure has considerable advantages. To start with, the thicker depletion region increases the quantum efficiency and the responsivity, which depend on the $(1 - e^{-\alpha d})$ factor in (13.4), due to the fact that photon absorption becomes more probable. Furthermore, as already observed, the thicker width d reduces the parasitic capacitance of the junction, even if the transit time increases. Last but not least, the ratio between the diffusion length and the depletion region length is reduced, so that the drift process becomes faster than diffusion. The *response times* thus obtained are around 10 ps, which correspond to 50 GHz bandwidths.

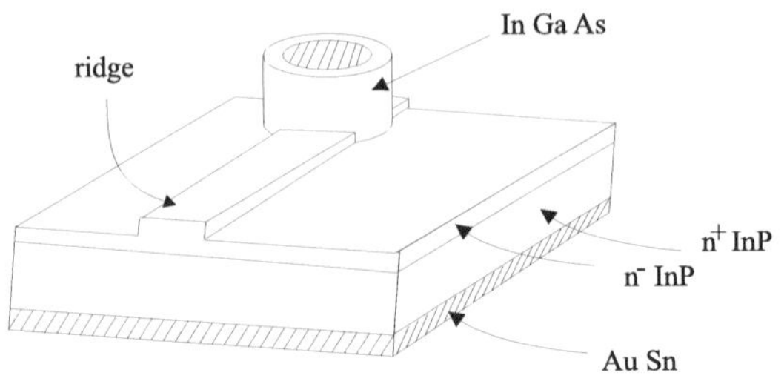

Fig. 13.10: Cross section of an InGaAs/InP photodetector.

The window area of PIN photodiodes should be large enough to trap as many photons as possible, but it should not be too large in order not to further increase the parasitic capacitance. Large area PIN photodiodes are used in advanced measurement applications.

13.2.4 Heterostructure Photodiodes

Heterostructure photdiodes have been created to solve the problem of surface absorption of the incident photons. Suppose to grow on a semiconductor layer another semiconductor layer, e.g. by molecular beam epitaxy. The two layers have similar lattice periods in order to avoid lattice defects, but they must have a different energy gap. This occurs with a ternary layer on a binary substrate, such as with $Al_xGa_{1-x}As$ on GaAs, or with $In_xGa_{1-x}As$ on InP. More specifically, suppose that the surface layer has a higher energy gap than the substrate, so that the surface layer is transparent to the incident light, which is not absorbed. The photon beam is thus able to reach the depletion region.

The advantages of this kind of structure are various. First of all, even if the junction is far below the surface, the structure can still be vertically illuminated. Secondly, it is possible to choose a variety of materials, so that the device can be *tuned* on a wider range of frequencies. Finally, the junction can be laterally illuminated if the field is directed through a waveguide, which must be made of a semiconductor material with higher E_g. Consider, for instance, the photodiode in figure 13.10. The field propagates through the guide and reaches the InGaAs region, which has a lower E_g than InP, and a higher refractive index. The field can thus propagate without being absorbed, and it finally reaches the depletion region.

13.2.5 Avalanche Photodiodes

Avalanche photodiodes operate in close proximity to the breakdown voltage, i.e. at high reverse biased voltage, in order to increase the electron-hole pairs which can be generated with a given

number of incident photons. The electric field of the depletion region accelerates the electrons and holes generated by a photon, called *primary carriers*, so that due to impact ionization other electron-hole pairs are generated, the so-called *secondary carriers*. Secondary carriers are then further accelerated, thus continuing the avalanche process. The result is a considerable increase of the device sensitivity, even if the process is intrinsically noisy due to the stochastic nature of secondary carrier generation.
The photodiode gain, G, is thus defined as:

$$G = \frac{N_T}{N_p}\,, \tag{13.12}$$

where N_T is the total number of electron-hole pairs, and N_p is the number of primary pairs. If the gain G is low, the photodetector has low sensitivity and its performance is limited by the thermal noise. Nevertheless, even if the thermal noise and the noise associated with the gain are eliminated, the performance is limited by the random nature of primary electron generation.
The most interesting aspect of avalanche photodiodes is that the avalanche process produces enough current to be detected also at weak light intensities.
The major drawbacks of avalanche photodiodes are related to the time required for the process, which also limits the bandwidth. Furthermore, it is a random process which generates noise. Finally, it is an unstable process which may cause breakdown.

13.2.6 Noise

In photodectectors, noise basically depends on the quantum nature of the material-field interaction, which in turn explains the non-deterministic process of photon transformation into electron-hole pairs. The major sources of noise are: photon noise, photoelectron noise, gain noise (typical of avalanche photodetectors), receiver circuit noise, and photodetector thermal noise.

Photon noise: although the power is constant, due to the quantum fluctuation of the electromagnetic field only the mean photon flux is constant. The number of detected photons follows a Poisson distribution with mean value $E_r/h\nu$, so that the probability of generating k electron-hole pairs becomes:

$$P(k) = \left(\frac{E_a}{h\nu}\right)^k \frac{e^{-\frac{E_r}{h\nu}}}{k!}\,, \tag{13.13}$$

where $E_r = \int_t^{t+T} P(t)\,dt$ is the energy of the received field.

Photoelectron noise: when a photon propagates through a semiconductor material, there is a probability η (defined as quantum efficiency) that it generates an electron-hole pair which does not recombine and thus contribute to the photocurrent. If we consider also η, the Poisson distribution

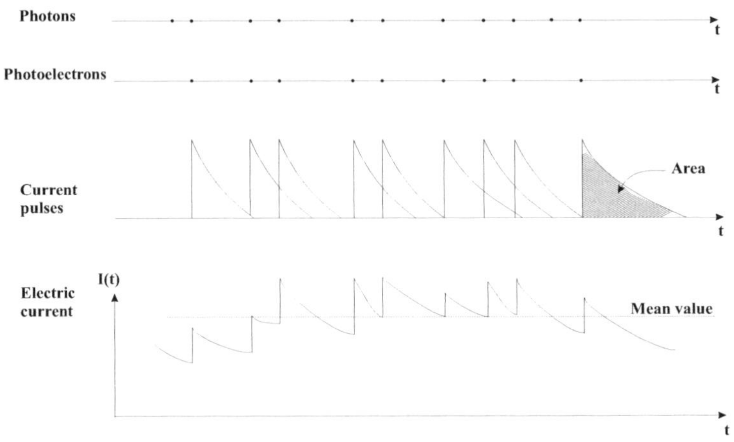

Fig. 13.11: In a photodetector circuit, the photocurrent derives from the superposition of electrical pulses, e.g. exponential current pulses, each associated with a detected photon.

in (13.13) becomes:

$$P(k) = \left(\eta \frac{E_r}{h\nu}\right)^k \frac{e^{-\eta \frac{E_a}{h\nu}}}{k!} \,. \tag{13.14}$$

The equation in (13.14), therefore, describes the probability of generating k photoelectrons by applying an energy E_r/T, where T is the time interval.

The total result produced by the two sources of noise is a fluctuating current whose mean value is constant. The fluctuations are known as *shot noise.* Figure (13.11) highlights the pulse of electric current produced by each electron, $e^{-t/\tau}$ (where τ is the response time of the photodiode), which has an area corresponding to its charge, q, i.e. $\int_0^T I(t)dt = q$.

Gain noise: it occurs only where there is carrier-multiplication, e.g. in avalanche photodiodes. Figure 13.12 shows that the number of electrons produced by each photon is random, and together with the photon noise it generates current fluctuations.

Receiver circuit noise: it is the thermal noise produced by the charged carriers in resistors, together with the noise of the first stage of the amplifier, which usually follows the photodiode in order to amplify the output signal.

Thermal noise: both in the depletion layer and in the quasi-neutral layers some electron-hole pairs can be generated due to thermal agitation, thus producing a current, called *dark current*, which might be present even in the absence of incident photons. Dark current does not depend on frequency, and it therefore constitutes a so-called *white noise.*

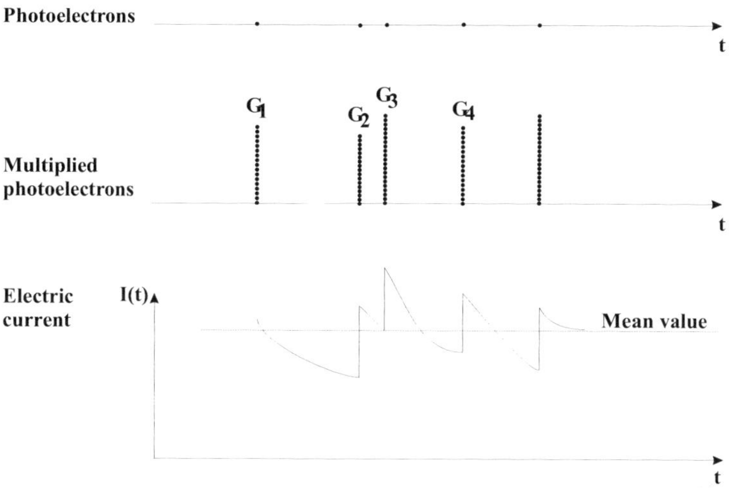

Fig. 13.12: In a photodetector with gain, each electron-hole pair generates a random number G_i of electron-hole pairs, each of which produces an electric current pulse of area qG_i in the detector circuit. The total electric current $I(t)$ is the overlap of these pulses.

Before concluding the observations on noise, it is interesting to define the *quantum limit.* Even if all the thermal sources of noise are eliminated, theoretically reaching the absolute zero, there might still be some probability of transmission error. Consider, for instance, an ON-OFF keying in which a 0 signal is transmitted, which means that the energy E_r incident on the photodiode is zero. According to the equation in (13.13), the error probability, i.e. the probability of detecting 1, is zero. On the other hand, by transmitting a 1 signal ($E_r \neq 0$), there is a non-zero probability that electrons are generated, i.e. that $k = 0$. In fact, according to (13.13), we might write:

$$P(0) = e^{-\eta \frac{E_r}{h\nu}} . \qquad (13.15)$$

the equation in (13.15) thus represents the probability that a beam of energy E_r generates zero photoelectrons. It is thus possible to determine the minimum incident energy required for a selected probability of error. If, for example, $P(0) = 10^{-9}$, the result is $\eta \frac{E_r}{h\nu} \simeq 21$, which means that in order to obtain an error probability of 10^{-9}, 21 photoelectrons must be generated in the time interval T. If the transmitted pulse is rectangular and has duration T, it follows that $E_r = PT$, which means that the minimum incident power to transmit a 1 signal is $P = 21 \frac{h\nu}{\eta T}$.

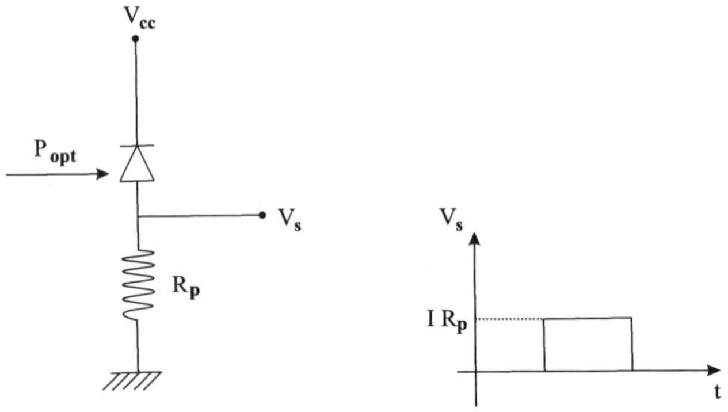

Fig. 13.13: Biasing circuit for a photodiode and output voltage (V_s) as a function of time (t). V_{cc} is the supply voltage, P_{opt} is the incident optical power, R_p is the polarisation resistance, and V_s the output voltage.

13.2.7 Biasing Circuit

Figure 13.13 shows the most simple structure of a biasing circuit, which is a reverse biased photodiode with a bias resistor of resistance R_p. In absence of the signal, when the incident optical power,P_{opt}, is zero, the polarisation resistance, R_p, is zero too because no current flows. In all other cases, the current flowing in the photodiode corresponds to:

$$I = \Re P_{opt} \ , \tag{13.16}$$

where $\Re$ is the responsivitiy, which on R_p determines a voltage drop in the output voltage, V_s:

$$V_s = R_p I = \Re R_p P_{opt} \ . \tag{13.17}$$

Consider, for example, $\Re = 0.5\ A/W$, $R_p = 50\ \Omega$ and $P_{opt} = 20\ \mu W$, corresponding to $-17\ dBm$. The output voltage is thus $V_s = 500\ \mu V$, which is too low to be processed directly, so that amplifiers are usually followed by a voltage preamplifier of gain A_V. The output voltage is thus $V_0 = \Re R_p A_V P_{opt}$, so that the responsivity $\Re_V$ of the whole circuit (photodiode and preamplifier) becomes:

$$\Re_V = \Re R_p A_V \ . \tag{13.18}$$

This kind of biasing circuit is non-inverting because the output electric signal increases when there is an incident optical signal, but then immediately goes back to zero when there is no incident beam. Figure 13.14, on the other hand, shows an inverting configuration. The output voltage becomes:

$$V_s = V_{CC} - R_p I \ . \tag{13.19}$$

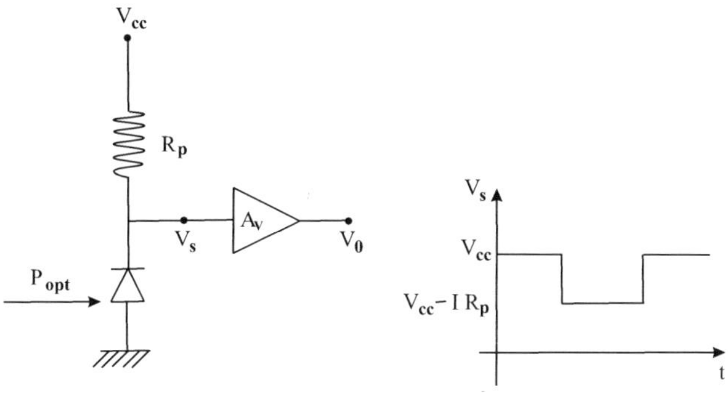

Fig. 13.14: Inverting configuration of a biasing circuit and its output voltage (V_s) as a function of time (t).

Due to the dynamic nature of this kind of biasing circuits, these devices are usually modified by adding two parallel capacitors betweeen the photodiode and the preamplifier, so that they alternatively operate as bypass capacitors for low or high frequencies.

Chapter 14

Fiber Dispersion

The major problem of propagation along a guided medium is that signals are distorted, thus limiting data transmission speed due to the intersymbol interference. This chapter examines the major causes of distortion and identifies possible solutions to the matter. Research on this issue is based on some hypotheses which need to be analysed accurately in order to avoid rough or simplistic generalisations. We will therefore follow a scientific method of description and we will simplify some hypotheses in order to make the analytical expressions easier, thus being able to emphasise the most significant aspects of the discussion.

14.1 Propagation of Modulated Signals

To start with, suppose that the input field excites a single mode. According to Fourier analysis, a modulated signal is made up of the sum of its harmonic components. Each harmonic contributes to the Fourier series representation weighed according to the value of the Fourier transform of the signal calculated at its angular frequency. The input signal can thus be written as:

$$\overline{e}_{in}(u_1, u_2, \omega) = \int_{-\infty}^{+\infty} \overline{\mathcal{E}}_{in}(u_1, u_2, t)e^{-j\omega t}dt, \tag{14.1}$$

where $\overline{e}_{in}$ is the Fourier transform of the vector $\overline{\mathcal{E}}_{in}$ in a point with coordinates (u_1, u_2) on the transverse plane. In the present discussion, $\mathcal{F}$ and $\mathcal{F}^{-1}$ will refer to the Fourier transform and the inverse Fourier transform, respectively. For each angular frequency, the mode is excited by an input field of $A(\omega)$ amplitude which depends on the modulating signal and corresponds to:

$$\overline{e}_{in}(u_1, u_2, \omega) = A(\omega)\overline{E}(u_1, u_2, \omega). \tag{14.2}$$

$A(\omega)$ represents the contribution of the mode at angular frequency ω to the representation of the input field. Nevertheless, $A(\omega)$ is not the Fourier transform of the input signal, due to the fact that ω depends on $\overline{E}$.

Similarly, in a generic section z of a guide, the field depends on the contribution of various frequencies of the single mode which propagates. As shown in figure 14.1, the contributions might be subject to attenuation or dispersion at different levels:

$$\overline{e}(u_1, u_2, z, \omega) = A(\omega)\overline{E}(u_1, u_2, \omega)e^{-\alpha(\omega)z}e^{-j\beta(\omega)z}. \tag{14.3}$$

The inverse Fourier transform of 14.3 represents the temporal variation of the field at the generic z axis:

$$\overline{\mathcal{E}}(u_1, u_2, z, t) = \frac{1}{2\pi}\int_{-\infty}^{+\infty} A(\omega)\overline{E}(u_1, u_2, \omega)e^{-\alpha(\omega)z}e^{-j\beta(\omega)z}e^{j\omega t}d\omega. \tag{14.4}$$

The equations in (14.1) and (14.4) link the input and output fields. Since Fourier analysis can only be applied to linear systems, the equations so far defined can be considered only if the nonlinear effects of propagation (e.g. the refractive index dependence on the amplitude of the electric field) can be ignored. Note that the mode functions $\overline{E}$ and $\overline{H}$ do not depend significantly on the angular

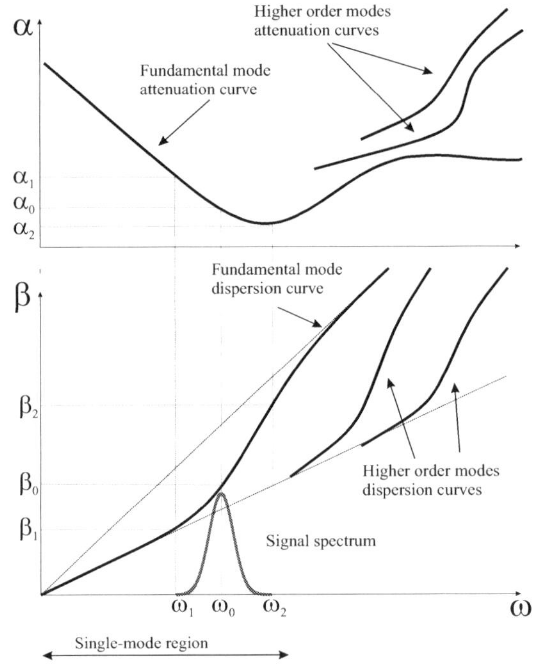

Fig. 14.1: Propagation of a modulated signal. The fundamental mode, which is the only mode which can propagate, is dephased and attenuated while propagating as a function of its angular frequency. The red lines mark the dispersion and attenuation curves.

frequency, so that in practical applications their variations can be ignored, thus simplifying the discussion. In fact, we can assume that within the signal bandwidth:

$$\overline{E}_0(u_1, u_2) \simeq \overline{E}(u_1, u_2, \omega_0) \qquad \overline{H}_0(u_1, u_2) \simeq \overline{H}(u_1, u_2, \omega_0) \tag{14.5}$$

where ω_0 is the central angular frequency.[1] Consequently, the mode contributions for different frequencies at position z differ due to the different attenuation and phase constants, but not due to the different distribution in the transverse plane. We can thus rewrite (14.2) and (14.3) as follows:

$$\overline{e}_{in}(u_1, u_2, \omega) = A(\omega)\overline{E}_0(u_1, u_2);$$

$$\overline{e}(u_1, u_2, z, \omega) = A(\omega)\overline{E}_0(u_1, u_2)e^{-\alpha(\omega)z}e^{-j\beta(\omega)z}.$$

The equation in (14.5) thus allows separating the spatial variables from the spectral one (and thus from the temporal one as well).

Consider the transverse component of the $\overline{E}_t$ electric field, in which the temporal variations at the input and along the guide are:

$$\overline{\mathcal{E}}_{t\,in}(u_1, u_2, t) = a_{in}(t)\overline{E}_{t0}(u_1, u_2) \qquad \overline{\mathcal{E}}_t(u_1, u_2, z, t) = a(z, t)\overline{E}_{t0}(u_1, u_2) \tag{14.6}$$

where

$$a_{in}(t) = \mathcal{F}^{-1}\left[A(\omega)\right] \qquad a(z, t) = \mathcal{F}^{-1}\left[A(\omega)F(\omega)\right] = \mathcal{F}^{-1}\left[A(\omega)e^{-\alpha(\omega)z}e^{-j\beta(\omega)z}\right].$$

$\overline{E}_{t0}$ gives information about the field variation on the transverse plane, whereas $a_{in}(t)$ and $a(z, t)$ give information on how the field evolves in time and space. Since we mainly deal with narrow band signals, it is easier to analyse the relation between $a_{in}(t)$ and $a(z, t)$ in terms of the equivalent lowpass signals. The relation between $a_{in}(t)$ and its equivalent lowpass signal $u_{in}(t)$ is:

$$u_{in}(t) = 2\mathcal{F}^{-1}\left[A_+(\omega + \omega_0)\right],$$

$$a_{in}(t) = Re\left\{u_{in}(t)e^{j\omega_0 t}\right\} = |u_{in}(t)|cos\left(\omega_0 t + \phi_{in}(t)\right)$$

where $\phi_{in}(t) = arg(u_{in}(t))$ and A_+ refers to the positive frequencies of the $a_{in}(t)$ spectrum, as shown in figure 14.2. In defining the equivalent lowpass signal of $a(z, t)$ it is useful to isolate the term $\beta_0 = \beta(\omega_0)$:

$$u(z, t) = 2\mathcal{F}^{-1}\left[A_+(\omega + \omega_0)e^{-\alpha(\omega+\omega_0)z}e^{-j\Delta\beta(\omega+\omega_0)z}\right] \tag{14.7}$$

where $\Delta\beta(\omega) = \beta(\omega) - \beta(\omega_0)$. Consequently:

$$a(z, t) = Re\left\{u(z, t)e^{-j\beta_0 t}e^{j\omega_0 t}\right\} = |u(z, t)|cos\left(\omega_0(t - \frac{z}{v_f}) + \phi(t)\right).$$

[1] The present discussion considers signal propagation in optical fibers. Note, however, that for TEM modes (used, for instance, in coaxial cables) the equations in (14.5) cease to be approximations, because the mode functions are independent of angular frequency and coincide with the electric and magnetic fields in the static case.

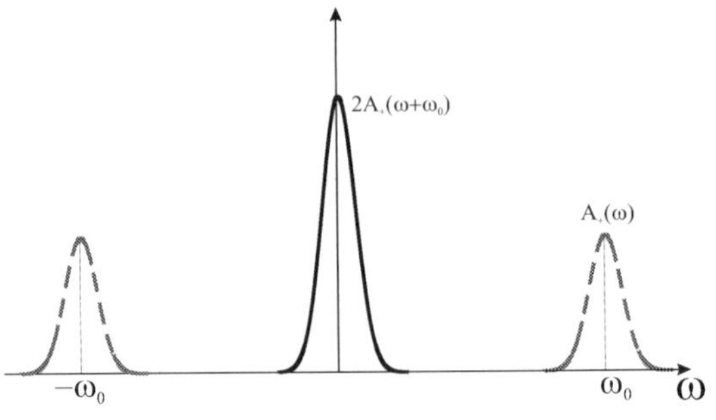

Fig. 14.2: The blue line represents the spectrum of signal $a_{in}(t)$, whereas the green line represents its equivalent lowpass signal $u_{in}(t)$.

We can now solve the equations in (14.6) and follow the same process for the longitudinal component. After that, by exploiting the equation in (1.8), which is valid for low loss, we obtain the final equation of electric field as a function of time and space:

$$\begin{aligned}\overline{\mathcal{E}}(u_1,u_2,z,t) &= \overline{E}_{t0}(u_1,u_2)|u(z,t)|\cos\left(\omega_0(t-\frac{z}{v_f})+\phi(z,t)\right) \\ &+ E_{z0}(u_1,u_2)|u(z,t)|\sin\left(\omega_0(t-\frac{z}{v_f})+\phi(z,t)\right)\hat{z}.\end{aligned} \tag{14.8}$$

Note that the carrier signal propagates at velocity v_f. Nothing can be said about $u(z,t)$, because its properties depend on how α and β vary according to the frequency (for more details, see the following sections). For constant time in (14.8), we obtain the spatial profile of the field along the waveguide at the selected instant. On the other hand, for constant z we obtain the temporal variation for that value of z. More specifically, if we consider a waveguide of length L, the temporal variation of the field at the output of the waveguide is:

$$\begin{aligned}\overline{\mathcal{E}}_{out}(u_1,u_2,t) = |u_{out}(t)|\Big(&\overline{E}_{t0}(u_1,u_2)\cos\left(\omega_0(t-\tau_f)+\phi_{out}(t)\right) \\ &+E_{z0}(u_1,u_2)\sin\left(\omega_0(t-\tau_f)+\phi_{out}(t)\right)\hat{z}\Big).\end{aligned} \tag{14.9}$$

At the output of the waveguide, the carrier is delayed by the phase delay, and its amplitude and phase modulations correspond to: $|u_{out}(t)| = |u(L,t)|$ and $\phi_{out}(t) = arg(u(L,t))$.

14.2 Conditions for the Absence of Distortion

According to the principles of signal processing, channels do not cause distortion provided that, within the signal bandwidth, the transfer function has a constant magnitude and its phase varies linearly with the angular frequency. Therefore, based on (1.6), the *conditions for the absence of distortion* are:

$$\alpha(\omega) = \alpha_0$$
$$\beta(\omega) = \beta(\omega_0) + \frac{d\beta}{d\omega}|_{\omega_0}(\omega - \omega_0) = \beta_0 + \beta_0^{'}(\omega - \omega_0).$$

If we define the group velocity as $v_g = 1/\beta_0^{'}$, in the absence of distortion we obtain $\Delta\beta = (\omega - \omega_0)/v_g$, so that (14.7) becomes:

$$u(z,t) = 2\int_{-\infty}^{+\infty} A_+(\omega + \omega_0)e^{-\alpha_0 z}e^{-j\frac{\omega z}{v_g}}e^{j\omega t}d\omega = e^{-\alpha_0 z}u(t - \frac{z}{v_g})$$

from which we obtain (14.8):

$$\overline{\mathcal{E}}(u_1, u_2, z, t) = e^{-\alpha_0 z}|u|(t - \frac{z}{v_g})\left\{\overline{E}_{t0}(u_1, u_2)cos\left(\omega_0(t - \frac{z}{v_f}) + \phi(t - \frac{z}{v_g})\right)\right. \tag{14.10}$$
$$\left. + \; E_{z0}(u_1, u_2)sin\left(\omega_0(t - \frac{z}{v_f}) + \phi(t - \frac{z}{v_g})\right)\hat{z}\right\}$$

As shown in figure 14.3, the information contained in the envelope and phase of the wave travels without distortion at the group velocity v_g, whereas the carrier propagates at the phase velocity. If we consider a fiber of length L, $\tau_g = L\beta_0^{'}$ is the group delay and represents the time needed by the envelope to travel through the waveguide. In real waveguides, however, the conditions for the

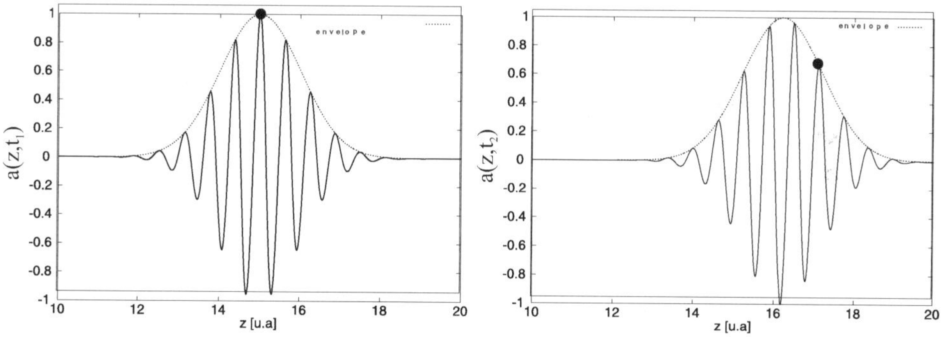

Fig. 14.3: Spectrum of $a(z,t)$ (continuous line) and its envelope (dashed line) along the propagation direction at instants $t = t_1$ and $t = t_2 > t_1$ for $v_f > v_g$. The envelope propagates at velocity v_g, whereas the equiphase points (black dots) at velocity v_f.

absence of distortion cannot be achieved, so that u_{out} is not a delayed equivalent of u. Distortion, in turn, limits the performance of communication systems. The effects of distortion caused by the channels are related to propagation, so that they increase on longer distances and for larger bandwidths, thus setting an upper limit to the product between the propagation velocity and the fiber length. In transmission lines, distortion is mainly caused by the fact that the the attenuation constant α depends on angular frequency (α increases according to $\sqrt{\omega}$). In optical fibers, on the other hand, there is a very large bandwidth (approximately 23 THz) centred around $\lambda = 1550$ nm wavelength where α is constant. This basically causes a phase distortion, because the phase constant does not depend linearly on angular frequency. Figure 14.4 shows the spectrum of the propagation constant of the fundamental mode. If ω tends toward zero, the mode is near cut-off, so that the phase constant tends asymptotically towards $n_{cl}k_0 = n_{cl}\frac{\omega}{c}$. If ω tends towards infinity, the mode is increasingly confined into the core, so that β tends asymptotically towards $n_{co}k_0 = n_{co}\frac{\omega}{c}$. Figure 14.4 takes into account also material dispersion, which causes the dependence of n_{co} and n_{cl} on angular frequency.

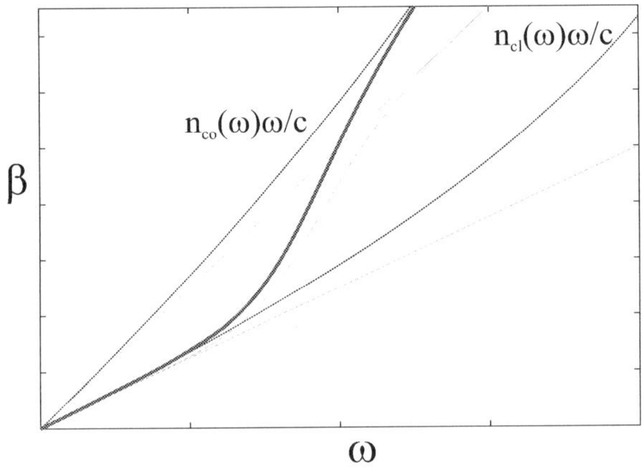

Fig. 14.4: Spectrum of the phase constant, β, as a function of angular frequency, ω. The green lines illustrate what happens in the absence of material dispersion, i.e. $n_{cl} = const.$, $n_{co} = const.$

14.3 Intramodal Dispersion

If the dependence of the propagation constant on angular frequency is nonlinear, the phase distortion caused by the waveguide ultimately produces pulse broadening. Nonlinear dependence, in turn, is caused by two combined phenomena: waveguide and material dispersion. The effect of material dispersion, i.e. the dependence of the refractive index of the material on angular frequency, is always present. On the other hand, the waveguide effect only affects the modes for which the phase constant depends on the spatial distribution of the field. In this kind of modes, which includes all except for TEM, the boundary and continuity conditions limit the spatial distribution of the field, thus modifying the *transverse propagation constant*, k_t, which in turn modifies the phase constant. The influence of waveguide and material dispersion is more evident if we write β as a function of both the material and k_t:

$$\beta = \sqrt{\omega^2 \frac{n}{c} - k_t^2}. \tag{14.11}$$

Only if n is constant ans k_t depends linearly on ω, β varies linearly and the envelope is not distorted. In order to understand the effects of distortion on the envelope, we need to consider an hypothetical case. First of all, we represent the spectrum of β in the signal bandwidth as a Taylor series:

$$\beta(\omega) = \beta_0 + \beta_0^{'}(\omega - \omega_0) + \frac{1}{2}\beta_0^{''}(\omega - \omega_0)^2 + \frac{1}{6}\beta_0^{'''}(\omega - \omega_0)^3 + \cdots.$$

We only write until the second-order term of the Taylor series, so that:

$$\beta(\omega) = \beta_0 + \beta_0^{'}(\omega - \omega_0) + \frac{1}{2}\beta_0^{''}(\omega - \omega_0)^2. \tag{14.12}$$

There is a term which is proportional to the parameter

$$\beta_0^{''} = \left.\frac{\partial^2 \beta}{\partial \omega^2}\right|_{\omega_0} = \left.\frac{\partial \beta'}{\partial \omega}\right|_{\omega_0}$$

which is called *GVD parameter* or *GVD coefficient* (Group Velocity Dispersion). It is in fact this term, which is absent when no distortion occurs, which causes the dependence of the group velocity on angular frequency:

$$v_g = \frac{1}{\beta_0' + \beta_0^{''}(\omega - \omega_0)}. \tag{14.13}$$

Suppose now that the type of modulation is *OOK*. Suppose to transmit a "1" and that the modulating signal is a Gaussian pulse with standard deviation T:

$$a_{in}(t) = A_0 e^{-\frac{1}{2}(\frac{t}{T})^2} cos(\omega_0 t). \tag{14.14}$$

This is typical of fiber optic communication systems which use an *RZ* (Return to Zero) modulation format. Under these conditions, the Fourier transform of the equivalent lowpass signal is:

$$A_+(\omega + \omega_0) = T/2\sqrt{2\pi} A_0 e^{-\frac{\omega^2 T^2}{2}}$$

and (14.7) becomes:

$$\begin{aligned} u(z,t) &= \frac{T}{\sqrt{2\pi}} A_0 e^{-\alpha z} \int_{-\infty}^{+\infty} e^{-\frac{\omega^2 T^2}{2}} e^{-j\frac{z}{v_g}\omega} e^{-j\frac{1}{2}\beta_0^{''}\omega^2 z} e^{j\omega t} d\omega \\ &= \frac{T}{\sqrt{2\pi}} A_0 e^{-\alpha z} \int_{-\infty}^{+\infty} e^{j(t-\frac{z}{v_g})\omega} e^{-\frac{1}{2}(T+j\beta^{''} z)\omega^2} d\omega. \end{aligned}$$

If we consider the definite integral:

$$\int_{-\infty}^{+\infty} e^{-\frac{a^2 x^2}{2}} e^{jbx} dx = \frac{\sqrt{2\pi}}{a} e^{-\frac{b^2}{2a^2}},$$

where $a = \sqrt{T^2 + j\beta_0'' z}$ e $b = t - z/v_g$, it follows that:

$$u(z,t) = \frac{A_0 T e^{-\alpha z}}{\sqrt{T^2 + j\beta_0'' z}} e^{-\frac{(t-\frac{z}{v_g})^2}{2}\frac{T^2}{T^4+(\beta_0'' z)^2}} e^{j\frac{(t-\frac{L}{v_g})^2}{2}\frac{\beta_0'' z}{T^4+(\beta_0'' z)^2}}$$

and if we consider (14.10) we obtain:

$$\overline{\mathcal{E}}(u_1,u_2,z,t) = \frac{A_0 T e^{-\alpha z}}{\sqrt[4]{T^4 + (\beta_0'' z)^2}} e^{-\frac{(t-\frac{z}{v_g})^2}{2}\frac{T^2}{T^4+(\beta_0'' z)^2}} \tag{14.15}$$

$$\left\{\overline{E}_{t0}(u_1,u_2)\cos\left(\omega_0(t-\frac{z}{v_f}) + \phi(z,t)\right) + E_{z0}(u_1,u_2)\sin\left(\omega_0(t-\frac{z}{v_f}) + \phi(z,t)\right)\hat{z}\right\},$$

where

$$\phi(z,t) = -\frac{1}{2}\,atan(\frac{\beta_0'' z}{T^2}) + \frac{(t-\frac{z}{v_g})^2}{2}\frac{\beta_0'' z}{T^4 + (\beta_0'' z)^2}.$$

The presence of $\phi(z,t)$ causes phase distortion and, consequently, a frequency modulation. The instantaneous frequency varies throughout the pulse. The envelope remains a Gaussian function even though, due to distortion, its peak value decreases and the pulse broadens. Figure 14.5 shows the spectrum of $a(z,t)$ along z at three different instants. Note that the pulse broadening effect happens as if there were a higher propagation velocity of the rising edge of the pulse with respect to the falling edge.

In optical systems, the receiver converts the optical signal into an electrical signal by means of a photodiode, in which current is proportional to the power of the incident electromagnetic field. According to the previous hypotheses, the incident power is proportional to:

$$I(L,t) = u(L,t)u^*(L,t) = \frac{A_0^2 T^2}{\sqrt{T^4 + (\beta_0'' L)^2}} e^{-(t-\tau_g)^2 \frac{T^2}{T^4+(\beta_0'' L)^2}}. \tag{14.16}$$

If we compare the time-domain profile with that of the input signal taken from (14.14):

$$I_{in}(t) = A_0^2 e^{-(\frac{t}{T})^2}$$

we note that the pulse duration increases for higher values of β_0'' and longer lengths of the waveguide. Its broadening factor (BF), i.e. the ratio between its standard deviation and that of the input pulse, is:

$$\frac{\sigma(L)}{\sigma_0} = \sqrt{1 + \left(\beta_0'' \frac{L}{T^2}\right)^2} = \sqrt{1 + \left(\frac{L}{L_D}\right)^2},$$

where L_D is the *dispersion length*:

$$L_D = \frac{T^2}{|\beta_0''|}.$$

The dispersion length is useful to estimate the distance through which the pulse can travel before dispersion effects become not negligible. For $L = L_D$ the pulse duration is $\sqrt{2}$ times the initial one,

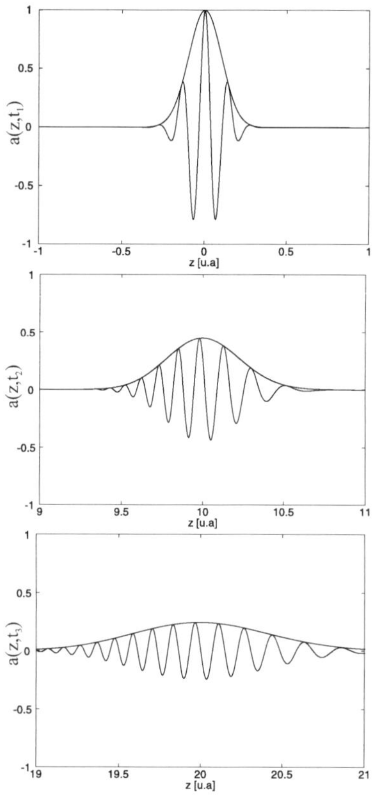

Fig. 14.5: Spectrum of $a(z,t)$ (solid line) and its envelope (dashed line) along the propagation direction at instants $t = t_1$, $t = t_2 > t_1$ and $t = t_3 > t_2 > t_1$. Note the temporal broadening of the envelope and the frequency modulation of the carrier.

which corresponds to a 40% pulse broadening. If the waveguide is long enough ($L >> L_D$) the pulse duration increases linearly at a rate which depends on both the dispersion caused by the waveguide, $\beta_0^{''}$, and the signal bandwidth, $1/T$. In fact, the higher $\beta_0^{''}$ and the signal bandwidth, the higher the rate at which pulse duration increases. If, for example, $\beta_0^{''} = 21.6ps^2/km$ (a typical value for standard fibers at $1550nm$) and $T = 100ps$, we obtain $L_D = 463km$. However, if we increase the transmission speed by a factor of four[2] ($T = 25ps$) the lenght is only $L_D = 29km$. This implies that in the absence of compensation, the signal is completely distorted after a few kilometers.
When the modulating signal is not a Gaussian pulse, no analytical solution is available. Nevertheless, it can be noticed that the pulse broadening is still proportional to the parameters above described.

[2]According to the ITU (International Telecomunication Union) standard for time division multiplexing (TDM), the transmission speed in communication systems must always be increased by a factor of four.

In optical communication, dispersion is usually calculated by means of the *dispersion parameter*, D:

$$D = \left.\frac{\partial\beta'}{\partial\lambda}\right|_{\lambda_0} = -\frac{2\pi c}{{\lambda_0}^2}\beta_0'' \qquad [D] = \frac{ps}{nm\ km}$$

from which:

$$v_g = \frac{1}{\beta_0' - D\frac{\lambda_0}{\lambda}(\lambda - \lambda_0)}, \tag{14.17}$$

$$L_D = \frac{2\pi c}{\lambda_0}\frac{T^2}{D}.$$

The relevance of D can be understood if we consider the group delay:

$$\tau_g = \frac{L}{v_g} = L\beta_0' - LD\frac{\lambda_0}{\lambda}(\lambda - \lambda_0) \simeq L\beta_0' - LD(\lambda - \lambda_0).$$

In fact, D measures how much the group delay (in ps) accumulated over a 1 km propagation varies for a spectral shift of 1 nm. In regions with anomalous dispersion, $\beta_0' < 0$, $D > 0$, and the group velocity increases with increasing frequency. In regions with normal dispersion, on the other hand, where $\beta_0' > 0$ and $D < 0$, group velocity and frequency are in inverse relationship, i.e. the group velocity decreases with increasing frequency.
In order to avoid unbalanced DWDM systems, the fiber should operate on different channels in a similar way. The *dispersion slope*, S, helps to define the relation between the dispersion parameter and the wavelength:

$$S(\lambda) = \frac{dD}{d\lambda}$$

from which

$$D(\lambda) = D(\lambda_0) + S(\lambda_0)(\lambda - \lambda_0).$$

In standard fibers at $1550nm$, S is typically around $0.06\,ps/(nm^2km)$. Consequently, in a DWDM system in which channels have a spacing of $0.4\,nm$ the dispersion coefficient varies by 0.15% between adjacent channels.

14.3.1 Walk-off

The dependence of the group delay on the wavelength has not only drawbacks. In fact, it can be exploited to reduce the negative effects of some nonlinear phenomena occurring in DWDM systems, such as cross-phase modulation (XPM) and four wave mixing (FWM). For the sake of simplicity, consider two channels and their nonlinear interaction, which occurs only when their pulses overlap. The non-zero GVD coefficient determines different group velocities for the channels, thus reducing the interaction length. The larger the difference in group velocity between the channels, the shorter the interaction length, usually referred to as walk-off, will be. The walk-off parameter, d_{12}, corresponds to:

$$d_{12} = \frac{1}{v_g(\lambda_1)} - \frac{1}{v_g(\lambda_2)}.$$

If

$$\beta_0'(\lambda_2) \simeq \beta_0'(\lambda_1) + \frac{\partial \beta_0'}{\partial \lambda}(\lambda_2 - \lambda_1) = \beta_0'(\lambda_1) D(\lambda_2 - \lambda_1),$$

the walk-off length for pulses of pulse width T is:

$$L_W = \frac{T}{|d_{12}|} = \frac{T}{|D||\lambda_2 - \lambda_1|}.$$

Consequently, if we want to reduce the distance between channels without increasing the nonlinear effects, we need to use fibers with higher GVD coefficients.

14.4 Dispersion Compensation Techniques

This section describes the most popular techniques to compensate GVD in optical communication systems. Before doing so, however, it is important to underline some characteristics of dispersion-shifted fibers, which reach the zero-dispersion wavelength at $1550\,nm$, where the minimum attenuation is achieved, too. In fact, this is possible because the dispersion curve depends on the refractive index profile. Nevertheless, the use of dispersion-shifted fibers in DWDM systems is usually avoided due to some problems which might occur. To start with, it is difficult to achieve zero-dispersion in a wide wavelength range, so that the number of channels is limited. Secondly, the very long walk-off length (approaching infinity in the limit) naturally increases the nonlinear distortions between channels. Therefore, DWDM systems usually exploit non-zero dispersion-shifted fibers and cascaded dispersion compensators to compensate GVD. In precompensation, the compensator is placed before the fiber, whereas in postcompensation the device is placed after it. In linear regimes, the precompensation and postcompensation schemes coincide. If we consider the nonlinear effects, however, there are significant differences. The *dispersion map* represents the profile of dispersion along the transmission line.

The pulse is broadened due to phase distortion because the rising edge propagates faster than the falling edge. Starting from this considerations, researchers have developed compensation techniques for channel dispersion which are based on the group velocity dependence on frequency. Therefore, if we consider (14.13) and (14.17), it is possible to compensate the dispersion by changing the instantaneous frequency of the rising and falling edges, a technique used in pre-chirping and soliton transmission. Another possibility for dispersion compensation are devices used in cascade which have a dispersion coefficient with opposite sign and an adequate length, such as dispersion compensating fibers or chirped fiber Bragg gratings.

14.4.1 Pre-chirping

If we apply to the transmitted signal a parabolic phase modulation $\phi(t) = C_0(t/T)^2$, where C_0 is the *chirp coefficient*, the instantaneous frequency of the pulse varies linearly, so that: $\nu(t) = \frac{C_0 t}{\pi T}$. If

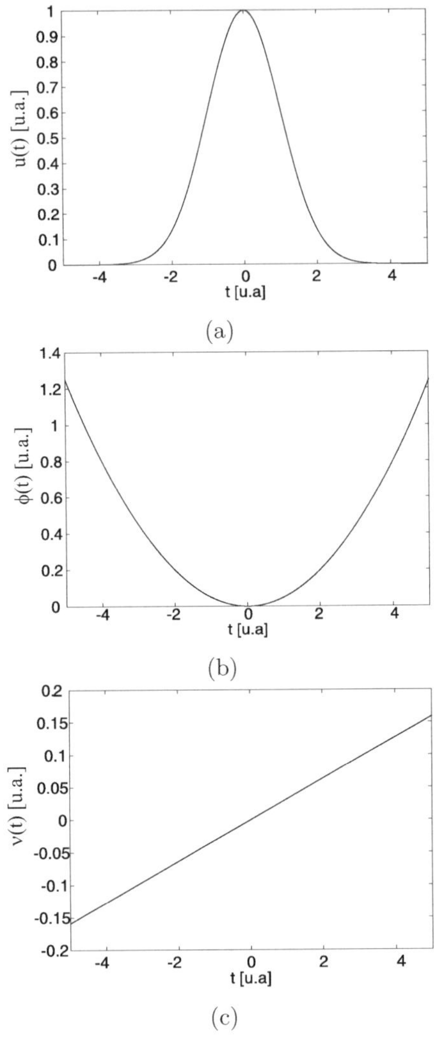

Fig. 14.6: Pre-chirping in the anomalous dispersion regime ($C_0 > 0$). (a) pulse envelope (rising edge in the negative time half-axis); (b) instantaneous phase as a function of time; (c) change of instantaneous frequency due to phase modulation.

we select the origin of the time axis so that the pulse is centred at $t = 0$, the rising edge of the pulse will be for $t < 0$, whereas the falling edge is for $t > 0$. Consequently, if $C_0 > 0$, the instantaneous frequency on the rising edge is lower than the instantaneous frequency on the falling edge, whereas if $C_0 < 0$ the opposite situation occurs. Figure 14.6 shows the case in which $C_0 > 0$. In the anomalous dispersion regime, the group velocity increases at higher frequencies, so that to slow down the rising

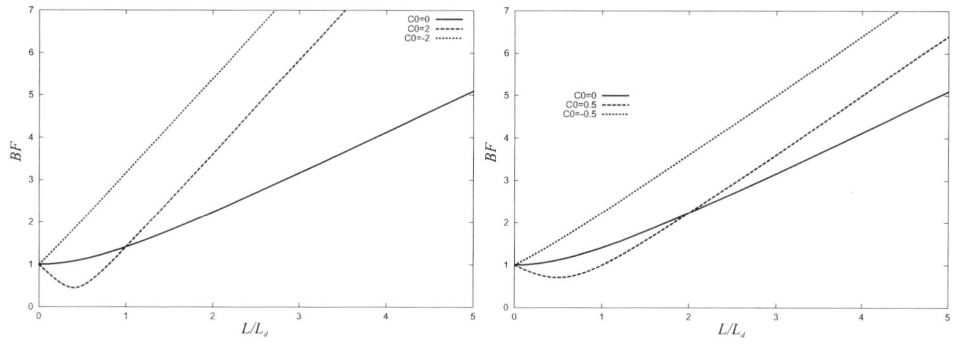

Fig. 14.7: Variation of the Broadening Factor for $\beta_0'' < 0$

edge we need to reduce the instantaneous frequency ($C_0 > 0$). On the other hand, in the normal dispersion regime we need that $C_0 < 0$. It is important to balance between the effects of chirping and dispersion, in order not to distort the signal. If we exceed in modulation, for instance, the rise and fall times become, respectively, too slow and too fast, so that the pulse becomes narrower at first, and than broadens at a much higher rate than in the absence of modulation. If we consider a Gaussian pulse, we can solve the equation analytically, thus relating quantitatively the GVD and the chirp parameter. Consider the complex envelope of a Gaussian pulse with a parabolic phase modulation:

$$u(t) = A_0 e^{-\frac{1}{2}(\frac{t}{T})^2} e^{jC_0(\frac{t}{T})^2}. \tag{14.18}$$

By solving the equations as we did in the previous section, we obtain the signal envelope along the fiber:

$$u(z,t) = \frac{A_0 T}{\sqrt{T^2 + j\beta_0'' z(1 - jC_0)}} e^{-\frac{(t-\tau_g)^2}{2}\frac{T^2}{T^4\left[(1+\frac{\beta_0'' z}{T^2}C_0)^2+(\frac{\beta_0'' z}{T^2})^2\right]}} e^{j\frac{(t-\tau_g)^2}{2}\frac{\beta_0'' z(1+C_0^2)+C_0T^2}{T^4\left[(1+\frac{\beta_0'' z}{T^2}C_0)^2+(\frac{\beta_0'' z}{T^2})^2\right]}};$$

so that the broadening factor becomes:

$$BF = \frac{\sigma(z)}{\sigma_0} = \sqrt{(1 + C_0\beta_0''\frac{zL}{T^2})^2 + \left(\beta_0''\frac{z}{T^2}\right)^2}.$$

Figure 14.7 shows the variation of the BF for different values of C_0. As it was mentioned before, in order to achieve compensation the GVD and chirp parameters must have opposite sign. If this situation occurs, the BF along the fiber first decreases, and reaches its minimum:

$$BF_{min} = \frac{1}{\sqrt{1 + C_0^2}}$$

at a distance

$$L_c = -\frac{C_0}{\beta_0''}\frac{T^2}{1+C_0^2}.$$

After the minimum, the BF increases at a higher rate than for an unchirped pulse. If $L = L_c$, the phase pre-modulation wholly compensates the modulation introduced by the channel. On the other hand, if C_0 has the same sign of β_0'', the pre-modulation is added to the modulation introduced by the channel, thus immediately and rapidly degrading the signal. Chirping is therefore fundamental in designing modern optical modulators. Due to the non ideality of modulators, sometimes the output signal may be unintentionally chirped. The resulting chirp must be considered in developing the compensation strategies. If the chirp and the GVD parameters have opposite sign, chirping can be exploited (at least partially) for dispersion compensation. If, on the other hand, the two parameters have the same sign, an additional compensation is required. The chirp parameter for modulators is usually defined as:

$$\alpha_m(t) = \frac{\frac{d\phi(t)}{dt}}{\frac{1}{2I(t)}\frac{dI(t)}{dt}}$$

where $I(t)$ and $\phi(t)$ represent the output field intensity and phase of the modulator. By substituting into α_m the equations of $I(t)$ and $\phi(t)$ found in (14.18), it follows that:

$$\alpha_m(t) = -2C_0.$$

The table illustrates some examples of the chirp parameters of some modulators.

Chirp parameters of some transmitter

Type of transmitter	Chirp parameter
Direct modulation DFB laser	−1
DFB laser with an electroabsorption modulator	−0.3
DFB laser with a Mach-Zehnder modulator	0
DFB laser with an MZ modulator for pre-chirping	0.8

If we consider dispersion, the best modulators are DFB laser modulators which operate in continuous wave regime integrated with a cascaded Mach-Zehnder modulator.

14.4.2 Solitons

Intramodal dispersion can be compensated by exploiting the nonlinearity of the material forming the waveguide. In silica, for instance, which is used to produce optical fibers, the refractive index and the dielectric permittivity both depend on the squared magnitude of the electric field:

$$n_{silica} = n_0 + n_2|E|^2,$$

where $n_2 = 6.1 \cdot 10^{-23} m^2/V^2$. This kind of nonlinearity is called Kerr nonlinearity or third order nonlinearity. The extremely small nonlinear coefficient allows to apply a perturbation approach, so that the equations so far introduced for linear materials can be employed. We can thus rewrite (14.11) in a way which emphasises that n depends on the field:

$$\beta = \sqrt{\omega^2 \frac{n(|\overline{\mathcal{E}}|^2)}{c} - k_t^2}, \tag{14.19}$$

It becomes evident that in a nonlinear material the phase constant of the mode (and thus the phase of the propagating wave) depends on the intensity of the wave itself. This effect is called Self Phase Modulation (SPM) and it can be exploited for dispersion compensation. At first, suppose that the nonlinearity does not alter the spatial distribution of the mode in the transverse plane, so that in (14.19) k_t is considered constant. If $n_2 > 0$, the nonlinearity is referred to as self-focusing nonlinearity. In this case, the refractive index increases on the rising edge and, according to (14.19), the phase constant increases too. The increase of the phase constant determines a lower instantaneous phase [3], which in turn determines a lower instantaneous frequency. On the falling edge, conversely, the instantaneous frequency increases. Figure 14.8 illustrates an example of self phase modulation. As it was seen in section 14.4.1, due to the signs of the instantaneous frequency variations, we immediately understand that dispersion compensation through SPM is possible only in the anomalous dispersion regime. Furthermore, due to the variation of instantaneous frequency, SPM wholly compensates the dispersion only with selected pulse shapes and at selected intensities. The selected pulses are known as *solitons*, and they have a hyperbolic secant shape which does not vary during propagation, so that:

$$u(z,t) = A_0 e^{j\frac{\beta_0'' z}{2T^2}} sech(\frac{t - \beta_0' z}{T})$$

where

$$|A_0|^2 = \frac{n_0 \beta_0''}{n_2 \beta_0 T^2}.$$

If we consider A_0 it is clear that the shorter the pulse (high bit rate), the higher the peak power will be. Moreover, it shows that, in a focusing nonlinear material, solitons exist only in regions with anomalous dispersion. Figure 14.9 shows the spectral profile of $sech(t/T)$. The major disadvantage of long distance transmission of solitons is the loss introduced by the fiber, which reduces the power of the pulse. If the power decreases below a certain value, the pulse is distorted by GVD. Consequently, it is necessary to reduce the distance between amplifiers from approximately $100 \div 120\,km$ to $20 - 40\,km$, and to use low loss fibers. The cost can be decreased and the performance enhanced if we use distributed Raman amplifiers combined with EDFAs (discrete amplification). Nevertheless, this technique is quite complex and expensive, so that it is generally used only in specialised systems, e.g. submarine systems, where the signal propagates along thousands of kilometres without regeneration.

[3] The relation between the phase constant and the instantaneous phase is: $\phi(t) = \omega t - \beta(t)z$.

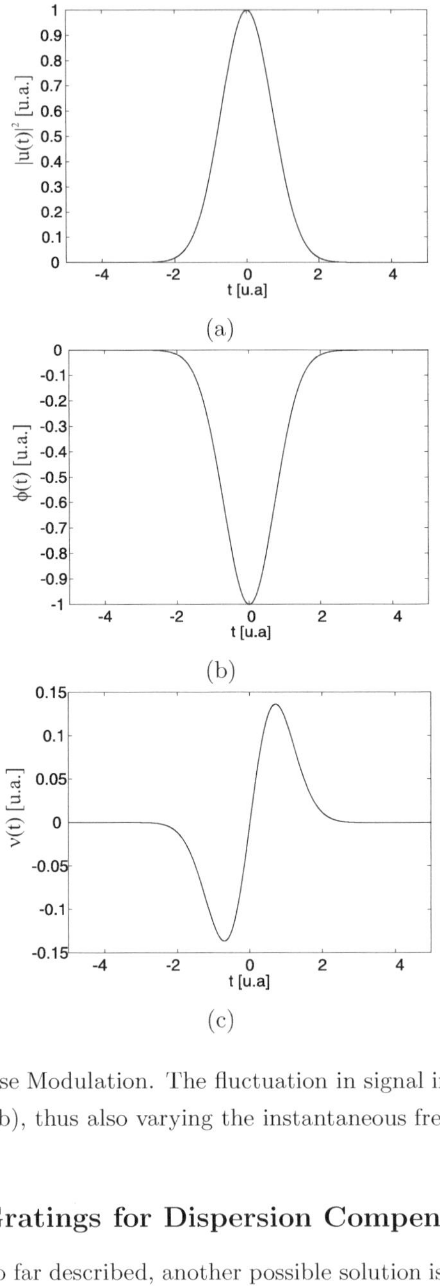

Fig. 14.8: Effect of Self Phase Modulation. The fluctuation in signal intensity (a), causes a temporally varying phase change (b), thus also varying the instantaneous frequency (c).

14.4.3 Fibers and Gratings for Dispersion Compensation

Apart from the techniques so far described, another possible solution is based on group delay equalization within the signal bandwidth. In the anomalous dispersion regime, the lower frequencies have lower group velocity, and thus a higher delay. In order to avoid dispersion, however, all the frequencies must have the same delay. Equalization can thus be obtained by placing at the end of

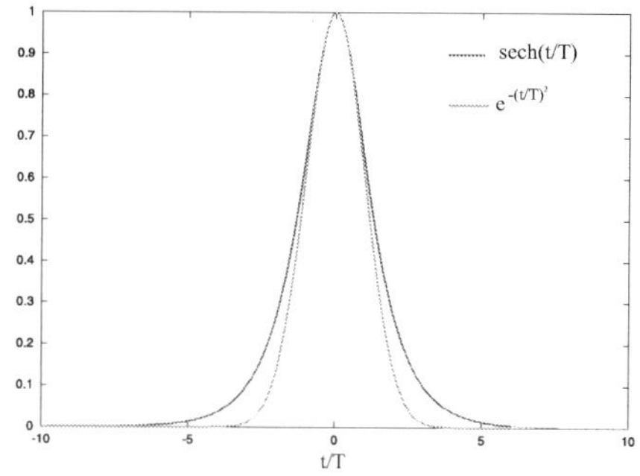

Fig. 14.9: Field amplitude of a soliton (red line) and of a Gaussian pulse (green line).

a segment of fiber a dispersion compensation fiber (DCF) with opposite dispersion characteristics. The compensation will be total at the output of the DCF if $DL + D_{DCF}L_{DCF} = 0$. In this case, the transfer function corresponds to:

$$\begin{aligned} F_{tot}(\omega) &= F(\omega)F_{DCF}(\omega) = e^{-\alpha L}e^{-j\beta_0 L}e^{-j\beta_0'(\omega-\omega_0)L}e^{-j\beta_0''(\omega-\omega_0)^2L/2} \\ & e^{-\alpha_{DCF}L_{DCF}}e^{-j\beta_{0DCF}L_{DCF}}e^{-j\beta_{0DCF}'(\omega-\omega_0)L_{DCF}}e^{-j\beta_{0DCF}''(\omega-\omega_0)^2L_{DCF}/2} \\ &= e^{-(\alpha L+\alpha_{DCF}L_{DCF})}e^{-j(\beta_0 L+\beta_{0DCF}L_{DCF})}e^{-j(\beta_0' L+\beta_{0DCF}' L_{DCF})(\omega-\omega_0)} \end{aligned}$$

and satisfies the conditions for the absence of signal distortion. In DCFs, the GVD coefficient is typically $-80 \div -110 ps/nmkm$, whereas the dispersion coefficient, D, between 1500 and 1600 nm varies linearly according to the equation $D(\lambda) = D_0 + S_0(\lambda - \lambda_0)$, where $\lambda_0 = 1550\ nm$, $D_0 = -100ps/(nmkm)$ and $S_0 = -0.22\ ps/(nm^2km)$. These values can be obtained by using fibers with specific refractive index profiles and a higher doping concentration than standard fibers (typically, depressed inner cladding fibers). Both the high doping concentration and the specific refractive index profile significantly increase the loss introduced by the fiber. Since the typical values of the attenuation coefficient are between 0.4 and $0.6dB/km$, it is fundamental to minimise the length of the DCF. In order to compensate a segment of $120km$ made of a G.652 fiber with $D = 19\ ps/(nmkm)$ with a DCF fiber having $D_{DCF} = -100ps/nmkm$, for instance, the DCF length must be $L_{DCF} = DL/D_{DCF} = 23km$. If we suppose that the attenuation coefficient is $\alpha = 0.5dB/km$, the DCF introduces a $11.5dB$ attenuation, which is compensated by inserting an EDFA amplifier at the fiber end. Recently, non-zero dispersion-shifted fibers, the G.655, in which $D = 8ps/nmkm$, have been developed. In this case, the length of the compensation fiber must be $L_{DCF} = 10km$, so that the loss is only $4.8dB$. It is thus evident that both the length and the loss are reduced. The drawback, however, is that these fibers have larger walk off lengths, so that they are more sensitive to other kinds of distortions related to the nonlinearity of their refractive index profile, such as Cross Phase Modulation (XPM) and Four Wave Mixing (FWM), which seriously damage DWDM systems.

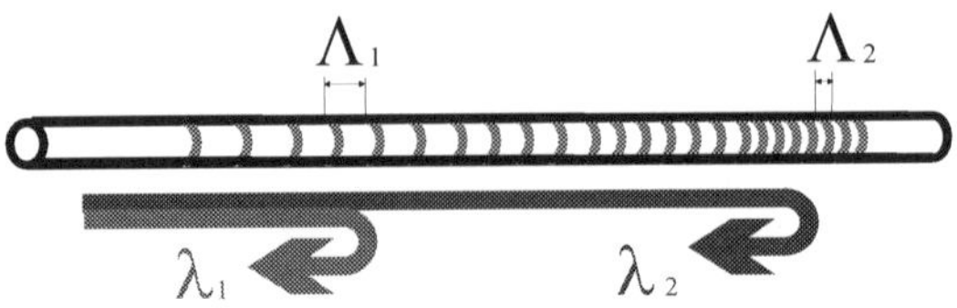

Fig. 14.10: Effect of a chirped grating on the two wavelengths λ_1 and λ_2 (where $\lambda_2 < \lambda_1$). Since λ_2 travels for a longer distance, it has a higher delay.

The same effect can be obtained by using a fiber Bragg grating, which also reduces the length of the fiber. In fiber Bragg gratings, the refractive index varies periodically along the propagation direction. Given Λ as the period of the refractive index, the grating operates like a mirror only when the phase constant of the mode satisfies the Bragg's condition, i.e. when $\beta(\lambda) = \frac{\pi}{\Lambda}$. If Λ varies along the propagation direction, the gratings are known as *chirped*. In this case, at each wavelength of the spectrum the mode travels along a different segment of the fiber before being reflected, as shown in figure 14.10. The grating period must be varied so that the wavelengths with higher group velocity travel along a longer segment than wavelengths with slower velocity. In GVD, the group delay depends linearly on frequency, so that the periodic refractive index variation must be $\Lambda(z) = \Lambda_0 + C_{Bragg}z$. The value and the sign of C_{Bragg} depend on the value and the sign of D, and on the travelled distance. Figure 14.11 shows an example of compensation scheme using a chirped fiber Bragg grating. More specifically, the scheme is based on a grating and a circulator

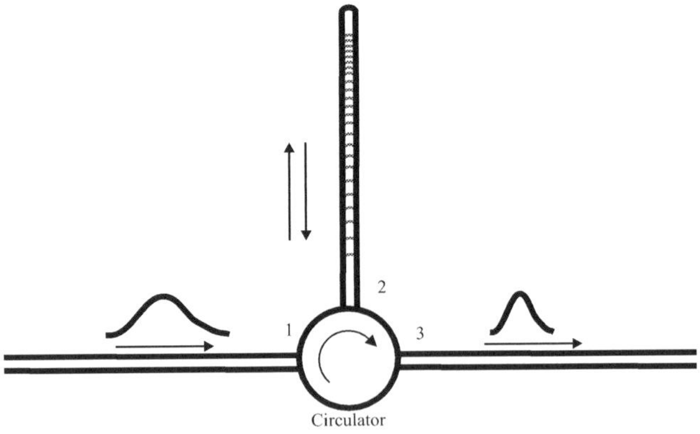

Fig. 14.11: Example of a chirped grating for dispersion compensation. The signal travels through a fiber connected to port 1 by a circulator, and then through port 2, which is connected to the grating. Finally, the compensated signal exits from port 3, which is connected to the output fiber.

connected as shown in the figure. As it was seen in chapter 6, the circulator is an optical device in which the signal enters from port 1 and exits from 2, then it enters from port 2 and exits from port 3 following the direction of the arrow. The major disadvantage of this technique is that compensation is allowed only within limited wavelength bandwidths, so that in DWDM systems it is necessary to use a cascade of chirped gratings of the type shown in figure 14.11.

14.5 Higher Order Dispersion

The relation in (14.12) is an approximation. Nevertheless, although in most cases the contribution of higher orders can be ignored, it is sometimes necessary to consider them for a correct analysis. Consider, for instance, a compensated system in which the effects of the GVD coefficient are ultimately cancelled, or an ultrashort pulsed system in which the pulses measure a few ps or fractions of them (thus having an extremely large bandwidth). If we consider only the effects of $\beta_0^{'''}$, we obtain a Third Order Dispersion (TOD), in which the phase constant can be expressed as:

$$\beta(\omega) = \beta_0 + \beta_0^{'}(\omega - \omega_0) + \frac{1}{2}\beta_0^{''}(\omega - \omega_0)^2 + \frac{1}{6}\beta_0^{'''}(\omega - \omega_0)^3.$$

Its dispersion length, on the other hand, is:

$$L_D^{'} = \frac{T_0^3}{|\beta_0^{'''}|}.$$

which permits to evaluate if the influence of $\beta_0^{'''}$ on the signal distortion can be ignored (unlike $\beta_0^{''}$). Nevertheless, the effects of TOD cannot be ignored if $L_D^{'} \leq L_D$, i.e. if:

$$T|\frac{\beta_0^{''}}{\beta_0^{'''}}| \leq 1.$$

As for Gaussian pulses, it must be mentioned (without further investigation) that their shape is also not constant during propagation. More specifically, the deformation shows an asymmetry with respect to the centre of the pulse itself and depends on the sign of $\beta_0^{'''}$. If $\beta_0^{'''} > 0$, the pulse develops a tail on the rising edge, whereas if $\beta_0^{'''} < 0$ the tail is on the falling edge. Consider the standard deviation, σ:

$$\sigma = \sqrt{< t^2 > - < t >^2}$$

where:

$$< t^p >= \frac{\int_{-\infty}^{\infty} t^p |u(z,t)|^2 dt}{\int_{-\infty}^{\infty} |u(z,t)|^2 dt}.$$

Suppose to transmit a Gaussian pulse with an intial chirp C_0, so that the broadening factor corresponds to:

$$BF = \frac{\sigma(z)}{\sigma_0} = \sqrt{(1 + \frac{C_0\beta_0^{''} z}{T^2})^2 + (\frac{\beta_0^{''} z}{T^2})^2 + (1 + C_0^2)^2(\frac{\beta_0^{'''} z}{2T^3})^2}.$$

Contrary to what occurred for GVD, the contribution of TOD does not depend on the sign of C_0. Therefore, it is not possible to contrast the effect of TOD by means of phase modulation. On the other hand, a possible solution is to insert cascaded optical devices such that:

$$\beta_0^{''} L + \beta_{0DC}^{''} L_{DC} = 0; \quad \beta_0^{'''} L + \beta_{0DC}^{'''} L_{DC} = 0.$$

over the whole signal bandwidth. The main difficulty, however, is to achieve both conditions over a large wavelength interval. Nevertheless, for pulses with a $1\,ps$ duration, it is enough to satisfy these conditions over a wavelength interval of about $4 - 5\,nm$.

14.6 Influence of the Light Source

So far, the hypothesis has been that the light source which creates the electromagnetic field before it is modulated is monochromatic, i.e. it has perfect temporal coherence. Consequently, the output signal generated by the light source, $p(t)$, is perfectly sinusoidal:

$$p(t) = \cos(\omega_0 t).$$

As a matter of fact, it is impossible to generate a perfectly monochromatic electromagnetic field due to spontaneous emission, which gives an incoherent contribution to the output signal. In this case, $p(t)$ can be written as:

$$p(t) = A_{sp}(t)\cos(\omega_0 t + \phi_{sp}(t))$$

where $A_{sp}(t)$ and $\phi_{sp}(t)$ are stochastic processes related to spontaneous emission. The situation thus becomes more complex. Although we will not demonstrate it, the power spectral density of the stochastic process, which corresponds to:

$$g(t) = A_{sp}(t)e^{j\phi_{sp}(t)}$$

is a Gaussian function:

$$G(\omega) = \frac{G_0}{\sqrt{\pi}W} e^{-\frac{\omega^2}{W^2}}$$

where W is the *linewidth*. If the transmitted signal is Gaussian, the output signal will be Gaussian too, and the Broadening Factor will be:

$$BF = \frac{\sigma(z)}{\sigma_0} = \sqrt{(1 + \frac{C_0\beta_0^{''} z}{T^2})^2 + (1 + (TW)^2)(\frac{\beta_0^{''} z}{T^2})^2 + (1 + C_0^2 + (TW)^2)^2(\frac{\beta_0^{'''} z}{2T^3})^2}.$$

The less coherent the light source, the larger is the linewidth and the higher is the BF growth rate. It is thus fundamental, in transmission, to have light sources with narrow linewidth. If $T << 1/W$ the influence of the light source on the BF can be ignored. LEDs (Light Emitting Diodes), for instance, are very cheap devices, but they have a spectral linewidth of tens of nm. If the linewidth is

$10nm$ in $\lambda_0 = 1550\,nm$, it follows that $1/W = -\lambda_0^2/c\,1/\Delta\lambda = 0.8\,ps$. Lasers, on the other hand, are higly coherent devices and their linewidth is around $0.05\,nm$, which corresponds to $1/W = 160\,ps$. Therefore, at $10Gbit/s$ T is some tens of ps, a value which cannot be achieved by LEDs, which would degrade the signal too much.

14.7 Intermodal Dispersion

When the input field excites more than one mode propagating for long distances, due to the linearity of Maxwell equations it is possible to apply the superposition principle. In the absence of distortion, the equation in (14.10) can be used for each propagating mode. If two modes propagate, we thus obtain:

$$\begin{aligned}\overline{\mathcal{E}}(u_1,u_2,L,t) = a_1|u|(t-\tau_{g1})\Big(&\overline{E}_{t1}(u_1,u_2)cos\Big(\omega_0(t-\tau_{f1})+\phi(t-\tau_{g1})\Big)\\ &+E_{z1}(u_1,u_2)sin\Big(\omega_0(t-\tau_{f1})+\phi(t-\tau_{g1})\Big)\Big)+\\ a_2|u|(t-\tau_{g2})\Big(&\overline{E}_{t2}(u_1,u_2)cos\Big(\omega_0(t-\tau_{f2})+\phi(t-\tau_{g2})\Big)\\ &+E_{z2}(u_1,u_2)sin\Big(\omega_0(t-\tau_{f2})+\phi(t-\tau_{g2})\Big)\Big);\end{aligned} \tag{14.20}$$

where:

$$a_1 = \frac{1}{2}\int_S \left(\overline{e}_{t0}\times\overline{H}_{t1}^*\right)\cdot\hat{z}dS$$
$$a_2 = \frac{1}{2}\int_S \left(\overline{e}_{t0}\times\overline{H}_{t2}^*\right)\cdot\hat{z}dS.$$

It is important to remember that the coefficients a_1 and a_2 do not depend on frequency, due to the assumption that within the signal bandwith we can neglect mode profile variations related to frequency. The two modes propagate independently because they are orthogonal. Their different group velocity, however, causes a pulse broadening which limits the transmission speed. The pulse broadening is proportional to:

$$\Delta\tau = \tau_{g1}-\tau_{g2} = \frac{L}{v_{g1}}-\frac{L}{v_{g2}} = \left(\beta_{01}^{'}-\beta_{02}^{'}\right)L,$$

which implies it is proportional to the propagation distance too. This kind of distortion is related to the different propagation characterisitcs of the modes, and it is called *intermodal dispersion*. Unlike *intramodal* distortion, intermodal dispersion can be easily eliminated if the guide operates in a single-mode regime. Nevertheless, in real guides it is difficult to obtain a real single-mode regime. As a matter of fact, in most guides the symmetric fundamental mode is formed by two degenerated and orthogonally polarised modes, which therefore share the same propagation constant. This occurs, for example, in metal waveguides with a square cross section, where the TE_{01} and TE_{10} modes have the same propagation constant. Theoretically, this kind of distortion does not create problems, because thanks to degeneracy it is possible to obtain the same phase and envelope velocities for the two kinds of polarization. Nevertheless, the perfect symmetry of the ideal guide cannot be actually achieved, and some small imperfection is always present. Therefore, the two modes lose the degeneracy and are affected by intermodal dispersion. Due to modern fabrication techniques it is possible to obtain waveguides with extremely small imperfections, so that the distortion caused by them can be ignored. Once all the other causes of distortion have been eliminated, however, intermodal dispersion becomes the only significant obstacle in transmission systems. Since this kind of dispersion affects two degenerate modes with different polarisation, it is usually referred to as *polarisation mode dispersion* rather than intermodal dispersion. The characteristics and effects of polarisation mode dispersion will be described in the following section.

14.8 Polarisation Mode Dispersion

Polarisation Mode Dispersion (PMD) mainly affects long distance, high speed optical fiber transmission. The distance considered can be up to hundreds of kilometres, and the bit rate achieves some Gb/s. This kind of systems usually exploits single-mode optical fibers, in which the cylindrical symmetry of the transverse mode pattern determines the presence of two degenerate orthogonally polarised modes. At the same time, the low index difference makes polarisation of the electric field constant and linear within the whole transverse plane. Due to degeneration, we normally refer only to one fundamental mode with two alternative polarisations, LP_{01}^{x} or LP_{01}^{y}, where the superscript refers to the direction of the electric field.

In real fibers, the characteristics of the modes are slightly different because of the imperfections related to the manufacturing process and the deformations linked to installation. The core-cladding interface, for instance, might not be perfectly circular. Furthermore, the processes previously mentioned may also cause mechanical stress, so that the silica constituting the core and the cladding becomes anisotropic. The final result is a general loss of degeneracy between the two polarisations of the fundamental mode. The optical fiber is in fact intrinsically multimode, even if the the propagation characteristics of the two modes are so similar that the phenomenon can be ignored if compared to other causes of dispersion, mainly GVD. Modern optical communication systems eliminate intermodal dispersion by means of single-mode fibers and use compensation systems for intramodal dispersion. In these cases, the polarisation mode dispersion becomes the major effect, which is the most limiting factor to the bit rate increase in long distance transmission.

14.8.1 Polarisation Maintenance and Birefringent Fibers

Birefringent fibers show the most clear example of polarisation mode dispersion. In order to simplify the discussion, we suppose that the polarisations of the two modes are both linear and tranverse

to the direction of propagation. Nevertheless, we can then generalise the discussion to all kinds of polarisation.
If the two modes are linear and transverse to the direction of polarisation, we obtain:

$$\overline{E}^{x} = \psi(u_1, u_2)e^{-\alpha_x z}e^{-j\beta_x z}\hat{\imath} \qquad \overline{E}^{y} = \psi(u_1, u_2)e^{-\alpha_y z}e^{-j\beta_y z}\hat{\jmath};$$

where α_x, α_y and β_x, β_y are the attenuation and phase constants of the two modes. The electric

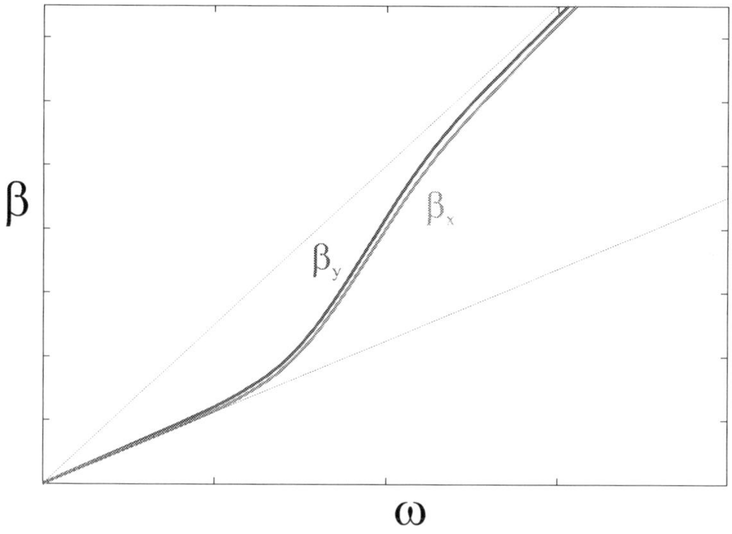

Fig. 14.12: Dispersion curves of a birefringent fiber. For the sake of simplicity, the effect of material has been neglected.

fields of the two modes are oriented along two directions, called the *major axes of the fiber*, which coincide with the x and y axes in our specific case. Figure 14.12 shows the dispersion curves of the two modes. Suppose to operate in the harmonic regime and to excite the fiber with a field which does not excite the radiation modes, so that:

$$\begin{aligned}\overline{e}_{in}(u_1, u_2) &= a_1\overline{E}^{x}(u_1, u_2) + a_2\overline{E}^{y}(u_1, u_2) = \\ &= \psi(u_1, u_2)\left(a_1\hat{\imath} + a_2\hat{\jmath}\right).\end{aligned}$$

The coefficients used to excite the modes, a_1 and a_2, depend on the polarisation state of the input field, which along the waveguide corresponds to:

$$\begin{aligned}\overline{e}(u_1, u_2, z) &= a_1\overline{E}^{x}(u_1, u_2)e^{-\alpha_x z}e^{-j\beta_x z} + a_2\overline{E}^{y}(u_1, u_2)e^{-\alpha_y z}e^{-j\beta_y z} \\ &= \psi(u_1, u_2)\left(a_1 e^{-j\beta_x z}\hat{\imath} + a_2 e^{-j\beta_y z}\hat{\jmath}\right) \\ &= \psi(u_1, u_2)a_2 e^{-j\beta_y z}\left(\frac{a_1}{a_2}e^{-j\Delta n_{eff}k_0 z}\hat{\imath} + \hat{\jmath}\right)\end{aligned} \tag{14.21}$$

where $\Delta n_{eff} = |\beta_x - \beta_y|/k_0$. Birefringence determines the beat length:

$$L_B = \frac{2\pi}{|\beta_x - \beta_y|} = \frac{\lambda}{\Delta n_{eff}},$$

which corresponds to the distance after which the polarisation state goes back to the original input state. The beat length increases as the modes approach the degeneracy point, so that it becomes a parameter to measure the fiber birefringence. Due to birefringence, the fiber has a different effect on the two modes and the channel transfer function is no longer a scalar function, but becomes a matrix:

$$\overline{e}_{out}(\omega) = [T]\,(\omega)\overline{e}_{in}(\omega). \tag{14.22}$$

$[T]$ is called *tranfer matrix* and in the case described, according to (14.21), it corresponds to:

$$[T] = \begin{bmatrix} T_{11} & T_{12} \\ T_{21} & T_{22} \end{bmatrix} = \begin{bmatrix} e^{-\alpha_x(\omega)z}e^{-j\beta_x(\omega)z} & 0 \\ 0 & e^{-\alpha_y(\omega)z}e^{-j\beta_y(\omega)z} \end{bmatrix}. \tag{14.23}$$

If we consider a second-order approximation of the two phase coefficients and we suppose that the attenuation coefficents are constant with angular frequency, (14.23) becomes:

$$[T] = \begin{bmatrix} e^{-\alpha_x z}e^{-j(\beta_{x0}+\beta'_{x0}\omega+\frac{1}{2}\beta''_{x0}\omega^2)z} & 0 \\ 0 & e^{-\alpha_y z}e^{-j(\beta_{y0}+\beta'_{y0}\omega+\frac{1}{2}\beta''_{y0}\omega^2)z} \end{bmatrix}. \tag{14.24}$$

When signals are modulated, according to what we explained in section 14.1, the carriers and the envelopes propagate at a different phase velocity ($v_{fx} = \omega_0/\beta_{x0}$ and $v_{fy} = \omega_0/\beta_{y0}$ respectively) and at a different group velocity ($v_{gx} = 1/\beta'_{x0}$ and $v_{gx} = 1/\beta'_{x0}$). Furthermore, the envelopes are affected by a different pulse broadening (due to GVD) and a different attenuation. The different attenuation also causes the *polarisation dependent loss* (PDL). Figure 14.13 schematically shows the input and output envelopes of the two modes in the absence of intramodal dispersion and loss. The difference between the group delays of the two polarisations is called *differential group delay* (DGD) and corresponds to:

$$\Delta\tau = \left(\beta'_{x0} - \beta'_{y0}\right) z. \tag{14.25}$$

On the receiver side, the current circulating in the photodiode is proportional to the intensity of the field and is not affected by its polarisation state. The differential group delay between the two modes determines a broadening of the pulse, and usually contributes to it more than GVD, so that transmission velocity is significantly reduced. On the other hand, if the polarisation of the input field excites only a single mode, the limit is once again set by GVD. In birefringent fibers the group delay increases linearly with distance. In long-distance transmission systems, the major axes of the fibers affected by PMD are not constant, but vary randomly along the propagation direction. In this case, as we will see afterwards, the group delay increases with the square root of the distance. The transfer matrix is usually expressed in a slightly different way from (14.24). In particular, we have to consider the propagation constants of the two modes:

$$\beta_m(\omega) = \frac{\beta_x(\omega) + \beta_y(\omega)}{2} \tag{14.26}$$

and substitute this equation and the one in (14.25) to (14.24). In the absence of polarisation dependent loss, we obtain:

$$[T] = e^{-\alpha z} e^{-j\beta_m(\omega)z} \begin{bmatrix} e^{-j(\frac{\Delta\tau}{2}\omega + \frac{1}{2}\Delta\beta''\omega^2)z} & 0 \\ 0 & e^{j(\frac{\Delta\tau}{2}\omega + \frac{1}{2}\Delta\beta''\omega^2)z} \end{bmatrix} = e^{-\alpha z} e^{-j\beta_m(\omega)z} [U]. \tag{14.27}$$

The terms which multiply the matrix describe the common behaviour between the two polarisations. The $[U]$ matrix is called the fiber *Jones matrix*. It describes the different polarisation behaviour and it is an additional contribution to (1.6). In the case considered, the transfer matrix is diagonal due to the fact that the modes are orthogonal to each other, so that there is no coupling. The coupling may

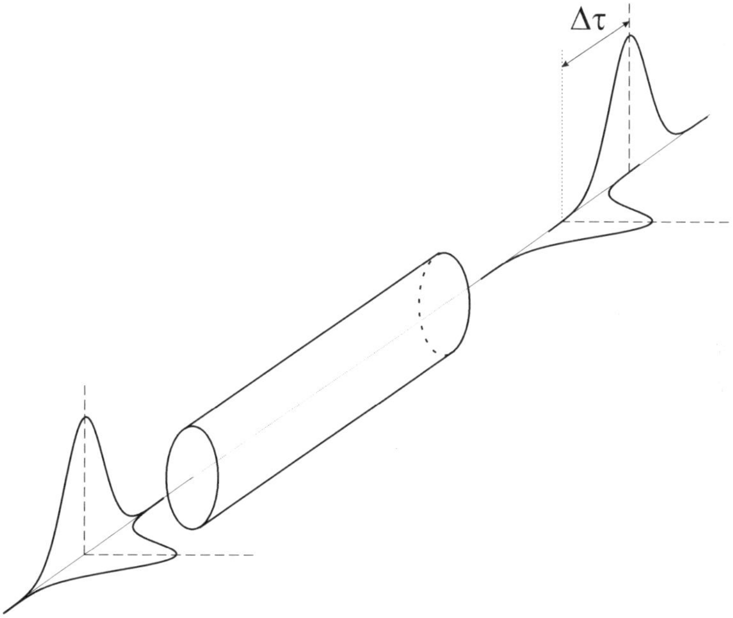

Fig. 14.13: Pulse propagation in a birefringent fiber. The major axes coincide with x and y. $\Delta\tau$ is the differential group delay (DGD).

be caused by some imperfections of the guide. If this happens, new boundary conditions must be set where the imperfections occur, because all the radiating and guided modes must be considered. The contribution of the radiating modes becomes zero after a fairly short distance, so that only guided modes are left. The amplitudes and phases of the guided modes will be different from the ones they had before they were perturbed, which implies that perturbation causes a subdivision between the modes of the transmitted power. This means that the transfer matrix will have nonzero off-diagonal terms (T_{12} e T_{21}), whose equation depends on the perturbation characteristics and can be calculated by means of the coupled mode theory.

14.8.2 Transfer Matrix of a Fiber Affected by PMD

In real fibers, due to the intrinsic and extrinsic perturbations, the two modes, LP_{01}^{x} and LP_{01}^{y}, are not degenerate and are coupled with each other. A real fiber is similar to a polarisation maintaining fiber, except for the fact that the orientation of the major axes and the beat length change randomly along the guide. Figure 14.14 shows an example of how the orientation of one axis changes in a real fiber. At first, we can imagine the fiber as a sequence of segments of polarisation maintaining fiber,

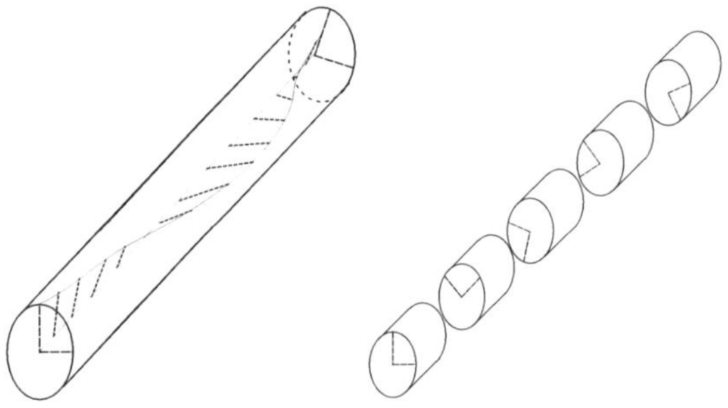

Fig. 14.14: Left: Orientation of one of the major axes along a fiber affected by PMD. Right: Waveplate model which divides the fiber into segments of polarisation maintaining fiber in which the major axes (dotted lines) have different orientations.

each having its own characteristics (see figure 14.14, right). This kind of model is called *waveplate model.* The length of the segments is constant, and must be shorter than the *correlation length*, L_c, which represent the distance to cover along the fiber so that the birefringence characteristics are no more related to the initial characteristics. The correlation length of a polarisation maintaining fiber is infinite and its tranfer matrix corresponds to (14.27). In real fibers, on the other hand, the random orientation of the chain segments causes polarisation mode coupling and the tranfer matrix corresponds to a generalisation of (14.27), which can be written as:

$$[T] = e^{-\alpha z} e^{-j\beta_m(\omega) z} [U] ,$$

where $[U]$ includes all the nonzero elements. The $[U]$ matrix has peculiar properties. To start with, it is unitary, which implies that its inverse coincides with the conjugate transpose: $[U] = [U]^{t*}$. This implies, in turn, that $det([U]) = 1$, so that we can write:

$$[U] = \begin{bmatrix} u_1 & u_2 \\ -u_2^* & u_1^* \end{bmatrix}. \qquad (14.28)$$

Once we know the value of u_1 and u_2 within the signal bandwidth, therefore, we also know the PMD of the fiber. Unfortunately, since the PMD depends on intrinsic and extrinsic causes (especially environmental conditions), the value of u_1 and u_2 randomly evolves in time and cannot be known a priori. The analytical description of their statistics is in fact one of the most important fields of research in photonics.

Experimentally, it has been observed that if the bandwidth of the transmitted signal and the distance covered are not too high, the fiber behaviour is similar to that of polarisation maintaining fibers and the envelope of the output signal corresponds to the sum of the two pulses with a different delay. Two kinds of polarisations of the input field can be applied in order to eliminate one of the pulses, and they are called *principal states of polarisation* (PSP).

There are, however, some differences with PMFs. To start with, if we excite only one PSP, polarisation is not maintained. As a matter of fact, it is necessary to specify whether we refer to PSP of the fiber input or output. Figure 14.15 shows the input and output PSPs referred to a linear polarisation. The approximation of the transfer matrix of a fiber affected by PMD is called first-order PMD approximation. While in polarisation maintaining fibers the PMD and DGD are deterministic quantities, in standard fibers they vary randomly depending on both time and frequency. If the fiber is excited by a source with a polarisation constant in time, the power splitting (γ) at the output, i.e. the splitting of the total power between the two PSPs, and the differential time delay ($\Delta\tau$) randomly fluctuate in time. Research has shown that if the fiber length is much longer than the correlation length, the power splitting is described by a uniform random variable with values between 0 and 0.5. Statistically, this implies that the PSPs are uniformly distributed in all the polarisation states. On the other hand, the differential group delay follows a random Maxwell distribution, so that:

$$f_{\Delta\tau}(t) = \frac{32}{\pi^2} \frac{t^2}{< \Delta\tau >^3} e^{-\frac{4t^2}{\pi < \Delta\tau >^2}}.$$

$< \Delta\tau >$ is the mean DGD and it is proportional to the square root of the fiber length:

$$< \Delta\tau > = \delta\tau\sqrt{L}.$$

$\delta\tau$ is the *PMD coefficient* and it is measured in $ps/\sqrt{km}$. Note that the Maxwell distribution is completely defined once we know its mean value, $< \Delta\tau >$. Consequently, standard fibers are less affected by length than polarisation maintaining fibers. This can be explained if we consider that in standard fiber the random evolution of the birefringence characteristics tend to equalise the propagation delays of the two polarisations. Consider a $100km$ length and a fiber with a $0.5ps/\sqrt{km}$ PMD coefficient, so that the mean DGD corresponds to $5ps$. According to Maxwell distribution,

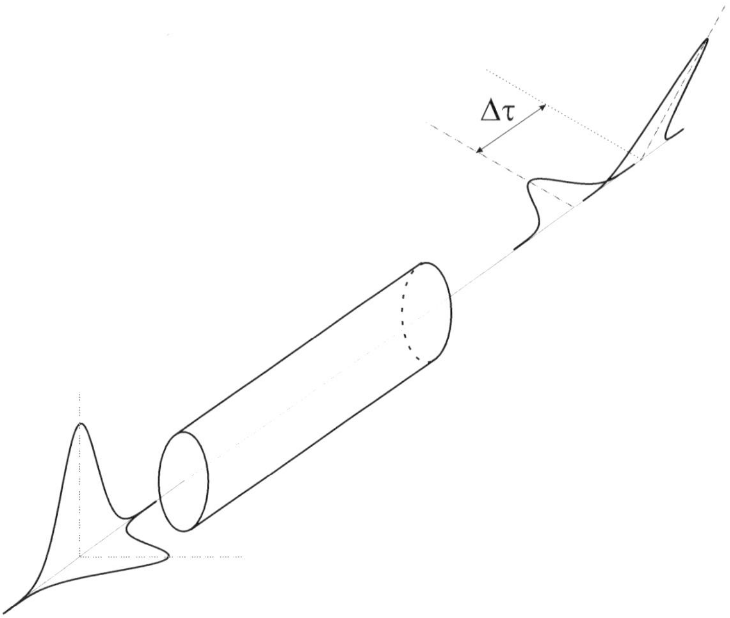

Fig. 14.15: Pulse propagation in a fiber affected by PMD. In a first-order PMD approximation, the behaviour of a real fiber corresponds to the behaviour of a birefringent fiber in which the PMD and DGD are random. The dotted lines indicate the input PSP (green line) and the output PSP (red line).

to this value of mean DGD corresponds a $4\,10^{-5}$ probability (21 minutes per year) that the DGD is higher than $25ps$, which corresponds to 25% of the bit time at $10Gbit/s$ and to 100% of the bit time at $40Gbit/s$. Due to the temporal fluctuations of DGD and PSPs the error probability at the receiver end is not constant. In such a situation, the system performance is evaluated in terms of the *outage duration*, i.e. the minutes per year in which the error probability is higher than the allowed probability. It has been shown that, in order to obtain an acceptable outage probability, the mean DGD must be lower than 10% of the bit time. This implies that the maximum length of the fiber can be $400km$ at $10Gbit/s$, which is reduced to $25km$ at $40Gbit/s$.

Appendix A

Optical Amplification

Amplification consists of an energy transfer from a primary source of high energy which does not carry information to a low energy signal which carries information. Figure A.1 schematically shows the specific case of optical amplification, in which the energy transfer from a pump source (primary source) to the signal is based on the interaction between matter and the electromagnetic field. A material which is able to transfer energy to an electromagnetic wave is called *active medium*. The

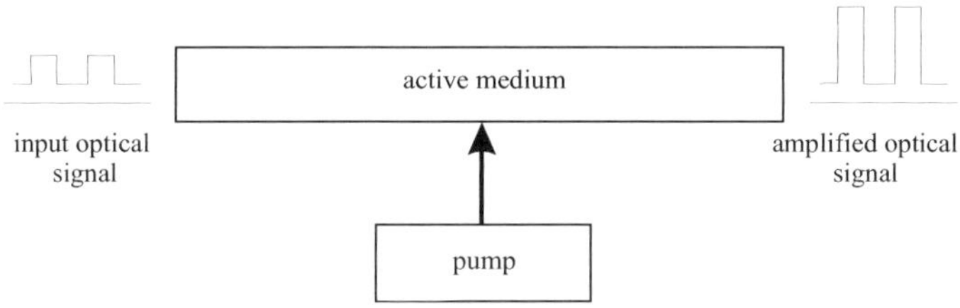

Fig. A.1: Schematic representation of optical amplification.

analysis of this kind of interaction requires a semiclassical model, according to which the electromagnetic field is treated classically, whereas the energy of matter is *quantized*, i.e. the material particle (electrons, atoms, molecules) acquire discrete values of energy, called *energy levels*. In the following section, we will revise the basic element of quantum mechanics according to which the EM field interacts with matter.

A.1 Photons and Energy Levels

According to quantum mechanics, a particle may be associated only with specific values of energy. In a rarefied medium, and when the distance between particles is sufficiently large to allow neglecting their interaction with other particles, the range of values is discrete, and they are called allowed *energy levels*. The energy levels can be associated with the energy of molecular vibration or rotation, or with the electron energy of isolated atoms or rarefied gases. If the density of the particles is high, the discrete levels tend to be separated into sublevels with such a small energy gap that they can be considered a continuum of energy levels, called *energy bands*. Figure A.2 shows the energy levels and energy bands of a generic material. Each energy level, E_i, is characterised by a *population*,

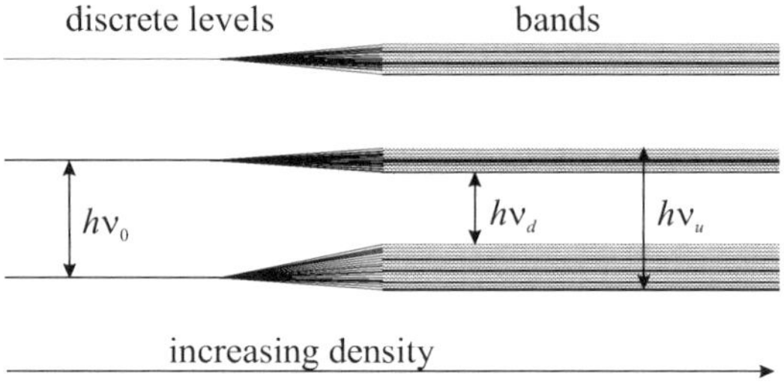

Fig. A.2: Energy levels and energy bands.

N_i (the number of particles per unit volume in the level considered), which is measured in $[1/m^3]$. In thermodynamic equilibrium, the population of the energy level depends on the probability of occupation, $P(E_i)$, of the level itself. The probability of occupation, in turn, statistically depends on the type of particle. Electron population in semiconductors, for instance, is defined by the *Fermi-Dirac* distribution:

$$P(E_i) = \frac{1}{e^{\frac{E_i - E_F}{KT}} + 1} \tag{A.1}$$

where $K = 1.381 \times 10^{-23} J/K$ is the Boltzmann constant, T is the absolute temperature, and E_F is the Fermi energy. As for more rarefied particles, they follow the *Boltzmann* distribution:

$$P(E_i) \simeq e^{-\frac{E_i}{KT}} \tag{A.2}$$

which can be described as an approximation of the Fermi-Dirac distribution valid when $E >> E_F$. The most important aspect to consider is the comparison of population between the energy levels. Consider two energy levels, E_1 and E_2, where $E_2 > E_1$ and $E_2 - E_1 >> KT$. According to both the Fermi-Dirac and the Boltzmann distribution, $N_2 << N_1$, i.e. most particles have low energy, and only very few particles have a much higher energy than the thermal energy KT of the system. The aim of this appendix is to analyse in detail the interaction between matter and the EM field. The EM field will not be classically described as the wave solution of Maxwell equations, but it will be rather defined in terms of its corpuscular nature. Consider a monochromatic electromagnetic wave of frequency ν and with a Poynting vector of magnitude I, which will be called *intensity* in the present discussion. The wave vector is $\overline{k}$ and the wave can be described as a beam of particles called *photons*, each one having its own energy:

$$E_{ph} = h\nu,$$

and moment:

$$\overline{p} = \hbar\overline{k} = \frac{h}{2\pi}\overline{k}.$$

The polarisation and initial phase of the photons, on the other hand, are identical to those of the EM wave. Figure A.3 shows a photon beam generating a photon flux Φ (i.e. a number of photons per unit time crossing the unit area) which corresponds to:

$$\Phi = \frac{I}{h\nu}. \tag{A.3}$$

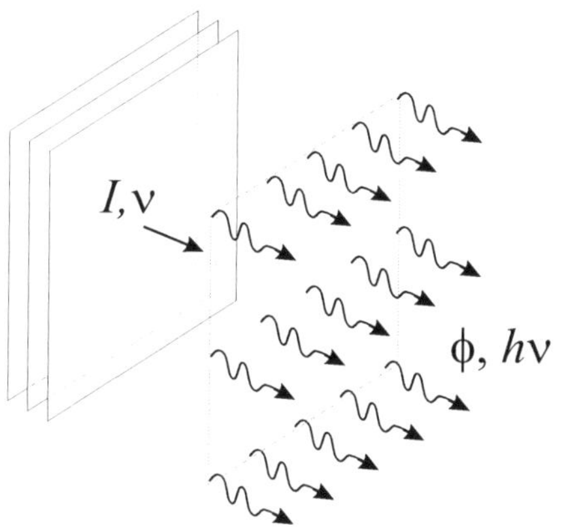

Fig. A.3: Electromagnetic wave and photon flux.

A.2 Electromagnetic Field - Matter Interaction

The interaction between the electromagnetic field and matter occurs by exchanging quanta of energy, each corresponding to the energy of a single photon. There are three types of interaction, which are schematically represented in figure A.4: *absorption*, *stimulated emission*, and *spontaneous emission*. The interaction occurs only if the photon energy is exactly the same as the energy gap between the two energy levels. For the sake of simplicity, consider a particle with only two allowed energy levels, which constitutes a *two-level system*. N_1 is the population of the lower energy level, E_1, which is normally referred to as the ground state. N_2, on the other hand, is the population of the excited state, whose energy corresponds to E_2. The electromagnetic wave propagating through the medium

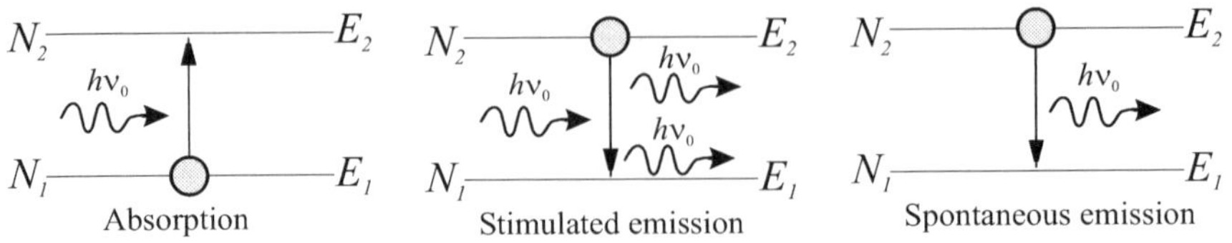

Fig. A.4: Absorption, stimulated emission and spontaneous emission in a two-level system.

has a frequency $\nu_0 = (E_2 - E_1)/h$. The three processes above mentioned are random, so that they are characterised by a probability to occur. More specifically, the probability W_{12} that a photon is absorbed by a particle, which therefore travels from level 1 to level 2, is proportional to the average number of photons interacting with the particle, i.e. to the flux Φ. The proportionality constant is written as σ_a, so that the probability becomes:

$$W_{12} = \sigma_a \Phi. \tag{A.4}$$

σ_a is usually referred to as the *absorption cross section* at frequency ν_0. It has the physical dimension of an area and it depends on the particle characteristics. By substituting the equation (A.3) into (A.4) we obtain:

$$W_{12} = \sigma_a \frac{I}{h\nu_0}. \tag{A.5}$$

If N_1 represents the number of particles at the ground state, the average number of photons absorbed per unit time and per unit volume, ΔN_a, corresponds to:

$$\Delta N_a = W_{12} N_1 = \sigma_a \frac{I}{h\nu_0} N_1. \tag{A.6}$$

Another hypothesis is that the photon interacts with a particle at the excited state 2. In this case, the photon may stimulate a transition of the particle to level 1. The transition is thus associated with the emission of another identical photon (same frequency, direction, polarisation and phase). The stimulated emission probability thus corresponds to:

$$W_{21} = \sigma_e \frac{I}{h\nu_0} \tag{A.7}$$

where σ_e is the *emission cross section*. The number of photons per unit time and per unit volume emitted due to stimulated emission is:

$$\Delta N_e = W_{21} N_2 = \sigma_e \frac{I}{h\nu_0} N_2. \tag{A.8}$$

Finally, spontaneous emission is a random phenomenon which occurs when a particle falls from the excited state to a lower energy state. If the decay is a radiative decay, the energy difference generates a photon with random direction, polarisation and phase. This phenomenon is characterised by a quantity called *lifetime* τ_{sp} of the excited state. The spontaneous emission probability, A_{21}, is inversely proportional to the lifetime, so that:

$$A_{21} = \frac{1}{\tau_{sp}}. \tag{A.9}$$

The variation of the number of photons caused by spontaneous emission corresponds to:

$$\Delta N_{sp} = A_{21}\, N_2 = \frac{N_2}{\tau_{sp}}. \tag{A.10}$$

A.2.1 Level Broadening

If the density of the material is particularly high, a single particle may be affected by all the others, and the energy levels are divided into sublevels which constitute energy bands. As figure A.2 shows, all the photons of $h\nu$ energy, which range between $h\nu_d$ and $h\nu_u$ might either cause stimulated emission or be absorbed. This phenomenon is an advantage if the material is used to produce an optical amplifier, because if the energy levels broaden, they allow the amplification of modulated signals, i.e. of signals characterised by a non-zero bandwidth. On the other hand, however, level broadening is a drawback if the active medium is used to produce lasers, because their linewidth must be as narrow as possible.
To each value of ν corresponds a specific value of the cross sections, σ_a and σ_e, so that:

$$W_{12}(\nu) = \frac{\sigma_a(\nu)}{h\nu} I(\nu), \qquad W_{21}(\nu) = \frac{\sigma_e(\nu)}{h\nu} I(\nu). \tag{A.11}$$

The cross sections represent the likelihood of the material to absorb or emit a photon at the frequency considered. If we consider two discrete energy levels, their cross sections are non-zero only for the photons at frequency $\nu_0 = (E_2 - E_1)/h$. Otherwise, the active material does not interact with the photon, and the cross sections correspond to zero. On the other hand, in the energy bands

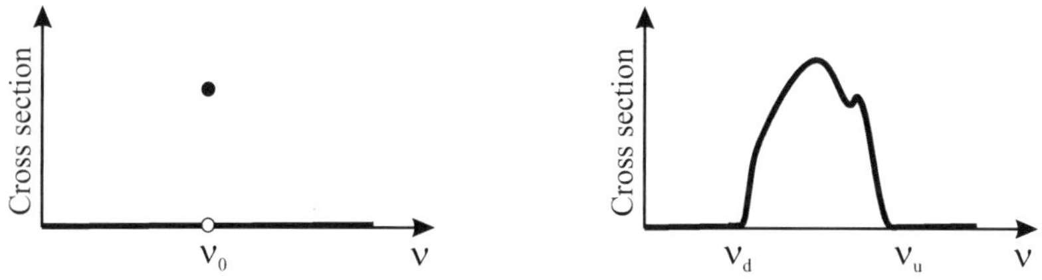

Fig. A.5: Cross sections of two discrete levels (left) and energy bands (right).

the cross sections vary considerably for different values of ν, even if they are still limited and continuous. Figure A.5 shows the cross sections of both discrete levels and energy bands. It has been demonstrated that for discrete levels $\sigma_a = \sigma_e$, whereas if we consider sublevels, $\sigma_a \neq \sigma_e$. Nevertheless, the values are often very similar, so that the approximation $\sigma_a \simeq \sigma_e$ can be adopted, as we will do in the following sections.

A.2.2 Population Rate Equations

In the previous section, we mentioned that the population at different levels depends on the probability of occupation. This is true in the absence of external electromagnetic fields, which stimulate emission or absorption and thus modify population. In this section, we will find the equations which link the incident electromagnetic radiation to the population of the energy levels. These equations are usually referred to as *population rate equations*. For the sake of simplicity, we will consider a

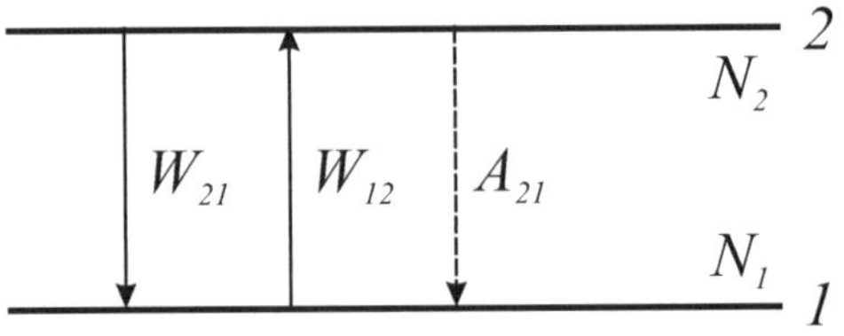

Fig. A.6: Transitions from one level to the other in a two-level system.

two-level system of discrete levels, as shown in figure A.6. We can then extend the discussion to energy bands by considering that the cross sections depend on frequency.
Consider a system with N_T particles per unit volume. According to the principle of conservation:

$$N_T = N_1 + N_2. \tag{A.12}$$

The phenomena which can vary the population rate are the transitions from level 2 to level 1 due to stimulated emission, spontaneous emission, or absorption. The probability that each occurs is, respectively: W_{21}, A_{21} and W_{12}. Both stimulated and spontaneous emission determine an increase of N_1 and a decrease of N_2, whereas the opposite occurs with absorption. In a stationary regime, the populations are constant in time, so that the values of N_2 and N_1 determine the same probability for favourable and unfavourable outcomes to happen. Since the phenomena are mutually exclusive, the probability of the union event corresponds to the sum of the single probabilities. If we consider level 2, we can thus write:

$$W_{12}N_1 = W_{21}N_2 + A_{21}N_2.$$

According to (A.12), if we isolate N_2 and consider (A.11), we obtain:

$$N_2 = \frac{W_{12}}{W_{12} + W_{21} + A_{21}} N_T = \frac{\sigma_a \frac{I}{h\nu}}{\sigma_e \frac{I}{h\nu} + \sigma_a \frac{I}{h\nu} + A_{21}} N_T. \tag{A.13}$$

Similarly, if we consider level 1, its population is:

$$N_1 = \frac{W_{21} + A_{21}}{W_{12} + W_{21} + A_{21}} N_T = \frac{\sigma_e \frac{I}{h\nu} + A_{21}}{\sigma_e \frac{I}{h\nu} + \sigma_a \frac{I}{h\nu} + A_{21}} N_T, \tag{A.14}$$

which corresponds to the product between N_T and the ratio of the favourable outcomes to the total number of possibilities. According to (A.13) and (A.14) we thus obtain:

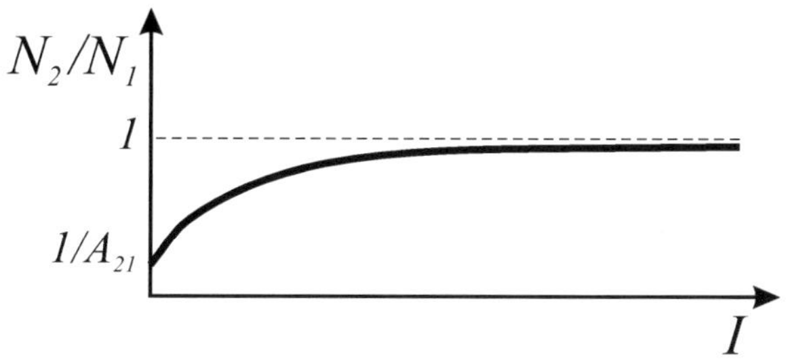

Fig. A.7: The ratio N_2/N_1 tends to saturate for increasing values of optical intensity.

$$\frac{N_2}{N_1} = \frac{\sigma_a I}{\sigma_e I + A_{21} h\nu}.$$

As shown in figure A.7, if we assume that $\sigma_e = \sigma_a$, this ratio is lower or equal to 1 for each value of I. Level 2 must necessarily have a lower population than level 1 and tends towards it only when I tends towards infinity. In conclusion, in a two-level system it is not possible to obtain population inversion, i.e. $N_2 > N_1$.

A.2.3 Propagation Equations

In this section, we consider the opposite side of the problem, and try to define the effects of absorption and stimulated emission on the incident radiation (for the sake of simplicity, we neglect spontaneous emission).

Consider once more a two-level system in which an electromagnetic radiation of intensity I hits a medium of volume $dV = dS\,dz$, as shown in figure A.8. While crossing the medium, the wave

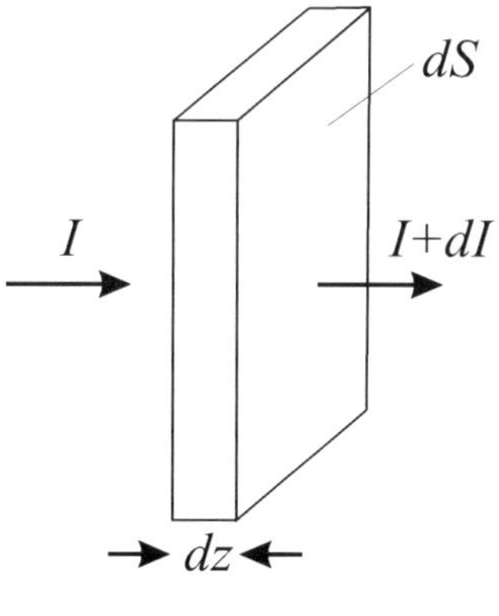

Fig. A.8: Medium with volume dV.

intensity will vary by dI. The variation increases or decreases depending on whether the number of stimulated emissions, ΔN_e, is higher or lower than the number of absorptions, ΔN_a. Since each photon is associated with an energy $h\nu$, the power balance equation corresponds to:

$$dI\,dS = [\Delta N_e - \Delta N_a] dV h\nu. \tag{A.15}$$

According to (A.8) and (A.6), if we consider that $dV = dS\,dz$ together with (A.11), (A.15) becomes:

$$\frac{dI}{dz} = [\sigma_e N_2(z) - \sigma_a N_1(z)]\, I(z). \tag{A.16}$$

The equation in (A.16) is called *propagation equation.* The intensity of the incident wave increases or decreases depending on the value of the *gain per unit length*:

$$g(z) = \sigma_e N_2(z) - \sigma_a N_1(z), \tag{A.17}$$

which can be higher or lower than zero. If $\sigma_a = \sigma_e$, in order for $g(z)$ to be positive it has to be:

$$N_2 > N_1.$$

In this case, *population inversion* occurs and the medium is defined an *active medium.* On the other hand, if $\sigma_a(\nu) \neq \sigma_e(\nu)$, the relation between N_2 and N_1 in order for the medium to be active is more complex: if $\sigma_e > \sigma_a$, the medium is active also if N_2 is slightly lower than N_1. Conversely, if $\sigma_e < \sigma_a$ a higher degree of population inversion is required.

A.3 Optical Pumping

A medium becomes active only by means of population inversion. Due to optical saturation, however, population inversion cannot be achieved in a two-level system by simply varying the intensity, I. Therefore, another technique is required, pumping, in order to move the particles from the ground state to the excited state. In optical amplifiers, population inversion can be obtained either electrically, as it occurs in semiconductor optical amplifiers (SOAs), or electromagnetically, as it occurs in doped fiber amplifiers (DFAs), in which the fibers are doped with rare-earth elements. DFAs use a pump scheme with a second optical beam, called *pump beam* and require a system with more than two levels. The most widespread system used is a three-level system.

A.3.1 Optical Pumping in Three-Level Systems

The pump scheme of a three-level system is shown in figure A.9, which is an extended representation of the pump scheme in a two-level system. The transitions represented are only those with a significant probability to occur in materials which can be used in optical amplifiers. Level 1 is the *ground state*, level 2 is the *metastable state*, and level 3 is the *excited state*. The transition from level

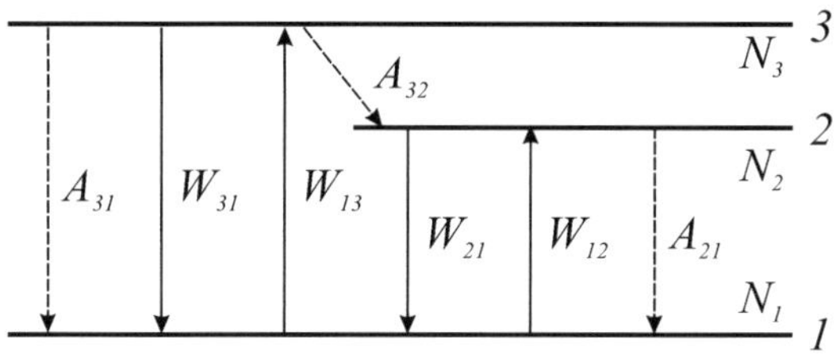

Fig. A.9: Transitions in a three-level system.

3 to level 2 is nonradiative, so that it does not affect the field characteristics. The $1-3$ transition is determined by the pump beam, which has a frequency $\nu_p = (E_3 - E_1)/h$. The probability of transition corresponds to:

$$W_{13} = \frac{\sigma_{13} I_p}{h\nu_p},$$

where σ_{13} is the absorption cross-section of the pump.
According to the law of conservation:

$$N_1 + N_2 + N_3 = N_T.$$

According to the same principles that we saw for the two-level system, in a stationary regime the *population rate equations* correspond to:

$$\begin{cases} -(W_{13}+W_{12})N_1 + (A_{21}+W_{21})N_2 + (A_{31}+W_{31})N_3 = 0 \\ W_{12}N_1 - (A_{21}+W_{21})N_2 + (A_{32})N_3 = 0 \\ W_{13}N_1 - (A_{31}+A_{32}+W_{31})N_3 = 0. \end{cases}$$

In erbium, however, it has been experimentally observed that the average lifetime is $\tau_{32} = 1/A_{32}$, which is extremely smaller than all the other parameters. This implies that an erbium ion can remain in E_3 level for a length of time effectively zero, so that the excited state cannot be populated. As a matter of fact, when a particle reaches the excited state, it immediately falls back to the metastable state. The relation which expresses this situation is found by dividing the second and third equation by A_{32}. We then make this parameter tend towards infinity. The two equations can thus be solved only if $N_3 = 0$, so that the final system becomes:

$$\begin{cases} -(W_{13}+W_{12})N_1 + (A_{21}+W_{21})N_2 = 0 \\ N_1 + N_2 = N_t \end{cases}$$

The situation is thus similar to the two-level system, but there is a different coefficient, W_{13}, to indicate the absorption from level 1 to level 3 (*pumping*). We thus call the system a *two-level approximation*, whose solution is:

$$N_2 = \frac{W_{12}+W_{13}}{W_{13}+W_{12}+W_{21}+A_{21}} N_t \tag{A.18}$$

$$N_1 = \frac{W_{21} + A_{21}}{W_{13} + W_{12} + W_{21} + A_{21}} N_t. \quad \text{(A.19)}$$

I_s and I_p correspond to the intensity of the signal beam and the pump beam respectively. The ratio between (A.18) and (A.19) is thus:

$$\frac{N_2}{N_1} = \frac{W_{12} + W_{13}}{W_{21} + A_{21}} = \frac{\dfrac{\sigma_a I_s}{h\nu_s} + \dfrac{\sigma_p I_p}{h\nu_p}}{\dfrac{\sigma_e I_s}{h\nu_s} + A_{12}}.$$

It follows that, in order to obtain $N_2 > N_1$:

$$I_p > \frac{\sigma_e - \sigma_a}{\sigma_p} \frac{\nu_p}{\nu_s} I_s + A_{21} \frac{h}{\sigma_p} \nu_p. \quad \text{(A.20)}$$

Therefore, it is the pump beam which provides energy to the system in order to transfer the population from N_1 to N_2. When the intensity is sufficiently high, inversion population occurs and the medium becomes active. The relation in (A.20) emphasises that the value of pump intensity increases for increasing values of signal intensity.

A.3.2 Electrical Pumping of Semiconductor Materials

In semiconductor materials, the energy levels are in fact energy bands divided by forbidden energy bands, called band gaps. The probability that bands are occupied is defined by the *Fermi-Dirac* distribution and depends on the energy reference level, called *Fermi level.* Emission occurs if there

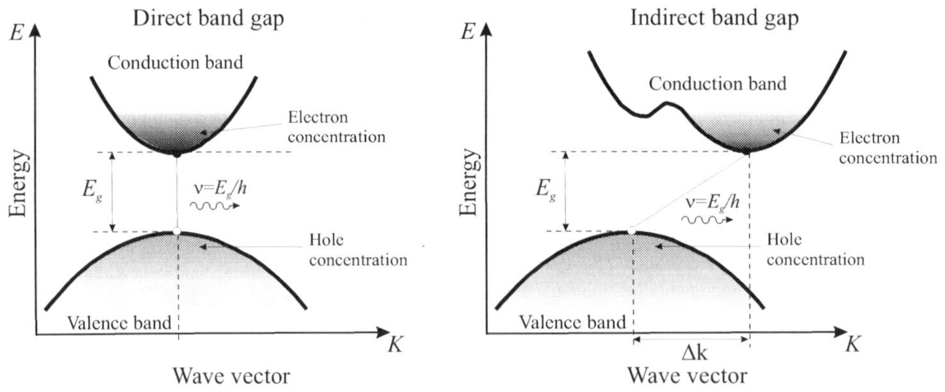

Fig. A.10: Valence and conduction bands in direct band gap semiconductors (left) and indirect band gap semiconductors (right). The different shades of grey are directly proportional to the electron concentration in the conduction bands and on the hole concentration in the valence band.

is an electron transition from the conduction band to the valence band. It is thus fundamental the presence of an electron in the conduction band (CB) and of a hole in the valence band (VB). Although emission is theoretically possible in all semiconductor materials, its probability highly depends on the types of band gaps, shown in figure A.10. In direct band gap semiconductors, such

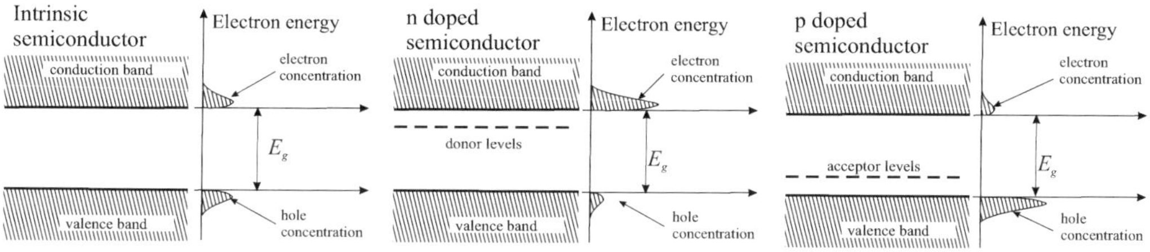

Fig. A.11: Carrier concentration (electron and holes) as a function of the doping type.

as GaAs, the minimum of the CB and the maximum of the VB occur at the same value of the electron wave vector, so that transition does not depend on the change in the momentum of the electron. On the other hand, in indirect band gap semiconductors, such as Si or Ge, the minimum and maximum of the two bands do not coincide, so that a change in the momentum of the electron is required. Since the momentum of a photon can be ignored when compared to the momentum of an electron, in direct band gap semiconductors, transition involves only a photon and an electron. Conversely, in indirect band gap semiconductors a phonon is required to compensate for the change in the momentum, so that the phenomenon is far less probable. Consequently, amplifiers and lasers are usually produced with direct band gap semiconductors.

Figure A.11 shows the concentration of carriers as a function of the doping type. In intrinsic semiconductors, the Fermi level lies in the middle of the band gap, and both the electron concentration in the CB and the hole concentration in the VB are extremely low. In a p-type or n-type semiconductor, the Fermi level shifts either towards the CB or the VB. In the first case, the electron concentration in the CB increases, whereas the hole concentration in the VB decreases. In the second case, the opposite occurs. Nevertheless, the stimulated emission of photons requires a high

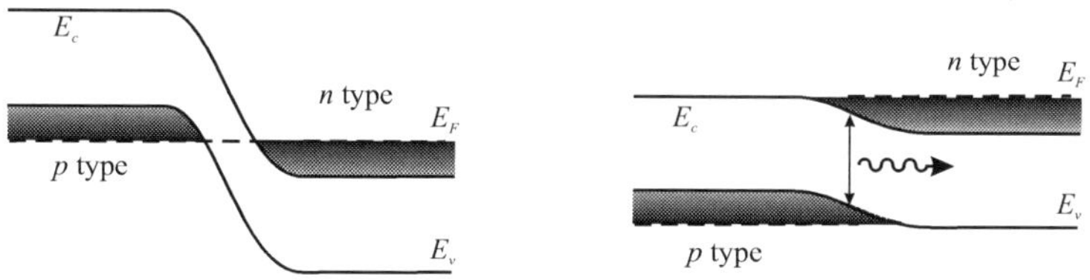

Fig. A.12: Carrier concentration across an unbiased p-n junction (left) and in a forward biased p-n junction (right).

concentration of both electrons in the CB and holes in the VB. This kind of population inversion is obtained in forward-biased p-n junctions, as shown in figure A.12. The energy band bending at the interface creates a region with high population inversion. The loss of carriers, which recombine to form photons, must be compensated by an external electrical current, called *injection current*, which operates as a pump.

In this case, determining the population rate equation is quite easy, because we can use the physical notions related to semiconductors. Nevertheless, this kind of topic will not be treated in the present discussion, so that we only provide the final result of population inversion, which is expressed through the Fermi inversion factor:

$$f_g = f_c(E_2) - f_v(E_1),$$

where:

$$f_c(E) = \frac{1}{e^{\frac{E-E_{F_c}}{KT}} + 1}, \qquad f_v(E) = \frac{1}{e^{\frac{E-E_{F_v}}{KT}} + 1} \tag{A.21}$$

are the Fermi functions of the conduction and valence bands respectively, whereas E_{F_c} and E_{F_v} are the quasi Fermi levels of the biased junction, which in turn depend on the bias voltage or on the pumping level of the junction.

A.4 Amplified Spontaneous Emission

The photons created due to spontaneous emission are not related to the signal beam, so that they represent a source of noise. Nevertheless, the photons which contribute more to noise are those created due to spontaneous emission at both ends of the active medium. As a matter of fact, this kind of photons propagates in the active medium and thus creates other photons due to stimulated emission. The photons created due to stimulated emission are associated with those created due to spontaneous emission and not with those belonging to the signal beam, so that they further increase

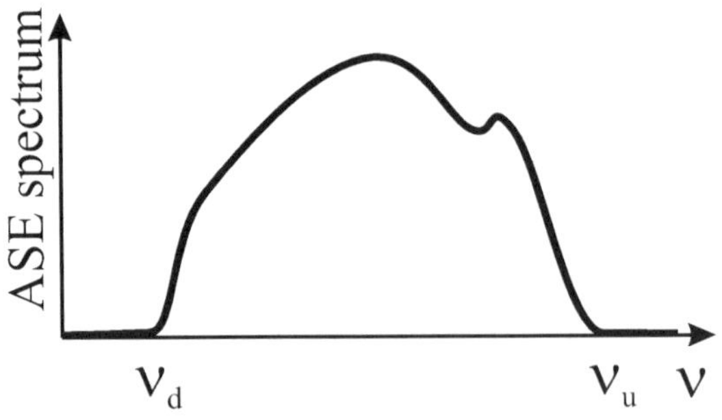

Fig. A.13: Example of ASE spectrum.

noise. This phenomenon is called *amplified spontaneous emission (ASE)*. Since amplification is progressive, ASE is mainly caused by the photons at both ends of the active medium. *Copropagating ASE*, or ASE^+, occurs if the ASE propagation direction is the same as the signal direction, whereas if the ASE propagation is the opposite of the signal propagation we refer to *counterpropagating ASE*, or ASE^-. Copropagating ASE reduces the signal-to-noise ratio, whereas the counterpropagating ASE has no effect on the signal-to-noise ratio because the emissions propagate in opposite directions. Nevertheless, counterpropagating ASE enhances the transitions from the metastable to the ground state, thus reducing population inversion and, consequently, the gain. ASE occurs at all the frequencies in which the gain coefficient is positive. Its spectrum is very similar to the emission cross section, as shown in figure A.13.

Appendix B

Coupled-mode Theory

The occurrence of mode coupling is sometimes caused by the dishomogeneity along waveguides or by the properties of the channel, e.g. nonlinearity. This coupling, which can be either voluntarily induced or can occur involuntarily, must be described in detail in order to understand and be able to exploit the energy transfer between modes during propagation.
Mathematically, the coupling of orthogonal modes can be described in different manners because, as it will be shown, there are different ways to obtain a system containing an infinite set of coupled equations, which is necessary to solve the problem. For the purposes of this study, we will consider the solutions for a uniform and homogeneous waveguide, where the perturbations created become sources for the waveguide modes. The perturbations cannot supply energy, but allow energy exchange between modes, which therefore can be coupled. That is why the theory is called *coupled-mode theory.*

Consider Maxwell's equations for an unperturbed waveguide:

$$\begin{aligned} \overline{\nabla} \times \overline{E}_I &= -j\omega\mu\overline{H}_I; \\ \overline{\nabla} \times \overline{H}_I &= j\omega\epsilon\overline{E}_I. \end{aligned} \tag{B.1}$$

Subscript I indicates the unperturbed guide, $\overline{E}$ and $\overline{H}$ are vector fields, respectively the electric and magnetic fields, ϵ is the dielectric permittivity of the medium, μ is the magnetic permeability and ω is the angular frequency. Consider a perturbation to the guide which is caused by a $\Delta\epsilon$ variation of permittivity. Using the subscript P to indicate the modes of the perturbed medium, Maxwell's equations can be written as follows:

$$\begin{aligned} \overline{\nabla} \times \overline{E}_P &= -j\omega\mu\overline{H}_P \\ \overline{\nabla} \times \overline{H}_P &= j\omega(\epsilon + \Delta\epsilon)\overline{E}_P. \end{aligned} \tag{B.2}$$

The second equation can then be written as:

$$\overline{\nabla} \times \overline{H}_P = j\omega\epsilon\overline{E}_P + j\omega\Delta\epsilon\overline{E}_P = j\omega\epsilon\overline{E}_P + \overline{J}_{eq}$$

It becomes evident that the perturbation can be described by means of equivalent sources, $\overline{J}_{eq}$. If then the sources are described through the corresponding polarization vector $\overline{P}(x, y, z)$, Maxwell's equations become:

$$\begin{aligned} \overline{\nabla} \times \overline{E}_P &= -j\omega\mu\overline{H}_P \\ \overline{\nabla} \times \overline{H}_P &= j\omega\epsilon\overline{E}_P + j\omega\overline{P} \end{aligned} \tag{B.3}$$

By applying Lorentz reciprocity theorem to the fields in (B.1) and (B.3) we thus obtain:

$$\overline{\nabla} \cdot \left(\overline{E}_P \times \overline{H}_I^* + \overline{E}_I^* \times \overline{H}_P\right) = -j\omega\overline{P} \cdot \overline{E}_I^*. \tag{B.4}$$

It is also useful to introduce the transverse nabla operator:

$$\overline{\nabla}_t = \overline{\nabla} - \frac{\partial}{\partial z}\hat{z}. \tag{B.5}$$

By integrating over the transverse plane we obtain:

$$\begin{aligned} &\iint_{-\infty}^{+\infty} \overline{\nabla}_t \cdot \left(\overline{E}_P \times \overline{H}_I^* + \overline{E}_I^* \times \overline{H}_P\right) dx\, dy + \\ &+ \iint_{-\infty}^{+\infty} \frac{\partial}{\partial z} \left(\overline{E}_P \times \overline{H}_I^* + \overline{E}_I^* \times \overline{H}_P\right) \cdot \hat{z}\, dx\, dy = \\ &= -j\omega \iint_{-\infty}^{+\infty} \overline{P} \cdot \overline{E}_I^*\, dx\, dy \end{aligned} \tag{B.6}$$

And if we apply the divergence theorem to the first integral in (B.6), we thus obtain:

$$\iint_{-\infty}^{+\infty} \overline{\nabla}_t \cdot \left(\overline{E}_P \times \overline{H}_I^* + \overline{E}_I^* \times \overline{H}_P\right) dx\, dy = \oint_C \left(\overline{E}_P \times \overline{H}_I^* + \overline{E}_I^* \times \overline{H}_P\right) \cdot \hat{l}\, dl \tag{B.7}$$

where the symbol "$\oint$" represents the line integral along the closed curve C, which in turn bounds the surface for which the double integral is calculated. All the solutions for (B.7) are fields which must satisfy the Sommerfeld's conditions. This implies that, in case of finite sources, at $r \to +\infty$ distance from the sources the following conditions must be satisfied:

$$\begin{aligned} \lim_{r \to \infty} r\overline{E} &= 0; \\ \lim_{r \to \infty} r\overline{H} &= 0. \end{aligned} \tag{B.8}$$

It follows that the line integral on a curve of infinite radius (see equation in (B.7)) must be zero:

$$\oint_C \left(\overline{E}_P \times \overline{H}_I^* + \overline{E}_I^* \times \overline{H}_P\right) \cdot \hat{l}\, dl = 0. \tag{B.9}$$

Therefore, the equation in (B.6) can be reformulated as follows:

$$\frac{\partial}{\partial z}\iint_{-\infty}^{+\infty}\left(\overline{E}_P \times \overline{H}_I^* + \overline{E}_I^* \times \overline{H}_P\right)\cdot \hat{z}\, dx\, dy = -j\omega \iint_{-\infty}^{+\infty} \overline{P}\cdot \overline{E}_I^*\, dx\, dy. \tag{B.10}$$

Note that the dot product $\left(\overline{A} \times \overline{B}\right)\cdot \hat{z}$ is non-zero only for components of $\overline{A} \times \overline{B}$ along $\hat{z}$, that is, by definition $\left(\overline{A} \times \overline{B}\right)\cdot \hat{z} = A_x B_y - A_y B_x$. Therefore, by introducing the transverse components ($\overline{E}_t$ e $\overline{H}_t$) and the longitudinal components ($\overline{E}_z$ and $\overline{H}_z$) of the electric and the magnetic field:

$$\begin{aligned}\overline{E} &= \overline{E}_t + \overline{E}_z\\ \overline{H} &= \overline{H}_t + \overline{H}_z\end{aligned} \tag{B.11}$$

the equation in (B.10) becomes:

$$\frac{\partial}{\partial z}\iint_{-\infty}^{+\infty}\left(\overline{E}_{Pt} \times \overline{H}_{It}^* + \overline{E}_{It}^* \times \overline{H}_{Pt}\right)\cdot \hat{z}\, dx\, dy = -j\omega \iint_{-\infty}^{+\infty} \overline{P}\cdot \overline{E}_I^*\, dx\, dy. \tag{B.12}$$

This equation describes a general relation between the solutions of an unperturbed guide and the solutions of a perturbed guide, which are functions of the perturbation itself. The relation is closely connected to the orthogonality properties of the guided modes. In particular, if we consider the case of $\overline{P} = 0$, that is, the absence of perturbations, (B.12) can be written as follows:

$$\iint_{-\infty}^{+\infty}\left(\overline{E}_{Pt} \times \overline{H}_{It}^* + \overline{E}_{It}^* \times \overline{H}_{Pt}\right) dx\, dy = 0 \tag{B.13}$$

which represents the case of orthogonality between different copropagating modes.

Since the modes form a complete set of orthogonal solutions, that is, they form an orthogonal basis, each field distribution within the waveguide can be described as the sum between a finite set of guided modes and a continuous spectrum of radiation modes. It follows that also the field of a perturbed guide can be described through a specific combination of modes:

$$\begin{aligned}\overline{E}_{Pt}(x,y,z) &= \sum_{\nu} \overline{E}_{t\nu}(x,y)\left[A_\nu(z)\, e^{-j\beta_\nu z} + B_\nu(z)\, e^{j\beta_\nu z}\right] + \int \overline{E}_{rad,\nu}\, d\nu\\ \overline{H}_{Pt}(x,y,z) &= \sum_{\nu} \overline{H}_{t\nu}(x,y)\left[A_\nu(z)\, e^{-j\beta_\nu z} - B_\nu(z)\, e^{j\beta_\nu z}\right] + \int \overline{H}_{rad,\nu}\, d\nu\end{aligned} \tag{B.14}$$

in which the summation refers to the guided modes, whereas the integral to the radiation modes. In the present discussion, we will consider only the coupling between guided modes, so that:

$$\overline{E}_{Pt}(x,y,z) = \sum_{\nu}\left[A_\nu(z)\, e^{-j\beta_\nu z} + B_\nu(z)\, e^{j\beta_\nu z}\right]\overline{E}_{t\nu}(x,y) \tag{B.15a}$$

$$\overline{H}_{Pt}(x,y,z) = \sum_{\nu}\left[A_\nu(z)\, e^{-j\beta_\nu z} - B_\nu(z)\, e^{j\beta_\nu z}\right]\overline{H}_{t\nu}(x,y). \tag{B.15b}$$

$A_\nu(z)$ refers to direct modes, that is, the modes propagating in the positive direction $+\hat{z}$, whereas $B_\nu(z)$ refers to reflected modes. Note that these coefficients must be functions of z because, otherwise, $\overline{E}_{Pt}(x,y,z)$ and $\overline{H}_{Pt}(x,y,z)$ would be solutions of an unperturbed waveguide. The minus sign in (B.15b) is linked to the magnetic fields of direct and reflected modes, which have an opposite sign, as explained by the duality theory.

It is now necessary to determine the coefficients A_ν and B_ν. Therefore, it is necessary to find the equations of which A_ν and B_ν are solutions in order for the field $(\overline{E}_P, \overline{H}_P)$ to be solution of the perturbed Maxwell's equations. There are two cases in which this occurs.

Coupling between copropagating modes

Consider an unperturbed guide and the solutions for the direct modes propagating in the positive direction:

$$\begin{aligned} \overline{E}_I &= \overline{E}_\mu e^{-j\beta_\mu z}; \\ \overline{H}_I &= \overline{H}_\mu e^{-j\beta_\mu z}; \end{aligned} \tag{B.16}$$

If we substitute the terms in (B.15) and (B.16) into the left side of the equation in (B.10), the equation becomes:

$$\sum_\nu \frac{\partial}{\partial z}\left(\iint_{-\infty}^{+\infty}\left[\left(A_\nu(z)\,e^{-j\beta_\nu z}+B_\nu(z)\,e^{j\beta_\nu z}\right)\overline{E}_{t\nu}\times\overline{H}_{t\mu}^{*}e^{j\beta_\mu z}+\right.\right.$$
$$\left.\left.\overline{E}_{t\mu}^{*}\times\left(A_\nu(z)\,e^{-j\beta_\nu z}-B_\nu(z)\,e^{j\beta_\nu z}\right)\overline{H}_{t\nu}e^{j\beta_\mu z}\right]\cdot\hat{z}\,dx\,dy\right)=$$

$$\begin{aligned}\sum_\nu \frac{\partial}{\partial z}\Bigg(e^{j\beta_\mu z}\Bigg[A_\nu(z)\,e^{-j\beta_\nu z}\iint_{-\infty}^{+\infty}\left(\overline{E}_{t\nu}\times\overline{H}_{t\mu}^{*}+\overline{E}_{t\mu}^{*}\times\overline{H}_{t\nu}\right)\cdot\hat{z}\,dx\,dy+\\ +B_\nu(z)\,e^{j\beta_\nu z}\iint_{-\infty}^{+\infty}\left(\overline{E}_{t\nu}\times\overline{H}_{t\mu}^{*}-\overline{E}_{t\mu}^{*}\times\overline{H}_{t\nu}\right)\cdot\hat{z}\,dx\,dy\Bigg]\Bigg).\end{aligned} \tag{B.17}$$

According to the orthogonality property of the modes, all the integrals in the previous equation are zero for $\nu \neq \mu$. In order to define the value of the other integrals, it is then necessary to set an arbitrary normalization which can vary from case to case. In fact, the formulations of the theory differ in the expression of the coupling coefficients due to normalization. In the present discussion, the normalisation adopted is:

$$\iint_{-\infty}^{+\infty}\frac{\overline{E}_\mu\times\overline{H}_\mu^{*}}{2}\cdot\hat{z}\,dx\,dy=\iint_{-\infty}^{+\infty}\frac{\overline{E}_\mu^{*}\times\overline{H}_\mu}{2}\cdot\hat{z}\,dx\,dy=1; \tag{B.18}$$

although the following one is more generally adopted:

$$\iint_{-\infty}^{+\infty} \left(\overline{E}_\mu \times \overline{H}_\mu^*\right) \cdot \hat{z}\, dx\, dy = \iint_{-\infty}^{+\infty} \left(\overline{E}_\mu^* \times \overline{H}_\mu\right) \cdot \hat{z}\, dx\, dy = \frac{1}{2}.$$

As it can be seen, the integrals in the second equation differ for a factor of 4 with respect to the first one. If then we substitute (B.18), (B.17) is reduced to $4\, dA_\mu/dz$ and the equation in (B.12) thus becomes:

$$\frac{dA_\mu}{dz} = -j\frac{\omega}{4} \iint_{-\infty}^{+\infty} \overline{P} \cdot \overline{E}_\mu^*\, dx\, dy\; e^{j\beta_\mu z}. \tag{B.19}$$

Note that the equation for copropagating modes only contains the coefficient A_μ, which refers to propagation in the positive direction.

Coupling between counterpropagating modes

In describing a μ-th counterpropagating mode, it is essential to remember that the exponent in $e^{j\beta_\mu z}$ has been taken with a positive sign. Furthermore, it is important to remember that the components of the mode functions can be described by means of those of the direct mode, so that:

$$\begin{aligned} \overline{E}_I &= (\overline{E}_{t\mu} - E_{z\mu}\hat{z})e^{j\beta_\mu z}; \\ \overline{H}_I &= (-\overline{H}_{t\mu} + H_{z\mu}\hat{z})e^{j\beta_\mu z}. \end{aligned}$$

Following a similar process to that used for the copropagating mode, for B_μ we obtain:

$$\frac{dB_\mu}{dz} = j\frac{\omega}{4} \iint_{-\infty}^{+\infty} \overline{P} \cdot \overline{E}_\mu^*\, dx\, dy\; e^{-j\beta_\mu z}. \tag{B.20}$$

It is now necessary to define the right side of both (B.19) and (B.20).

So far, the perturbation has been generically described using a polarization vector, $\overline{P}$. The perturbation, however, can also be described as a deviation $\Delta\varepsilon\,(x, y, z)$ of dielectric permittivity from its nominal value $\varepsilon\,(x, y)$, so that the polarization vector can be formulated as:

$$\overline{P} = \Delta\varepsilon\,(x, y, z)\, \overline{E}_P \tag{B.21}$$

The variation in permittivity results in a variation of the refractive index $n\,(x, y, z)$, so that:

$$n^2\,(x, y, z) = \frac{\varepsilon\,(x, y) + \Delta\varepsilon\,(x, y, z)}{\varepsilon_0}. \tag{B.22}$$

More generally, the anisotropy of the medium can be represented by a tensor, and the elements which cannot be found on the diagonal matrix describing the tensor can be associated to couplings between transverse-electric (TE) and transverse-magnetic (TM) modes. The present discussion will consider only the simplified case of perturbations along the z-axis of the guides, which implies that $\Delta\varepsilon = \Delta\varepsilon(z)$. In order to apply the orthogonality property, the polarization vector must be decomposed:

$$\overline{P} = \overline{P}_t + P_z \cdot \hat{z} \tag{B.23}$$

Where $\overline{P}_t$ corresponds to:

$$\overline{P}_t = \Delta\varepsilon \overline{E}_{Pt} = \Delta\varepsilon \sum_\nu \left(A_\nu(z)\, e^{-j\beta_\nu z} + B_\nu(z)\, e^{j\beta_\nu z}\right) \overline{E}_{t\nu}. \tag{B.24}$$

By considering Maxwell's equations:

$$\begin{gathered} j\omega(\varepsilon + \Delta\varepsilon)\, \overline{E}_{Pz} = \overline{\nabla}_t \times \overline{H}_{Pt} \\ j\omega\varepsilon E_z \hat{z} = \overline{\nabla}_t \times \overline{H}_t \end{gathered} \tag{B.25}$$

we obtain:

$$\begin{aligned} \overline{P}_z &= \Delta\varepsilon E_{Pz}\hat{z} \\ &= \frac{\Delta\varepsilon}{\varepsilon + \Delta\varepsilon} \frac{1}{j\omega} \overline{\nabla}_t \times \overline{H}_{Pt} \\ &= \frac{\Delta\varepsilon}{\varepsilon + \Delta\varepsilon} \frac{1}{j\omega} \sum_\nu \left(A_\nu(z)\, e^{-j\beta_\nu z} - B_\nu(z)\, e^{j\beta_\nu z}\right) \overline{\nabla}_t \times \overline{H}_{t\nu} \\ &= \frac{\varepsilon \cdot \Delta\varepsilon}{\varepsilon + \Delta\varepsilon} \sum_\nu \left(A_\nu(z)\, e^{-j\beta_\nu z} - B_\nu(z)\, e^{j\beta_\nu z}\right) E_{z\nu}\hat{z}. \end{aligned} \tag{B.26}$$

By using the equations in (B.23), (B.24) and (B.26), the right sides of the equations in (B.19) and (B.20) can be reformulated as follows:

$$-j\frac{\omega}{4} \iint_{-\infty}^{+\infty} \sum_\nu \left[\Delta\epsilon \left(A_\nu e^{-j\beta_\nu z} + B_\nu e^{j\beta_\nu z}\right) \overline{E}_{t\nu} \cdot \overline{E}^*_{t\mu} + \frac{\epsilon\Delta\epsilon}{\epsilon + \Delta\epsilon} \left(A_\nu e^{-j\beta_\nu z} - B_\nu e^{j\beta_\nu z}\right) E_{z\nu} E^*_{z\mu}\right] dx\, dy\; e^{j\beta_\mu z}$$

$$j\frac{\omega}{4} \iint_{-\infty}^{+\infty} \sum_\nu \left[\Delta\epsilon \left(A_\nu e^{-j\beta_\nu z} + B_\nu e^{j\beta_\nu z}\right) \overline{E}_{t\nu} \cdot \overline{E}^*_{t\mu} - \frac{\epsilon\Delta\epsilon}{\epsilon + \Delta\epsilon} \left(A_\nu e^{-j\beta_\nu z} - B_\nu e^{j\beta_\nu z}\right) E_{z\nu} E^*_{z\mu}\right] dx\, dy\; e^{-j\beta_\mu z}$$

Finally, the *coupled-mode equations* are obtained:

$$\begin{aligned} \frac{dA_\mu}{dz} &= -j \sum_\nu \left(K^+_{\nu\mu} A_\nu e^{-j(\beta_\nu - \beta_\mu)z} + K^-_{\nu\mu} B_\nu e^{j(\beta_\nu + \beta_\mu)z}\right); \\ \frac{dB_\mu}{dz} &= j \sum_\nu \left(K^-_{\nu\mu} A_\nu e^{-j(\beta_\nu + \beta_\mu)z} + K^+_{\nu\mu} B_\nu e^{j(\beta_\nu - \beta_\mu)z}\right); \end{aligned} \tag{B.27}$$

where:

$$
\begin{aligned}
K_{\nu\mu}^{+} &= K_{\nu\mu}^{t} + K_{\nu\mu}^{z} \\
K_{\nu\mu}^{-} &= K_{\nu\mu}^{t} - K_{\nu\mu}^{z}
\end{aligned}
\tag{B.28}
$$

and

$$
\begin{aligned}
K_{\nu\mu}^{t} &= \frac{\omega}{4} \iint_{-\infty}^{+\infty} \Delta\varepsilon \overline{E}_{t\nu} \overline{E}_{t\mu}^{*} \, dx \, dy \\
K_{\nu\mu}^{z} &= \frac{\omega}{4} \iint_{-\infty}^{+\infty} \frac{\varepsilon \Delta\varepsilon}{\varepsilon + \Delta\varepsilon} \overline{E}_{z\nu} \overline{E}_{z\mu}^{*} \, dx \, dy.
\end{aligned}
\tag{B.29}
$$

$K_{\nu\mu}^{t}$ and $K_{\nu\mu}^{z}$ are called *transverse and longitudinal coupling coefficients*, whereas $K_{\nu\mu}^{+}$ and $K_{\nu\mu}^{-}$ are the *coupling coefficients of the co- and counterpropagating modes.* Before applying these equations, it is important to remember some basic assumptions already mentioned. First of all, we have considered a cylindrical waveguiding structure, which is invariant to traslations along the z-axis. Secondly, the type of perturbation applied to the waveguiding structure has been a modulation of its dielectric permittivity. The perturbation must be small with respect to the unperturbed term, because the waveguide must be kept cylindrical. This is because, otherwise, the modes of a cylindrical waveguide could not be used to develop the coupled-mode theory.

Despite this relevant limitation, the coupled-mode theory represents a useful and important tool of analysis. As a matter of fact, the equations in (B.27) are general and can be applied to waveguiding structures whose features might appear considerably different, such as couplers, long- and short-period gratings.

B.1 Synchronism or Phase Matching Condition

Thanks to the solutions found so far, it is possible to solve the problem of perturbed structures by solving a system of infinite differential equations in infinitely many variables. Since the coupling actually occurs only between a limited number of modes, the results can then be considerably simplified in many applications, so that the advantages of such a process become more evident. The following paragraphs will try to explain why simplification is allowed.

First of all, notice that the behaviour of the coupling coefficients depends on a perturbation $\Delta\epsilon$. Therefore, provided that the permittivity of the unperturbed waveguide is independent of z, if the perturbation does not depend on the longitudinal coordinate, neither the coupling coefficients do. Conversely, if the perturbation periodically depends on the longitudinal coordinate, such as in gratings, the coefficients will be of the type:

$$
K_{\nu\mu}^{\pm} = K_{\nu\mu 0}^{\pm} cos\left(\frac{2\pi}{\Lambda} z\right) = k_{\nu\mu}^{\pm} \left(e^{-j\frac{2\pi}{\Lambda}z} + e^{j\frac{2\pi}{\Lambda}z}\right) \tag{B.30}
$$

where $k^{\pm}_{\nu\mu}$ is independent of z and Λ is the period of the perturbation. If we consider, for example, the first equation in (B.27) and integrate it with respect to z, we obtain:

$$A_\mu(z) = A_\mu(0) - j\sum_\nu \int_0^z k^+_{\nu\mu} A_\nu(\zeta)\left(e^{-j(\beta_\nu - \beta_\mu + \frac{2\pi}{\Lambda})\zeta} + e^{-j(\beta_\nu - \beta_\mu - \frac{2\pi}{\Lambda})\zeta}\right) d\zeta \; +$$

$$-j\sum_\nu \int_0^z k^-_{\nu\mu} B_\nu(\zeta)\left(e^{j(\beta_\nu + \beta_\mu - \frac{2\pi}{\Lambda})\zeta} + e^{j(\beta_\nu + \beta_\mu + \frac{2\pi}{\Lambda})\zeta}\right) d\zeta;$$

from which it is clear that $A_\mu(z)$ is a sum of terms of the type:

$$\int_0^z f(\zeta) e^{-j\delta\zeta} d\zeta,$$

with:

$$\delta = \beta_\nu - \beta_\mu \pm \frac{2\pi}{\Lambda}, \tag{B.31}$$

$$\delta = \beta_\nu + \beta_\mu \pm \frac{2\pi}{\Lambda}. \tag{B.32}$$

The first relation refers to the copropagating modes, whereas the second one refers to the counterpropagating modes. According to the principle of stationary phase, all contributions equal zero apart from those in which $\delta \cong 0$. Therefore, if the perturbation is independent of z, that is, if $\Lambda \to \infty$, the coupling only occurs between copropagating modes, either *direct* or *reflected*, for which:

$$\beta_\mu \cong \beta_\nu.$$

On the other hand, if the perturbation depends on z, the coupling can only occur between modes which satisfy the *synchronism or Bragg's condition*, usually referred to as *phase-matching condition*:

$$\beta_\nu - \beta_\mu \cong \pm\frac{2\pi}{\Lambda};$$

$$\beta_\nu + \beta_\mu \cong \pm\frac{2\pi}{\Lambda};$$

where the above equations refer to copropagating and counterpropagating modes, respectively. Therefore, the right sides of the equations in (B.27) will in fact be constituted by a limited range of terms, so that the problem is considerably simplified. The phase-matching condition can thus be generalized as follows:

$$\beta_\mu \pm \beta_\nu \cong m\frac{2\pi}{\Lambda}$$

where β_ν and β_μ always have positive values and the parameter m refers to the diffraction order considered. In fiber gratings, the standard is to apply a first order diffraction, in which $m = \pm 1$.

The phase-matching condition can be graphically explained as shown in figure (B.1). The line represents some possible values of the propagation constant for a given wavelength λ. The black circles represent the guided modes of the core ($n_{cl}\frac{2\pi}{\lambda} < \beta < n_{co}\frac{2\pi}{\lambda}$), the white circles represent the

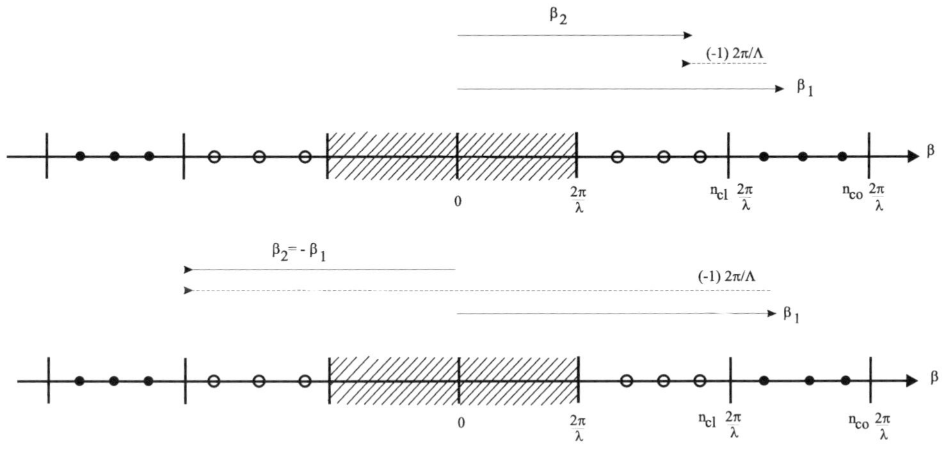

Fig. B.1: Diagrams of the *phase-matching* condition, which relate the propagation constant β_1 of a polarizing mode to the propagation constant β_2 of its related grating-induced mode. The constants are expressed as functions of the wavelength λ and the period Λ of the grating. Positive values of β represent modes propagating in the positive direction

cladding guided modes ($\frac{2\pi}{\lambda} < \beta < n_{cl}\frac{2\pi}{\lambda}$), and the marked area represents the spectrum of radiation modes. The negative values of β describe modes propagating in the negative direction. Consider, for example, a guided mode of the core which propagates in the positive direction. Short-period gratings allow coupling between counterpropagating modes, so that we obtain *Bragg gratings* or *reflection gratings*. On the other hand, a longer period allows coupling between copropagating modes, which is used to produce *long-period gratings* or *transmission gratings*. The scheme in figure (B.1), however, is partially simplified because it represents the coupling between specific modes without describing the coupling efficiency. As a matter of fact, the exchange of energy can sometimes be difficult or even impossible, but this can be seen only if the coupling coefficients defined in (B.29) are calculated.

B.2 Solution to the Grating Equations

Consider the equations in (9.4) and (9.5) which control the coupling between the direct and reflected modes of a grating:

$$\frac{dA}{dz} = -jkBe^{j\delta z};$$
$$\frac{dB}{dz} = jk^*Ae^{-j\delta z}.$$

The two equations can be decoupled if we work out the derivative of the first one with respect to z and then plug in the second equation:

$$\frac{d^2A}{dz^2} = -jk\left[\frac{dB}{dz} + j\delta B\right]e^{j\delta z} = |k|^2 A + j\delta\frac{dA}{dz};$$

If we then set:

$$S = 2\sqrt{|k|^2 - (\delta/2)^2}$$

we obtain:

$$A(z) = e^{-j\delta z/2}\left(a_1 \sinh(Sz/2) + a_2 \cosh(Sz/2)\right)$$

so that:

$$\begin{aligned} B(z) &= -\frac{1}{jk}\frac{dA}{dz}e^{-j\delta z} = \\ &= \frac{j}{k}\left[\left(\frac{S}{2}\cosh(Sz/2) + j\frac{\delta}{2}\sinh(Sz/2)\right)a_1 + \right. \\ &\qquad \left. + \left(\frac{S}{2}\sinh(Sz/2) + j\frac{\delta}{2}\cosh(Sz/2)\right)a_2\right]e^{-j\delta z/2}; \end{aligned}$$

By setting the following boundary conditions:

$$A(0) = A_0; \quad B(L) = 0$$

the first one becomes:

$$a_2 = A_0$$

and the second one:

$$a_1 = \frac{(S/2)\sinh(SL/2) + j(\delta/2)\cosh(SL/2)}{(S/2)\cosh(SL/2) + j(\delta/2)\sinh(SL/2)}A_0;$$

so that:

$$\begin{aligned} A(z) &= \left(\frac{A_0 e^{j\delta z/2}}{(S/2)\cosh(SL/2) + j(\delta/2)\sinh(SL/2)}\right) \times \\ &\times \left\{\left[(S/2)\cosh(SL/2) + j(\delta/2)\sinh(SL/2)\right]\cosh(Sz/2) \quad + \right. \\ &\qquad \left. - \left[(S/2)\sinh(SL/2) + j(\delta/2)\cosh(SL/2)\right]\sinh(Sz/2)\right\}; \end{aligned}$$

As for $B(z)$, in a few steps we can easily obtain:

$$\begin{aligned} B(z) &= \frac{j}{k}\left(\frac{|k|^2 A_0 e^{-j\delta z/2}}{(S/2)\cosh(SL/2) + j(\delta/2)\sinh(SL/2)}\right) \times \\ &\times \left[\cosh(SL/2)\sinh(Sz/2) - \sinh(SL/2)\cosh(Sz/2)\right]. \end{aligned}$$

B.3 Solution to the Coupler Equations

By decoupling the following coupling equations:

$$\frac{dA_a}{dz} = -jkA_b e^{-j\delta z}; \tag{B.33}$$

$$\frac{dA_b}{dz} = -jk^* A_a e^{j\delta z}. \tag{B.34}$$

we obtain a second order equation with a single variable, $A_a(z)$:

$$\frac{d^2 A_a}{dz^2} + j\delta \frac{dA_a}{dz} + |k|^2 A_a = 0.$$

$A_a(z)$ is usually written as:

$$\alpha^2 + j\delta\alpha + |k|^2 = 0 \ ;$$

With:

$$S = 2\sqrt{|k|^2 + (\delta/2)^2} \ ,$$

its roots can be written as:

$$\alpha_{1,2} = -j\frac{\delta}{2} \mp j\frac{S}{2}$$

so that $A_a(z)$ becomes:

$$\begin{aligned} A_a(z) &= d_1 e^{\alpha_1 z} + d_2 e^{\alpha_2 z} \\ &= d_1 e^{-j\frac{\delta}{2}z} e^{-j\frac{S}{2}z} + d_2 e^{-j\frac{\delta}{2}z} e^{j\frac{S}{2}z} \\ &= e^{-j\frac{\delta}{2}z}\left[d_1 e^{-j\frac{S}{2}z} + d_2 e^{j\frac{S}{2}z}\right] \\ &= e^{-j\frac{\delta}{2}z}\left[d_1(cos\frac{Sz}{2} - jsin\frac{Sz}{2}) + d_2(cos\frac{Sz}{2} + jsin\frac{Sz}{2})\right] \\ &= e^{-j\frac{\delta}{2}z}\left[(d_1 + d_2)cos\frac{Sz}{2} - j(d_1 - d_2)sin\frac{Sz}{2}\right]. \end{aligned}$$

By setting $c_1 = d_1 - d_2$ and $c_2 = d_1 + d_2$ we obtain:

$$A_a(z) = e^{-j\delta z/2}\left(-jc_1 \sin\frac{Sz}{2} + c_2 \cos\frac{Sz}{2}\right).$$

By means of this equation and the one in (B.33), we can write $A_b(z)$ as:

$$A_b(z) = -\frac{e^{j(\delta z/2)}}{2k}\left[j\left(\delta c_1 + Sc_2\right)\sin\frac{Sz}{2} - \left(\delta c_2 + Sc_1\right)\cos\frac{Sz}{2}\right].$$

Now we have to set the boundary conditions. Suppose that they involve only a single incident wave in the $z = 0$ region of the second waveguide, so that:

$$A_a(0) = 0 \ ; \ A_b(0) = A_0 \ .$$

By means of the first one, we then find that $c_2 = 0$. The second one can be substituted in:

$$\frac{1}{2}\frac{S}{k}c_1 = A_0 \; ;$$

so that we find:

$$c_1 = \frac{2A_0 k}{S} \; .$$

The two solutions can finally be written as:

$$A_a(z) = -jA_0\frac{2k}{S}e^{-j\delta z/2}\sin\frac{Sz}{2} \; ,$$

$$A_b(z) = A_0 e^{j(\delta z/2)}\left(\cos\frac{Sz}{2} - j\frac{\delta}{S}\sin\frac{Sz}{2}\right) .$$

Bibliography

[1] G.P. Agrawal, "Nonlinear Fiber Optics", Academic Press, New York, 2001.

[2] B.E.A. Saleh, M.C. Teich, "Fundamentals of Photonics", Wiley & Sons, New York, 1991.

[3] J.M. Senior, "Optical Fiber Communications", Prentice Hall, London, 1992.

[4] D.K. Mynbaev, L.L. Scheiner, "Fiber-Optic Communications Technology", Prentice Hall, Upper Saddle River, 2001.

[5] J.A. Buck, "Fundamentals of Optical Fibers", Wiley-Interscience, Hoboken NJ, 1995.

[6] C.D. Poole, J. Nagel, "Optical Fiber Telecommunications - IIIA", Academic Press, New York, 1997.

[7] D. Meschede, "Optics, Light and laser", Wiley-VCH Verlag GmbH & Co.KGaA, Weinheim, 2004.

[8] G. Lifante, "Integrated Photonics", Wiley & Sons, Chichester, 2003.

[9] F.G. Smith, T.A. King, "Optics and Photonics", Wiley & Sons, Chichester, 2000.

[10] W.B. Leigh, "Devices for Optoelectronics", Marcel Dekker, New York, 1996.

[11] F. Poli, A. Cucinotta, S. Selleri, "Photonic Crystal Fiber - Properties and Applications", Springer, Dordrecht, 2007.